First the seed

Science and Technology in Society

Daniel Lee Kleinman
Jo Handelsman
Series Editors

First the seed

The political economy of plant biotechnology, 1492–2000

Second Edition

JACK RALPH KLOPPENBURG, JR.
University of Wisconsin

The University of Wisconsin Press

The University of Wisconsin Press
1930 Monroe Street
Madison, Wisconsin 53711

www.wisc.edu/wisconsinpress/
3 Henrietta Street
London WC2E 8LU, England

First edition published in 1988 by Cambridge University Press

5 4 3 2

Printed in the United States of America

Library of Congress Cataloging-in-Publication Data
Kloppenburg, Jack Ralph.
First the seed: the political economy of plant biotechnology /
Jack Ralph Kloppenburg, Jr.—2nd ed.
p. cm. – (Science and technology in society)
Includes bibliographical references and index.
ISBN 0-299-19244-X (pbk. : alk. paper)
1. Seed industry and trade – United States. 2. Plant breeding – United States.
3. Plant biotechnology – United States. 4. Seed industry and trade. 5. Plant breeding.
6. Plant biotechnology. I. Title. II. Series.
SB117.3.K57 2004
631.5′2 – dc22 2004053597

For Thomas Robinson Kloppenburg

Contents

Tables

Figures

Preface to the second edition

Sixteen years after the initial publication of *First the Seed,* I am presented with the opportunity to revisit my work and evaluate it against the backdrop of subsequent events. I find I am both pleased and disappointed. I am pleased by the accuracy with which I limned the historical trajectories that were shaping the development and the deployment of that expanding set of knowledge, techniques and technologies that is still referred to generically as "biotechnology." I am disappointed that those trajectories have not been materially altered as much as they need to be.

Commodification has not just continued, it has been considerably accelerated. Given the recent flood of patents granted not just on genes (of humans as well as other species) but on short sequences and DNA fragments of unknown function, the initial reluctance in 1985 of the U.S. Patent and Trademark Office (PTO) to countenance the extension of patents to whole plants now seems almost quaint. Further, the advent of patents on research materials and techniques has reinforced long-standing trends in the social *division of labor* characteristic of the plant sciences. If in the 1980s public breeders were still struggling to maintain the capacity to release finished varieties, today they all too often must obtain a license from a corporate patent-holder in order to do their work at all. The clear asymmetries in the regime of *germplasm transfer* made "biopiracy" a central topic of debate during the 1990s in international fora from Rio to Seattle to Johannesburg. After twenty years of "seed wars," an International Treaty on Plant Genetic Resources for Food and Agriculture was signed in 2001, but its implications are uncertain. There is recognition of "Farmers' Rights" in principle, but of intellectual property rights in practice. The treaty is harmonized with the requirements of the World Trade Organization (WTO), but not with the interests either of peasant farmers or indigenous peoples.

Still, my disappointment that these trajectories so far have not been redirected much is tempered by a sense that conditions are developing in which they plausibly *could* be diverted onto substantially more progressive headings.

The world is a rather different place in 2004 than it was in 1988 when *First the Seed* was initially released. This is true both in regard to the social and natural circumstances immediately surrounding the development and deployment of biotechnology, and with reference to the larger political economy in which those processes are embedded. Seed nevertheless remains the vehicle in which many of the products of biotechnology – GMOs (genetically modified organisms) in the current argot – must be embodied and valorized. As both food and means of production, seed sits at a critical nexus where contemporary struggles over the technical, social, and environmental conditions of production and consumption converge and are made manifest.

The major difference between 1988 and 2004 is that over the last decade, opposition to the way in which private industry has chosen to develop biotechnology has emerged in robust, globally distributed, and increasingly well organized forms. In a new final chapter I show that opposition is expressed not just by advocacy and activist groups but has materialized within the public plant science community as well. I am grateful to the University of Wisconsin Press for giving me the occasion to explore how an alliance between civil society organizations and public agricultural research institutions might be made manifest, and how such a partnership could facilitate the realization of new, more just, and more sustainable trajectories for crop improvement. Truly, *still* the seed!

Jack R. Kloppenburg, Jr.

Madison
July 2004

Preface to the first edition

It is only March 8, but I planted today. It has been one of the mildest winters on record here in Wisconsin. Warmed by the heat reflecting off the stone facade of my house, the soil in my south-facing front garden has already thawed. When the temperature reached 71 degrees yesterday, I couldn't resist putting a spade to the soil to see how it turned. The earth fell cleanly away from the blade, and I crumbled the clods to a fine and receptive tilth. I prepared two beds. And then today I planted a few rows of "Easter Egg" radishes, "Red Sails" lettuce, and a line of the marvelous "Sugar Snap" peas. I never need much encouragement to – in the argot of seed packet instructions – "plant as soon as the soil can be worked." Really, it is too early for even these hardy species. There is no small bit of winter to go as yet, and these initial sowings probably will not survive. But they might; where there is seed and soil there is always hope of a harvest.

In the simple act of planting I was engaged in one of the most universal – and certainly one of the most important – of all human activities. I share the act of planting and my hope for a harvest with most of the world's population and with unnumbered previous generations. People must eat. And the chain of production processes that finally delivers food to our mouths – long for the New Yorker, short for the Thai peasant – begins everywhere with the sowing of seed. This is no less true of the animal products we consume, for milk or bacon is really nothing more than transformed grain. Crop production is the necessary foundation upon which the complex structures of human society have historically been raised. And the seed is the irreducible core of crop production. Truly, as the motto of the American Seed Trade Association has it, "First – the seed."

Despite the pivotal importance of the seed as the very stuff of the great American granary and as the fundamental input of the global "Green Revolution," the parallel development of plant breeding and the seed industry has received little attention from social scientists. This book seeks to redress this deficiency by providing a social history of both the scientific and commercial aspects of plant improvement. I trace the historical transformation of the seed from a public good produced and reproduced by farmers into a commodity that is a mechanism for the accumulation and reproduction of capital. While the development of scientific understanding in the plant sciences provides a narrative structure for the book, the central focus of my analysis is the interaction of scientific advances with three themes of political economy: (1) progressive commodification of the seed, (2) elaboration of a social division of labor between public and private plant breeding, and (3) asymmetries in global patterns of seed commerce and exchange between the less developed countries of the South and the advanced industrial nations of the North.

Further, I am interested in the history of plant breeding not only for what can be learned of the past but also for what the past can tell us of the present and even for what it can reveal of prospects for the future. As Russell Hoban expressed it in *Riddley Walker*, "What ben makes tracks for what wil be." Recent advances in genetics and molecular biology have given scientists access to the fundamental building blocks of life. The emergent "biotechnologies" constitute a crucial, perhaps epochal new technical form. One of the principal areas of application for the new biotechnologies is plant improvement, and the raw material of the plant genetic engineer is germplasm – the genetic information encoded in the seed. In this book I show that the weight of the past does indeed shape the present and bear upon future possibilities in concrete and specifiable ways. Contemporary issues such as the nature of global flows of germplasm, genetic erosion and vulnerability, the restructuring of research institutions, changing university-industry relations, and the development of patent rights for new crop varieties can be adequately understood only when viewed in historical perspective.

Such a historical perspective, as the title of this book implies, must encompass the year 1492. Contact between the Old and New Worlds touched off what has been called the "Columbian exchange," a dramatic and unprecedented movement of plants around the globe. In 1986, the celebration of the Statue of Liberty's centennial highlighted the central role that immigration has played in American history. Yet it is seldom recognized that our population of agricultural plants is as immigrant in character as the nation's human population. None of the crops that today make the United States an agricultural power is indigenous to North America. Had new crop plants not been introduced from other regions of the world, there would not have been a plant genetic base sufficient to provide an agriculture capable of sustaining the tide of human immigration. And the importation of "raw" plant genetic material for further processing in the test plots of American seed companies and agricultural colleges is a phenomenon of enduring importance. The pea, for example, originated in Asia Minor, and the germplasm that put the "sugar" in the "Sugar Snap" peas I planted today was derived from East Asian material. The evolution of global patterns of access to and control over plant genetic resources has materially conditioned the development of plant breeding in both the public and private sectors.

That plant breeding is not today an exclusively private endeavor is an interesting anomaly for American capitalism. The development of a private presence in plant breeding has historically faced two obstacles, one biological and the other institutional. First, the very reproducibility of seed made the farmer the commercial seed company's prime competitor and constrained private investment in plant improvement. Into this vacuum of investment moved the state to become an institutional obstacle to the expansion of a commercial seed industry. Whereas in other sectors of the economy the state may act indirectly to shape the character of a product through regulation, in plant breeding it has done so directly by actually creating the product – the new plant variety. Public breeders have thus significantly limited the possibilities for capital accumulation by private breeders by directly competing with them.

The social history of plant breeding in the twentieth century is essentially a chronicle of the efforts of private industry to circumvent these twin obstacles. These efforts have involved the elaboration of two distinct but intersecting solutions to the constraints facing seed companies. One involves the use of science to make the seed more amenable to commodification. The prime example of this technical solution is hybridization, a breeding technique that is capable of providing more productive plants but that eliminates the possibility of saving and replanting seed. The hybridization of corn, the archetypal success story of the plant breeder's art and science, served the interests of capital by bringing farmers into the commercial seed market every year. A second solution is the extension of property rights to plant germplasm by legislative fiat, as with passage of the Plant Variety Protection Act in 1970. The act conferred patent-like rights on breeders of new plant varieties. So, my Burpee's seed catalog not only describes my "Sugar Snap" peas as the "#1 All-Time All-America Vegetable," but also warns "Unauthorized propagation prohibited – U.S. Protected Variety." As the private seed trade has grown stronger, it has been able to continuously redefine the social division of labor in plant improvement, with public breeders becoming increasingly limited to activities complementary to rather than competitive with those of private capital.

It is clear that the new biotechnologies contain tremendous potential for increasing the productivity of agricultural crops. For private industry, they also offer the prospect of facilitating continued movement on the two paths of commodification along which capital has historically penetrated plant breeding. There is also the possibility that the new biotechnologies will produce significant externalities, for new technologies always carry liabilities as well as benefits. As a society, I think we would like to use our enhanced capabilities for manipulating the genetic code to develop and deploy new plant varieties in ways that are economically productive, socially equitable, and ecologically benign. Will we be able to do so?

To answer this question, we need to look both back into history and forward into tomorrow; hence the parameters 1492-2000 in the subtitle of this book. As plant breeder Norman Simmonds has noted, "There can be no better basis for a view of the future of a crop than a thorough understanding of its past." The extensive social impacts – both positive and negative – stemming from the introduction of hybrid corn were clearly evident in the United States *before* the inauguration of the international Green Revolution of the 1960s. Had social scientists been attentive to those impacts, they would not have been so surprised by the appearance of certain negative consequences associated with the introduction of Green Revolution plant types. Indeed, they might even have been in a position to have avoided or mitigated some of them.

If, as many believe, we are indeed on the threshold of a biorevolution, it would be both dangerous and socially irresponsible to move into the age of synthetic biology as blindly as we did into the Green Revolution. This book is written with the conviction that we need not do so.

Jack R. Kloppenburg, Jr.

Madison
8 March 1987

Acknowledgments

It is with mingled senses of relief and pleasure that I complete this book. It is the fruit of a project begun some four years ago as a graduate student at Cornell University. In my work at Cornell I was fortunate to have had the advice and guidance of Charles Geisler and Frederick Buttel. Both were instrumental in creating the intellectual and institutional space in which I could pursue the social analysis of seeds and biotechnology.

Having achieved another milestone in my intellectual journey, I also want to acknowledge the assistance of George Dalton. It was he who first set me on the path I am traveling, and the example he sets as scholar and teacher has helped me to stay there.

I want also to thank my friend and colleague, Martin Kenney. Continuous interaction with Martin over a four-year period shaped my intellectual growth in many ways. I will always be grateful that Providence (in the historical personage of Gordon Huckle) assigned me to an office with Martin upon my arrival at Cornell. I am also grateful to Daniel Kleinman, whose inquiring mind and diligent work have added much to my research effort.

For material support of the research that is reflected in this book, I would like to thank Cornell University, Resources for the Future, and the Mellon Foundation. Cornell University generously supported me with fellowships through every year of my graduate career in Ithaca. I have also been the grateful recipient of a Resources for the Future Dissertation Fellowship and a Mellon Foundation Student Research Grant. More recently, the College of Agriculture and Life Sciences and the Graduate School of the University of Wisconsin have provided assistance that has enabled me to extend the analysis I began as a graduate student. Finally, without the cooperation of the many people who generously gave of their time in interviews, my analysis would be much the poorer. I am not unmindful of my responsibility to justify the confidence these institutions and people have shown in my work. I hope that they will find this book worthy of their support.

Harriet Friedmann and Jean-Pierre Berlan both reviewed the original manuscript. Their insights and criticism resulted in many changes and a much improved book. I am especially grateful to Harriet Friedmann for helping me clarify the central issues and work through the sticky problem of integrating conceptual and chronological order. What deficiencies remain in the book are not the responsibility of these reviewers, but are mine alone. For their efficient management of the literary production process I thank my editors, Jim DeMartino, Frank Smith, and Louise Calabro Gruendel. Mary Lybarger is responsible for the fine illustrations and graphs. Pat Cartwright's editorial assistance eased the burden of multiple rereadings of my own prose.

Finally, my wife has given me far more assistance than I ever had a right to ask in more ways than I could ever have anticipated. And she gave it without having to be asked.

Without the contributions and succor of these people and organizations, I would not be in the happy position in which I find myself today. The poverty of my words stands in inverse proportion to the magnitude of my debt.

Abbreviations

AAA	Agricultural Adjustment Act
AAACES	Association of American Agricultural Colleges and Experiment Stations
ABA	American Breeders Association
AGS	Advanced Genetic Sciences, Inc.
ARI	Agricultural Research Institue
ARS	Agricultural Research Service
ASA	American Society of Agronomy
ASTA	American Seed Trade Association
B.t	*Bacillus thuringiensis*
CBD	Convention on Biological Diversity
CBI	Council for Biotechnology Information
CGIAR	Consultative Group on International Agricultural Research
CMS	Cytoplasmic Male Sterility
DNA	Deoxyribonucleic acid
DNAPT	DNA Plant Technology, Inc.
EPA	Environmental Protection Agency
EU	European Union
FAO	Food and Agriculture Orgnization of the United Nations
FDA	Food and Drug Administration
FTO	Freedom to operate
GAO	General Accounting Office
GMOs	Genetically modified organisms
GRAIN	Genetic Resources Action International
GURTs	Genetic Use Restriction Technologies
HYV	High yielding variety
IARC	International agricultural research center
IBPGR	International Board for Plant Genetic Resources
ICGB	International Cooperative Biodiversity Grants
ICIA	International Crop Improvement Association

LGU	Land-grant university
MNC	Multinational Corporation
MTAs	Materials Transfer Agreements
NAS	National Academy of Sciences
NASULGC	National Association of State Universities and Land-Grant Colleges
NBF	New biotechnology firm
NCCPB	National Council of Commercial Plant Breeders
NEPA	National Environmental Policy Act
NF	Nitrogen fixation
NIH	National institutes of Health
NRC	National Research Council
NSF	National Science Foundation
NSSL	National Seed Storage Laboratory
OSTP	Office of Science and Technology
OTA	Office of Technology Assessment
PBR	Plant breeders' rights
PGRs	Plant genetic resources
PIPRA	Public-Sector Intellectual Property Resource for Agriculture
PTO	United States Patent and Trademark Office
PVPA	Plant Variety Protection Act
RAC	Recombinant DNA Advisory Committee
rBGH	Recombinant bovine growth hormone
rDNA	Recombinant deoxyribonucleic acid
RTLAs	Reach Through Licensing Agreements
SAB	Scientific advisory board
SAES	State agricultural experiment station
TRIPS	Trade-Related Aspects of Intellectual Property Rights
TUA	Technology Use Agreement
UPOV	International Union for the Protection of New Varieties of Plants
USDA	United States Department of Agrigulture
WTO	World Trade Organization

1

Introduction

> Darwin has directed attention to the history of natural technology, i.e. the formation of the organs of plants and animals, which serve as the instruments of production for sustaining their life. Does not the history of the productive organs of man in society, of organs that are the material basis of every particular organization of society, deserve equal attention? And would not such a history be easier to compile, since, as Vico says, human history differs from natural history in that we have made the former, but not the latter? Technology reveals the active relation of man to nature, the direct process of the production of his life, and thereby it also lays bare the process of the production of the social relations of his life, and of the mental conceptions which flow from those relations.
>
> Karl Marx, *Capital 1* (1977)

This book is a political and economic history of what has been one of the most fundamental of humanity's "productive organs": plant biotechnology. Whatever the historical period, whatever the mode of production, plants and their products have been necessary components of the material base on which the complex structures of human societies have been raised. We must all eat, and what we eat is ultimately derived from plant material. What is a steak, after all, but embodied corn? As the prophet Isaiah phrased it: "All flesh is grass." And plants have provided us not only with food but also with the raw materials needed for the production of a multitude of useful goods ranging from cotton cloth to life-saving drugs. Moreover, the domestication and subsequent improvement of numerous plant species also represent instances in which, contrary to Vico, humanity has in some measure made natural history.

"Biotechnology" is broadly defined by the Congressional Office of Technology Assessment (OTA 1984) as "any technique that uses living organisms (or parts of organisms) to make or modify products, to improve plants or animals, or to develop microorganisms for specific uses." Though the term has only recently been added to our lexicon, it encompasses human activities of considerable antiquity. The fermentation of beer, the making of cheese, and the baking of bread can all be considered "biotechnological" processes given the use they make of yeasts.

More important than the manipulation of microorganisms, however, has been the breeding of plants and animals. Crop improvement is as old as agriculture itself, and the earliest agriculturalists were engaged in a simple form of biotechnology. There is a substantial amount of genetic diversity within species. And germplasm – the complement of genes that shapes the characteristics of an organism – differs from individual to individual. Out of each year's harvest, farmers selected seed from those plants with the most desirable traits. Over thousands of years the slow but steady accumulation of advantageous genes produced more productive cultivars. Following the rediscovery in 1900 of Mendel's work illuminating the hereditary transmission of traits, this global process of simple mass selection was augmented by the systematic "crossing" of plants by scientists with the express purpose of producing new varieties with specific characteristics.

The process of plant breeding can be thought of as "applied evolutionary science," because it encompasses all of the features of neo-Darwinian evolution (Simmonds 1983a:6). Plant breeders collect the genetic material provided by nature and recombine it in accordance with the parameters of speciation. In essence, they apply artificial selection to naturally occurring variance in the DNA[1] "messages" characteristic of different genotypes (Medawar 1977). On this basis humanity has enjoyed tremendous productive advances in plant agriculture.

But though modern breeding methods of considerable sophistication have been developed for the recombination of plant genetic material, the sense in which plant breeders can be said to have "made natural history" is somewhat limited. Breeders have had to work within the natural limits imposed by sexual compatibility. In their work, plant scientists have rearranged a given genetic vocabulary, but they have not been able to create new words or novel syntactical structures. As Marx might have phrased it, we have not historically had the power to alter "species being." That is, we have not had this capacity until very recently.

Outdoing evolution

With the appearance over the last decade of a set of new and uniquely powerful genetic technologies, we are poised on the edge of an era in which humanity *will* be "making natural history" in a much more complete sense of the phrase. The most prominent of these novel technologies are recombinant DNA (rDNA) transfer (gene "splicing"), protoplast fusion, and great improvements in the established technique of tissue culture ("cloning"). These technologies share a qualitative superiority over conventional methods of genetic manipulation in their potential for the directed alteration of living organisms.

This superiority has two principal dimensions. First, while conventional breeding operates on whole organisms, the new technologies operate at the cellular and even the molecular level. Second, while conventional breeding relies upon sexual means to transfer genetic material, rDNA transfer and protoplast fusion make it possible to bypass sexual reproduction and move genes between completely unrelated organisms. The new technologies permit the modification of living organisms with an unprecedented specificity and allow a qualitatively different degree of genetic transformation.

Under the impact of these *new* biotechnologies, the walls of speciation, heretofore breached only infrequently and with great difficulty, are now crumbling. At a recent National Research Council (NRC) Convocation on Genetic Engineering of Plants, Harvard botanist Lawrence Bogorad asserted that "We now operationally have a kind of world gene pool... Darwin aside, speciation aside, we can now envision moving any gene, in principle at least, out of any organism and into any organism" (NRC 1984:12). Profound transformations can be made in living organisms not over a period of millions, thousands, or even hundreds of years, but in a matter of days. Such transformations are not the result of randomly occurring changes but are the product of conscious and direct human intention, of human engineering and design. As God told Noah of the beasts of the earth, the birds of the air, and the fish of the sea: "Into your hand they are delivered... as I gave you the green plants, I give you everything." Or, as 1975 Nobel laureate and M.I.T. microbiologist David Baltimore put it, "We can outdo evolution" (quoted in Cavalieri 1981:32).

Now, outdoing evolution is not to be pursued as an end in itself, or simply for the purpose of satisfying the curiosity of scientists. Genetic engineering is an undertaking with far-reaching economic and social implications. Biotechnology promises to enhance significantly our power to create and reproduce the material conditions of our existence. The sectors upon which the new genetic technologies will have the greatest impact are fundamental components of the modern economy: medicine, energy generation, pollution and waste management, chemical and pharmaceutical production, food processing, and, of course, agricultural production. We are now witnessing a radical recharacterization of the nature of the link between the "productive organs of man in society" and the productive organs of living creatures. So profound an advance in the forces of science and technology can be expected to have broad and important effects on social and economic relations.

Many observers of what is often perceived as a "biorevolution" have emphasized the degree to which the global "biofuture" will break with the past (e.g., Hutton 1978; McAuliffe and McAuliffe 1981; Sylvester and Klotz 1983; Yoxen 1983). In his book *Algeny*, Jeremy Rifkin (1983) has gone so far as to greet the dawning of a new "biotechnical age." But however far-reaching the social impacts of the new biotechnologies ultimately may be, it is a premise of this book that they will be shaped in important ways by existing social relations. History is not a series of discontinuous events; the future is systematically connected to the past. New technologies are not deployed in a historical vacuum.[2] Rather, they are introduced into a particular set of social, economic, and ecological circumstances with established and knowable trajectories. The existing social formation conditions the manner in which a new technology is deployed, even as it may be changing under the influence of that deployment. An adequate approach to the assessment of the social impacts of novel technology is necessarily historical; an understanding of the "old" biotechnologies is a prerequisite to the understanding of the "new" biotechnologies.

I have produced a historical analysis of plant biotechnology in the United States out of a sociological interest in the past that is directly linked to a deep concern for the future. This book shows how concrete historical processes in the development of plant breeding and seed production have shaped the sector's present characteristics and thereby conditioned its prospects for the future. This understanding is used to illuminate the range of strategic choices we have in attempting to enhance our degree of social control over the deployment of biotechnology, and to support the continuing struggle to achieve a sustainable agriculture responsive to human needs.

First the seed

The plant is the irreducible core of crop production on the farm and the most fundamental agricultural input. As the motto of the American Seed Trade Association has it: "First – the Seed." But while scholars and political analysts representing a wide variety of theoretical positions have long recognized that technological advance is a principal factor contributing to structural change in agriculture, the role of new plant varieties in this process has gone largely unexamined.

Since 1940, and especially in the last decade, a substantial literature has treated the relationship of agricultural research to such phenomena as the disappearance of the "family farm," the concentration of farm ownership and production, the displacement of labor, the decline in the quality of food, the deterioration of the environment, the rise of agribusiness, the marginalization of the small producer, and the exacerbation of income inequalities in farming (e.g., Carson 1962; Hightower 1973; Barry 1977; Perelman 1977; Friedland et al. 1981; Vogeler 1981; Berardi and Geisler 1984; Levins and Lewontin 1985). This literature constitutes a sustained critical interpretation of the social impacts of agricultural science, but it has focused almost exclusively on mechanical and chemical technologies.

The social consequences of the introduction of new plant varieties have been only infrequently and inadequately addressed. When social scientific efforts have been applied to this area they have been narrowly directed to the highly visible case of tomato breeding for mechanized harvest (e.g., Webb and Bruce 1968; Schmitz and Seckler 1970; Friedland and Barton 1976; Schrag 1978). However useful and instructive the example of the tomato has been, it has not led to a more comprehensive assessment of the role of the plant sciences in the dramatic transformations that the structure of agriculture in the United States has undergone over the last half century.

The paucity of critical analysis devoted to plant breeding reflects prevailing perceptions that it is one of the most unambiguously beneficial of scientific endeavors. The product of plant breeding, the seed, is regarded as a uniquely benign input in both environmental and structural terms. As a natural product, seed is perceived as "ecologically positive" (Teweles 1976:66). And according to economists, the perfect divisibility of seed makes it scale neutral (Dorner 1983:77). Seed thus embodies yield-enhancing genetic improvements without damage to the environment and without a biasing effect on farm structure. In a widely used text, the well-known breeder N. W. Simmonds (1979:38) asserts that "plant breeding, in broad social terms, does indeed generate substantial benefits and is remarkably free of unfavourable side-effects (the economists' 'externalities')." Simmonds concludes, "As plant breeding, *per se*, is a wholly benign technology, any enhancement of it must be welcomed as being in the public good, no matter who does it."

That plant breeding might have managed to avoid "unfavourable side-effects" is all the more remarkable given the scale of what are regarded as its positive impacts. Since 1935, yields of all major crops in the United States have at least doubled, and at least half of these gains are attributable to genetic improvements. Indeed, plant breeders have been responsible for what the United States Department of Agriculture (USDA) considers the "food production story of the century": the development of hybrid corn (U.S. Congress, House of Representatives 1951:2).

In the twenty years following the commercial introduction of hybrid varieties in 1935, corn yields doubled. And in 1985 the average yield for corn stood at about six times the Depression-era figure of 20 bushels per acre. Certainly corn breeders themselves have done little to dispel the notion that they are indeed the "prophets of plenty." Testifying on science legislation before the Senate Committee on Military Affairs, L. J. Stadler credited the increased production attributed to hybrid corn varieties with paying for the development of the atomic bomb (Shull 1946:550). Paul Mangelsdorf went still further, asserting that hybrids had contained the spread of communism after World War II by ensuring an adequate food supply for a decimated Western Europe (Glass 1955:3).

The 700 percent annual social return on research investment that economist Zvi Griliches (1958) calculated for hybrid corn remains the paradigmatic example of the large benefits society enjoys from agricultural research. In his 1982 presidential address to the American Society of Agronomy's diamond jubilee convocation, C. O. Gardner (1983) still could find no more fitting example of the contributions made by plant scientists than to cite once more the "success story" of hybrid corn. Even now, in the brave new world of recombinant DNA transfer, the National Academy of Sciences (NAS) sees the "spectacular success" hybrid corn had in increasing yields as the model of achievement to which the new biotechnologists should aspire (NRC 1984:6).

But have the development and deployment of new crop varieties in the United States really been the unalloyed good they are made out to be? The superlatives attached to hybrid corn reflect an obsessive preoccupation with yield increases. Can such yield increases have been achieved without a complex constellation of far-reaching socioeconomic changes rippling throughout the agricultural sector? Is yield increase the only objective to which the agricultural plant sciences should be directed? What realities are masked by the language of "success" and the prevailing ideology of the benevolence of plant breeding?

That the role of new plant varieties in contributing to transformations in the structure of agriculture and in the natural environment has not been systematically addressed in the United States is curious, since it is this very connection that has so interested social scientists engaged in study of the international "Green Revolution" of the 1960s and 1970s. Both critics and defenders of the Green Revolution recognize that, whatever the benefits, the introduction of the "miracle" wheats and rices developed at the Ford- and Rockefeller-funded international agricultural research centers (IARCs) played a crucial role in galvanizing not just substantial yield increases, but a wide range of negative primary and secondary social and environmental impacts as well. These include the exacerbation of regional inequalities, generation of income inequalities at the farm level, increased scales of operation, specialization of production, displacement of labor, accelerating mechanization, depressed product prices, changing tenure patterns, rising land prices, expanding markets for commercial inputs, agrichemical dependence, genetic erosion, pest-vulnerable monocultures, and environmental deterioration (Cleaver 1972; Jennings 1974; Perelman 1977; Simmonds 1979; Pearse 1980; Plucknett and Smith 1982; Lipton and Longhurst 1986).

The introduction of hybrid corn in the 1930s touched off an American precursor of the international Green Revolution. Can we have passed through our own domestic Green Revolution without having experienced profound transformative social change? I think not. And, if one listens carefully, plant scientists occasionally admit as much.

In an unusually frank invitational paper read at the 1977 annual meeting of the American Society of Agronomy (ASA), University of California-Berkeley plant physiologist Boysie E. Day implicated the plant sciences as important contributors to social upheaval:

> I begin with the proposition that the agronomist is the moving force in many of the social changes of our time. I include under the title "Agronomist" all crop production scientists of whatever discipline. He has brought about the conversion of a rural agricultural society to an urban one. Each advance has sent a wave of displaced farm workers to seek a new life in the city and a flood of change throughout society. This is true in all of the developed nations but is particularly evident in the United States where the changes have been greater than elsewhere. Be assured that at the 1977 ASA annual meeting, as in the past, there were enough new findings disclosed to render many thousands of American farms economically superfluous and cause the displacement of many farm workers from the country to the city. Probably, no meeting in 1977 of politicians, bureaucrats, social reformers, urban renewers, modern-day Jacobins, or anarchists will cause as much change in the social structure of the country as the ASA meeting of crop and soil scientists. [Day 978:19]

Day is claiming for the plant sciences social impacts on a scale and of a type such as those widely associated with the international Green Revolution and with chemical and mechanical technologies. Day's assertions are not mere hyperbole.

At another ASA meeting, agronomist Werner Nelson (1970:1) voiced a perennial complaint of plant scientists: "We have a tremendous story to tell about past accomplishments and the exciting future . . . but who knows about it." I quite agree. This book is an attempt to go beyond a narrow preoccupation with yield increase and to explore the broad social impacts and the darker side of the "success story" of plant biotechnology.

Structuring the story

How then should the story be told? An analytical framework is required if, to paraphrase Wallace Stevens, the squirming facts are not to exceed the squamous mind. I begin with the premise that scientific plant improvement has developed in the historical context of capitalism. Given this premise, I follow Kautsky's admonition to ask:

> is capital, and in what ways is capital, taking hold of agriculture, revolutionizing it, smashing the old forms of production and of poverty and establishing the new forms which must succeed. [Banaji 1980:40]

In terms of the subject at hand, the question becomes: Have plant breeding and seed production become means of capital accumulation? If so, how has this been accomplished, and what have been its effects?

This book answers the first question in the affirmative. At the most general level, my argument is that the agricultural plant sciences have over time become increasingly subordinated to capital and that this ongoing process has shaped both the content of research and, necessarily, the character of its products. This is not to say that capital has achieved complete domination of the sector. Capital has encountered a variety of barriers in its attempts to penetrate plant breeding and make the seed a vehicle of accumulation. Such gains as it has enjoyed have been achieved only through struggle, a struggle that is still continuing.

The problem of explaining how this increasing subordination of agricultural science to capital has been achieved and what its effects have been is rather complex. Traditional histories of science have emphasized chronological order as an organizing principle. Such a structure corresponds to a predominantly "internalist" perspective that attributes an immanent technical logic to scientific discovery.[3] The secrets of nature are thought to be uncovered serially and in orderly fashion by a series of savants, each of whom begets his or her successor in a sequence of almost biblical linearity (Kuhn 1962). Any historical treatment needs to pay attention to chronology, of course. But in their concern with establishing what happened when, histories and sociologies of science have too often lost sight of the broader social context in which science is performed. Once analysis focuses on scientific practice to the exclusion of its social integument, it is but a small step to the technological determinism that has proved so persistent a component of contemporary interpretations of science and technology. And this is unfortunate, for as Noble (1984:xiii) points out:

> technology does not necessitate. It merely consists of an evolving range of possibilities from which people choose. A social history of technology that explores beneath the appearance of necessity to illuminate these possibilities which technology embodies, reveals as well the contours of the society that realizes or denies them.

So the task of explaining how capital has come to dominate plant breeding and seed production to the extent that it does demands attention to broad social dynamics as much as to the technical rationale behind the development of specific scientific techniques or technologies. More than this, it also demands sensitivity to lost possibilities or forgone alternatives to the technologies that ultimately emerged from the laboratories and test plots of plant breeders. What was not done may be as revealing as what was done.

While integrating chronological and conceptual order is a necessary task, it is also a delicate matter. In efforts to preserve the narrative thread, one may fail to do justice to fundamental processes that cut across time. On the other hand, by emphasizing such features, one runs the risk of losing the sense of historical continuity and perspective. In the chapters that follow, I have attempted to achieve a judicious balance of both approaches. For the most part, the structure of the book follows the temporal development of scientific understanding and technical capacity in the plant sciences. But this development is never presented alone. Rather, it always appears in relation to one or more of three conceptual themes that run through the entire book. Scientific and technical development in plant improvement can be properly understood only in the context of its interaction with these three parallel lines of historical development: (1) political economy-commodification, (2) institutions-division of labor, and (3) world economy-germplasm transfer. It is important that the reader keep these themes in mind throughout the book; hence, a brief summary of the themes and their substantive roles in shaping plant biotechnology follows.

Political economy – commodification

The capitalist mode of production is characterized by the existence of a class of direct producers who have been dispossessed of the means of production. Their labor power is sold to an opposed class that has monopolized the means of production. This necessarily implies the generalization of commodity production, because workers sell the commodity labor power for the means to purchase the goods by which they maintain their lives and which are available to them only as commodities. The fundamental historical processes associated with the political economy of capitalism are therefore those of *primitive accumulation*, the separation of the worker from the means of production, and *commodification*, the extension of the commodity form to new spheres. These processes are inextricably linked insofar as primitive accumulation necessarily implies the subordination of the dispossessed worker to the commodity relation.

The "capitalist mode of production" is an abstract construct that does not appear in society as a pure form. Rather, at any time there exists a particular social formation that represents the current shape of capitalist development. It is true that capitalism is quite advanced in many areas. But this advance is uneven and is conditioned by the struggle of those who are being dispossessed. This resistance may be reinforced by the structural or natural recalcitrance of certain productive sectors. The penetration of capitalist relations may be delayed or may occur incompletely. In consequence, the processes of primitive accumulation and commodification are still under way, even in advanced industrial societies.

There is no question that Marx saw agriculture as one of these recalcitrant sectors, and Mann and Dickinson (1978) have specified a number of the particular features of agricultural production that provide obstacles to capitalist penetration. Yet capitalism is nothing if not vitally expansionist. It constantly pushes against the barriers that restrain its advance, eroding them slowly, overwhelming them suddenly, or flowing past them and isolating them if their resistance is strong.

And capital has by no means been met with a complete rebuff in agriculture. Over the last century, farming has been converted from a largely self-sufficient production process into one in which purchased inputs account for the bulk of the resources employed. This transformation has been, in essence, a process of primitive accumulation characterized by the progressive separation of the farmer from certain (though not all) of the means of agricultural production (e.g., seed, feed, fuel, motive power), which come to confront him as commodities. The corollary to this process has been the rise of agribusiness: capitalist firms producing agricultural inputs with wage labor. No longer does the farmer autonomously reproduce most of his own means of production; these activities have moved off-farm into a capitalist production process yielding surplus value that is realized in the commodity-form.

This transformation has been undergirded by the advance of science and technology. A novel and useful way of thinking about agricultural research is to view it as the incorporation of science into the historical processes of primitive accumulation and commodification. As such, agricultural research can also be seen as an important means of eliminating the barriers to the penetration of agriculture by capital. This theoretical approach to agricultural research has special explanatory power when it is applied to the case of plant breeding and seed production.

Included in any compendium of "obstacles to the capitalist penetration of agriculture" should be the natural characteristics of the seed itself. Like the Phoenix of myth, the seed reemerges from the ashes of the production process in which it is consumed. A seed is itself used up (or, rather, transformed) as the embryo it contains matures into a plant. But the end result of that process is the manyfold replacement of the original seed. The seed thus possesses a dual character that links both ends of the process of crop production: It is both means of production and, as grain, the product. In planting each year's crop, farmers also reproduce a necessary part of their means of production. This linkage, at once biological and social, is antagonistic to the complete subsumption of seed (as opposed to grain) under the commodity-form. Thus, a farmer may purchase seed of an improved plant variety and can subsequently propagate the seed indefinitely for future use. As long as this condition holds, there is little incentive for capital to engage in plant breeding for the purpose of developing superior crop varieties, because the object in which that research is valorized – the seed – is unstable as a commodity-form. The natural characteristics of the seed constitute a biological barrier to its commodification.

Capital has pursued two distinct but intersecting routes to the commodification of the seed. One route is technical in nature and the other social. The technical approach has been provided by agricultural science in the form of hybrid corn. More important than enhanced productivity is the fission of the identity of the seed as product and as means of production that hybridization achieves. Because the progeny of hybrid seed cannot economically be saved and replanted, it has use-value and exchange-value only as grain, not as seed. Farmers using hybrid seed must return to the market every year for a fresh supply. Moreover, the peculiarities of breeding hybrid corn mean that the parent lines of any particular variety can be developed and maintained as trade secrets, thus making hybrid seed a proprietary product. Hybridization has proved to be an eminently effective technological solution to the biological barrier that historically had prevented more than a minimum of private investment in crop improvement. It opened to capital a whole new frontier of accumulation that commercial breeders moved rapidly to exploit. The initial breach in the barrier made in corn was widened as research efforts produced hybrid varieties in other species.

But not all crops yielded to this frontal assault by science. For technical reasons, hybridization has not been achieved in many economically important species. But there is also a second path to the encouragement of private investment in plant breeding. If property rights to privately developed plant varieties are established, the two social souls embedded in the seed can be split by institutional as well as technical force. The seed can be rendered a commodity by legislative fiat as well as by biological manipulation. The Plant Patent Act of 1930 had granted patent protection to breeders of novel varieties of asexually reproducing plants (principally fruit species and ornamentals), and the seed industry had long lobbied for provision of similar legislation for plants that reproduce sexually (i.e., by seed). In 1970 this ambition was partially realized with the passage of the Plant Variety Protection Act (PVPA), and capital has continuously sought to extend the reach of private property in germplasm.

Institutions – division of labor

Because of the difficulty of securing proprietary control over plant germplasm, little systematic private effort was devoted to the improvement of agricultural crops prior to the development of hybrid corn in 1935. Yet investment in plant improvement was crucially important to the development of American capitalism. Settlement of the continent, provision of a cheap food supply for the working class, the generation of foreign exchange, and, ultimately, the establishment of industrial capitalism were fundamentally dependent on a productive agricultural sector. A productive agricultural sector was in turn contingent upon the development of improved and adapted crop varieties. Because private capital was not made available, social capital was called forth. That is, the state undertook the task of plant improvement.

The history of plant improvement in the United States until 1935 or so is essentially that of the continuous growth and elaboration of publicly performed research and development in a virtual vacuum of private investment. Global plant germplasm collection was initiated by the U.S. Patent Office in 1839. Thus was established a powerful tradition of state commitment to agriculture in general and the plant sciences in particular. This commitment was explicitly institutionalized in 1862 with creation of the Department of Agriculture (USDA) and passage of the Morrill Act, which authorized federal support for agriculturally oriented "land-grant universities" (LGUs). The Hatch Act of 1887, which provided assistance to state agricultural experiment stations (SAESs), established a framework for the systematic application of science to agricultural problems.

Industrial and financial interests supported these state activities more strongly than did farmers themselves. For though farming proved relatively resistant to the extension of capitalist relations of production – with important exceptions – there was still much profit to be gained from expanded flows of agricultural inputs and outputs resulting from the "rationalization" of farming. Thus, by the turn of this century there was in place at both federal and state levels an extensive *public* sector devoted to a scientific approach to agricultural development and ideologically committed to a "mission" of serving the farmer, even as it was sensitively responsive to the political and financial leverage of non-farm capital.

A substantial proportion of this public research effort was channeled to plant improvement. Plant breeding began to move from a craft foundation to a truly scientific basis with the rediscovery of Mendel's work in 1900. Simple mass selection of chance crosses was replaced with the purposeful recombination of varieties for the transfer of specific traits. As breeding methodologies became more powerful, the flow of improved cultivars from public breeding plots quickened considerably. In order to distribute efficiently the seed of the new "college-bred" varieties to farmers, the SAES/LGUs encouraged the establishment of crop improvement associations and seed certification programs. These were established in most states between 1900 and 1930.

Because public varieties were not released unless they were clearly superior to existing cultivars, they established a standard against which privately developed varieties could be measured. Nonexclusive release policies ensured that seed of new varieties was available to all interested parties, and the mechanism of the crop improvement association functioned to provide individual farmer-producers with seed stock for multiplication. These policies tended to keep the structure of the seed market unconcentrated. Certification also acted to moderate prices, because farmers associated the "blue tag" of certified seed with a consistent level of quality and therefore purchased from lowest-priced sellers. Finally, the Federal Seed Act mandated that new varieties had to be merchandised under their original names, thus reducing the scope for product differentiation and reducing advertising incentives.

This constellation of administrative structures and policies associated with public breeding programs constituted an institutional barrier to the systematic penetration of plant breeding and seed marketing by entrepreneurial capital. In the first place, the development of "finished"[4] varieties by public agencies meant that the products of public research competed directly with those of private breeders. Whereas in other sectors of the economy the state may act indirectly to shape the nature of a product via regulation, in plant breeding it has done so directly by actually creating the product (the new plant variety) itself. Public breeders discipline the market as to quality, price, and structure, and so have significantly limited the ability of private seedsmen to accumulate capital.

So, even when capital found that hybrid corn provided it with an entry point, it immediately encountered the activities of the state itself as a fresh obstacle. The state, which in capitalist society is charged with providing the conditions for profitable accumulation, had become a barrier to such accumulation. Capital undertook the resolution of this contradiction by fostering the development of a particular division of labor between the public and private sectors. The "proper" role of public research is held to be the support of "basic" investigations while, private enterprise pursues "applied" problems. The terms "basic" and "applied" are misleading. The pivotal question has nothing to do with a particular type of science, but with proximity to and degree of control over the seed as a commodity-form. The parameters of this division of labor must continually be redefined as technological advance occurs and as private enterprise grows stronger.

The history of plant breeding since 1935 is a reversal of the previous pattern of institutional development. It is a chronicle of the loss of public leadership and the ascendancy of private industry. There has been continuous friction between the public sector and an expansive private sector. With a good deal of success, the seed industry has constantly attempted to direct state-subsidized plant breeding research toward activities that are complementary to rather than competitive with its own efforts. Public agencies have not been wholly acquiescent regarding the circumscription and redefinition of their historical functions. Though the resulting conflict is better characterized as endemic border skirmishing than as a war, any analysis that fails to take this struggle into account misses a crucial dimension of social and political dynamics.

World economy – germplasm transfer

While this book focuses its analysis on the United States, a full understanding of the development of American plant science and the rise of the seed industry must incorporate a global perspective. The export of primary and raw materials is widely considered to be a defining feature of the position of much of the Third World in the international division of labor (Lenin 1939; Emmanuel 1972; Castells 1986). Germplasm, the genetic information encoded in the seed, is the raw material used by the plant breeder. The development of agriculture in the advanced capitalist nations has involved the systematic acquisition of this raw material from the "gene-rich" periphery. And agricultural productivity in the capitalist core remains fundamentally dependent on constant infusions of plant materials from the Third World.

It happens that natural history rendered those areas of the world that now contain the advanced capitalist nations "gene-poor." On the other hand, nearly every crop of significant economic importance – and indeed, agriculture itself – originated in what is now called the Third World. American crop plants are as immigrant in character as its population, and Europe's crops are only slightly less so. The development of modern agriculture in these regions has necessarily been accompanied by the continuous appropriation of plant genetic resources from source areas of genetic diversity that lie principally in the Third World. This primitive accumulation of plant germplasm for processing in the scientific institutions of the developed world is one of the enduring features of the historical relationship between the capitalist core and its global periphery. The evolution of access to, utilization of, and control over plant genetic resources is a matter of fundamental importance.

Of particular interest are the institutional structures that have been created to facilitate and control the movement of plant genetic materials. Global collection of plant germplasm was initiated by the U.S. Patent Office as early as 1839, and it is in this beginning that the complex edifice that is now public agricultural science finds its origins. Similar motivations can be found at the root of the international Green Revolution. The creation of the Green Revolution research centers (e.g., the International Rice Research Institute, the International Center for the Improvement of Maize and Wheat) was the product not only of an effort to introduce capitalism into the countryside but also of the need to collect systematically the exotic germplasm required by the breeding programs of the developed nations. Western science not only made the seed the catalyst for the dissolution and transformation of pre-capitalist agrarian social formations, it also staffed an institutional network that has served as a conduit for the extraction of plant germplasm from the Third World.

The flow of plant germplasm between the gene-poor and the gene-rich has been fundamentally asymmetric. This asymmetry is expressed on at least two dimensions. First, in purely quantitative terms, the core has received much more material than it has provided to the periphery. Second, in qualitative terms the germplasm has very different *social* characters depending on the direction in which it is moving. The germplasm resources of the Third World have historically been considered a free good – the "common heritage of mankind" (Myers 1983; Wilkes 1983). Germplasm ultimately contributing billions of dollars to the economies of the core nations has been appropriated at little cost from – and with no direct remuneration to – the periphery. On the other hand, as the seed industry of the advanced industrial nations has matured, it has reached out for global markets. Plant varieties incorporating genetic material originally obtained from the Third World now appear there not as free goods but as *commodities*.

This book traces the development of technique in plant breeding, but the process of scientific advance is not seen to be the simple unfolding of immanent technical necessity. While there are always technical parameters as to what is physically possible, what gets done in the laboratory and what emerges from the test plot is determined by the interplay of technical possibility with political economy, institutional structure, and the exigencies of the world economy.

First the Seed is an attempt to tease out the cross-cutting warp and woof of these interacting moments in the complex weave of historical process. And if there is indeed a dynamic continuity between the past and the future, it should be possible to use an understanding of these historical trajectories to analyze current conjunctures in the development and deployment of the *new* plant biotechnologies.

New biology, new seedsmen

Plant improvement is a particularly appropriate avenue along which to explore the political economy of the emerging biotechnologies. As the president of Agrigenetics – a company formed expressly to apply the new genetic technologies to commercial crop improvement – has observed, "The seedsman, after all, is simply selling DNA. He is annually providing farmers with small packages of genetic information" (Padwa 1983:10). There is no question that the development of biotechnology has made selling packets of DNA to farmers a most enticing prospect even for firms not historically associated with the seed industry.

Since 1970, an astonishing wave of mergers and acquisitions has swept virtually every American seed company of any size or significance into the corporate folds of the world's industrial elite. Many of these acquisitions have been made by transnational petrochemical and pharmaceutical firms with substantial agrichemical interests and strong commitments to the commercialization of biotechnology in a variety of sectors. The seedsmen of today are the Monsantos, Pfizers, Upjohns, Ciba-Geigys, Shells, and ARCOs of the world. In addition, the last decade has seen the founding of over one hundred genetic engineering firms sporting such evocative names as Agrigenetics, Advanced Genetic Sciences, DNA Plant Technology Corp., Hybritech, Molecular Genetics, and Repligen. Born of the passionate marriage of academia and venture capital, these companies are devoted to the commodification of the research process itself. Both transnationals and the "genetic research boutiques" are gearing up to enter a market for seed that is projected to be worth some $7 billion in the United States alone by the year 2000 (*Business Week* 1984a:86).

The appeal of biotechnology for commercial plant breeding and the seed industry is twofold. At one level, the new genetic technologies promise heretofore unattainable improvements in the agronomic characteristics of crop varieties and thereby will improve the competitive position of individual firms. At a deeper level, biotechnology offers the prospect of the further elaboration of the twin vectors of commodification along which capital has historically penetrated plant breeding. Biotechnology may well allow hybridization of those crops that have not yet succumbed to conventional breeding methods for the production of hybrids. Moreover, as a result of a September 1985 decision of the Board of Patent Appeals and Interferences, plants are now considered patentable subject matter. Biotechnology pushes the commodification of the seed forward along both technical and juridical paths. It thus offers opportunities for private sector profits in an enlarged market.

But full realization of this promise is yet constrained by the refractory presence of the state in agricultural research. The current economic crisis and the high stakes and sense of urgency associated with biotechnology have compelled private interests to push unambiguously for a major restructuring of public agricultural research efforts (Kenney and Kloppenburg 1983). This overt, even heavy-handed, pressure has brought the contradictions of public plant breeding closer to the surface than they have ever been. There is a significant measure of resistance, among both state managers and bench scientists, to the wholesale reorientation of publicly funded agricultural research along lines dictated by commercial interests.

The contradictory position of the LGU complex is being highlighted in a period of established popular concern with issues relating to agriculture and, now, biotechnology. The effect of agricultural technology on the environment has been a topic of widespread debate ever since the publication of Rachel Carson's path-breaking *Silent Spring* (1962). Over the last decade, a series of critics (e.g., Hightower 1973; Perelman 1977; Berry 1977; Vogeler 1981) have torn away the veil that concealed the coupling of public research and agribusiness. This apparent cuckolding of the farmer called the very legitimacy of the LGUs into question and forced the USDA to confront the "structure issue" (USDA 1981).

The activities of the United Farm Workers and other labor groups have focused attention on the social impacts of mechanization. California Rural Legal Assistance has brought suit against the regents of the University of California, charging that the university's agricultural research has failed in its legislative mandate to respond to the needs of all rural residents and has instead come to serve the interests of an agricultural and industrial elite. Even the seed industry came under fire in 1980 as efforts to extend the Plant Variety Protection Act met with opposition from environmental and consumer organizations that feared the impact of increasing oligopoly in the "genetic supply" industry.

Nor have the new biotechnologies been immune from criticism. The actual and projected deployments of the first products of genetic engineering by capital have generated a variety of oppositional pressures from groups concerned about the possible socioeconomic and environmental impacts of the deliberate release of modern biological chimeras. Such opposition has an international component as well. As the new forces of production focus attention on the value of genetic resources, Third World nations have come to recognize the asymmetries in current patterns of plant germplasm exchange. Through the medium of the Food and Agriculture Organization of the United Nations (FAO), developing nations are insisting that a "new international genetic order" be a part of the New International Economic Order.

The development and deployment of the new genetic technologies in agriculture in general, and plant breeding in particular, are occurring in a particular historical context of struggle. As Edward Yoxen points out,

> New technologies, processes and products have to be dreamt, argued, battled, willed, cajoled and negotiated into existence. They arise through endless rounds of conjecture, experiment, persuasion, appraisal and promotion. They emerge from chains of activity, in which at many points their form and existence is in jeopardy. There is no unstoppable process that brings inventions to the market. They are realized only as survivors. [Yoxen 1983:27]

Biotechnology is still an embryonic technical form. The public agricultural research agencies are currently in a phase of flux and transition. We have before us both a tremendous opportunity and an enormous challenge. We have the opportunity of taking hold of a powerful new productive force in its formative stages and bending it to the satisfaction of human needs in a rational manner. At the same time, we face the challenge of wresting control over the new genetic technologies from those who would allow profit to be the prime determinant of the manner in which biotechnology is applied. In the agricultural sectors the terrain of this struggle as to who will determine how evolution is outdone will center on the role of the public research complex.

2

Science, Agriculture, and Social Change

> But e.g. if agriculture rests on scientific activities – if it requires machinery, chemical fertilizer acquired through exchange, seeds from distant countries, etc., and if rural, patriarchal manufacture has already vanished – which is already implied in the presupposition – then the machine-making factory, external trade, crafts, etc. appear as *needs* for agriculture... Agriculture no longer finds the natural conditions of its own production within itself, naturally, arisen, spontaneous, and ready to hand, but these exist as an industry separate from it... This pulling-away of the natural ground from the foundations of every industry, and this transfer of the conditions of production outside itself, into a general context – hence the transformation of what was previously superfluous into what is necessary, as a historically created necessity – is the tendency of capital.
>
> Karl Marx, *Grundrisse* (1973)

Before moving on to the historical matter that constitutes the greater part of *First the Seed*, it is useful to treat a number of thematic elements more completely than was possible in Chapter 1. This chapter provides an elaboration of the theoretical framework that informs my interpretation of the historical and contemporary records.

I begin with an examination of Marx's writings on science as he saw it developing within the capitalist mode of production. This is followed by a section on that most basic building block of capitalism, the commodity-form. The next two sections link the development of science and the extension of the commodity-form to agriculture and explore the special characteristics of that sector of production. The problematic articulation of the seed itself to the circuits of capital is then described. There follows a critical evaluation of the true social significance of the distinction between "basic" and "applied" science. Finally, an examination of the nature of "plant genetic geography" provides a framework for understanding the role of global germplasm flows in the world economy.

Science and capitalism

Marx wrote eloquently of the technological dynamism that characterizes capitalism. In the *Communist Manifesto* we find that

> The bourgeoisie cannot exist without constantly revolutionizing the means of production, and thereby the relations of production, and with them the whole relations of society. Conservation of the old modes of production in unaltered form, was, on the contrary, the first condition of existence for all earlier industrial classes. Constant revolutionizing of production, uninterrupted disturbance of all social conditions, everlasting uncertainty and agitation distinguish the bourgeois epoch from all earlier ones. [Tucker 1978:476]

Capitalist industry never views the existing form of a production process as definitive, but moves constantly toward technical transformation.

In Marx's understanding, this tendency is not simply a function of the inherent potential of science and technology. Rather, it stems from the interaction of these forces of production with the social relations of production.[1] In part, the pace of technical innovation is quickened by competition between capitalists and by accumulation and investment. But principally, the bourgeoisie eschews repose and continually solicits what Joseph Schumpeter called "gales of creative destruction" because of what was, in Marx's view, an absolute contradiction between the potential of the forces of production and the social matrix within which they are utilized. Technological innovation is called forth in response to the continuous struggle between competing capitalists, and between capital and other classes, as workers and petty commodity producers strive to gain a larger share of the social product and to maintain what control they have over the shape and duration of the labor process. As Marx notes in the *Grundrisse* (1973:706), "Capital is the moving contradiction . . . it calls to life all the powers of science and of nature, as of social combination and of social intercourse, in order to make the creation of wealth independent (relatively) of the labour time employed on it." But science, as opposed to technology, was not quickly or easily called into the service of a nascent bourgeoisie. The wedding of science to the useful arts of industry came only in the latter half of the nineteenth century (Braverman 1974; Noble 1977), long after the emergence of capitalism. Nathan Rosenberg (1974) suggests that Marx recognized at least two factors constraining the application of science to problems of production.

First, production based on handicraft or manufacture absorbs only the simplest practical advances of science, because innovations must be limited to those that can be encompassed by the limited physical capacity of the individual worker who carries out each particular process with manual implements. In capitalist manufacture, the worker has been appropriated by the process, but the process must still be adapted to the worker. The systematic incorporation of science into the productive process had to be preceded by technological advances that replaced "not some particular tool but the hand itself" (Marx 1977:507). This condition was fulfilled with the development of machinery and large-scale industry[2] and reinforced with the subsequent production of machines by machines. With the principle of machine production established, "the instrument of labor assumes a material mode of existence which necessitates the replacement of human force by natural forces, and the replacement of the rule of thumb by the conscious application of natural science" (Marx 1977:508).

But though the technological advances associated with the growth of machine-based industry increasingly emancipated production from the parameters set by the organic limits of human labor power, thereby opening the way to the "conscious application of natural science," there was no assurance that science was ready to contribute. A second historical obstacle to the union of science and industry is the uneven rate at which various areas of scientific knowledge have developed. In *Dialectics of Nature*, Engels (1940) postulated a hierarchy of disciplines ordered by increasing complexity and, by implication, by the sequence in which natural laws associated with a particular discipline could be usefully appropriated: mechanics to physics to chemistry to biology. Engels' schema fits well with historical reality. In his monumental four-volume *Science in History*, J. D. Bernal (1965:49) records "a definite succession of the order in which regions of experience are brought within the ambit of science. Roughly it runs: mathematics, astronomy, mechanics, chemistry, biology, sociology."

This is not to say that Marx and Engels conceive of the course of scientific advance as independent of the influence of industrial development. Indeed, they emphasize the manner in which the needs of industry focus scientific effort on particular problems and the way in which technology provides the mass of empirical and experiential data on which the sciences have been raised up.[3] The immediate requirements of economy directly stimulate the pursuit of particular avenues of scientific research. But Marx and Engels add that the availability of science is not purely a function of demand, it is also governed by the differential difficulty of comprehending certain features of the natural world.

Marx notes that it was not chemistry, geology, or physiology that had reached a "certain degree of perfection during the eighteenth century, but mechanics" (quoted in Rosenberg 1974:726). And it was mechanics on which the Industrial Revolution of the eighteenth and nineteenth centuries was founded. Science itself initially had little to do with the technical transformations of this period (Landes 1969:61; Noble 1977:6-7). Braverman (1974:157) notes that science

> did not systematically lead the way for industry, but often lagged behind and grew out of the industrial arts. Instead of formulating significantly fresh insights into natural conditions in a way that makes possible new technologies, science in its beginnings under capitalism more often

> formulated its generalizations side by side with, or as a result of, technological development.

Only late in the nineteenth century, standing on the backs of inventive craftsmen, did scientists begin to contribute significantly to the process of commodity production. Though Marx lived only long enough to see the merest beginnings of the methodical introduction of the findings of institutionalized science onto the shop floor, he could see that as this occurred. "Invention then becomes a business, and the application of science to direct production itself becomes a prospect which determines and solicits it" (Marx 1973:704).

It is precisely this characteristic – scientific invention become business – that distinguishes the technical base of contemporary capitalism from that of its previous forms. The contrast between the nineteenth-century Industrial Revolution and the twentieth-century scientific-technical revolution is that between "science as generalized social property incidental to production and science as capitalist property at the very center of production" (Braverman 1974:156). In using production of machines by machines to stand on its own feet technically, capital also created for itself a social foundation, a standing place from which it subsequently used the lever of science to move the world.[4] As the technological possibilities of the Industrial Revolution played themselves out (Landes 1969:237), so, in corresponding measure, did a maturing capitalism turn to science to maintain the momentum of accumulation.

Primitive accumulation and imposition of the commodity-form

An adequate understanding of the development of science under capitalism must take into account the distinctive social and economic characteristics of that mode of production. And if we are to understand how science becomes "capitalist property," as Braverman put it, we must understand the commodity-form and its genesis.

When Marx comes in *Capital* to the dissection of a historically specific mode of production, capitalism, he begins his analysis with the simplest social form of the product of labor characteristic of capitalist society: the commodity. His first lines run:

> The wealth of societies in which the capitalist mode of production prevails appears as an 'immense collection of commodities'; the individual commodity appears as its elementary form. Our investigation therefore begins with the analysis of the commodity. [Marx 1977:125]

The commodity – an article that is produced for exchange rather than use – is not unique to capitalism. What is distinctive about capitalism is that it is characterized by a system of *generalized* commodity production in which *labor power* also figures as something that is bought and sold; in a sense, the production of commodities by commodities.

Capitalism did not spring forth fully formed out of feudalism like Athena from the head of Zeus. Rather, capitalist relations were extended in proportion to the progressive generalization of the commodity form, especially its application to labor power. This, in turn, was not given but was achieved in a process of "primitive accumulation," which Marx (1977:875) defines as "nothing less than the historical process of divorcing the producer from the means of production." This was accomplished in the first instance through the expulsion of peasants from the land and by the dissolution of bands of feudal retainers. With legislation and bloody discipline was created a class of "free and rightless proletarians" who, being doubly free in that they neither were part of the means of production themselves (as slaves would be) nor owned means of production, had "nothing to sell except their wn skins" (Marx 1977:873), their own capacity to work.[5]

The commodity form was thus *imposed* upon the activity of human labor by forcible separation of the worker from the land. From this germinal act follows a series of far-reaching transformations, for, as Marx quotes Shylock, "You take my life when you do take the means whereby I live." Separated from the means of production with which they were accustomed to sustain themselves, the newly proletarianized workers had to sell labor power in order to obtain the necessities of life. These necessities now also generally assumed the commodity-form, because there was a growing class of persons who had to purchase food, clothing, and other basics not as a matter of occasional need but as a constant and ineluctable condition of their lives.[6] A corollary to the creation of the working class was the genesis of an opposed class of capitalists who monopolized the means of production. It is they who purchased the labor power for the production of the commodities that, when realized in the market, constituted the accumulating wealth of the capitalist mode of production.

So, to borrow Daniel Bell's (1973:378) imagery, the commodity is the "monad" that contains not just the "imago" of capitalist society, but its fundamental material, political, and social bases as well. As the simplest form of the product of labor, the commodity is capitalism's elementary material phenomenon. As the object in which labor is accumulated and valorized, it is its elementary form of wealth. As the material incarnation of a particular relation of power between classes, it expresses the basic political dynamic of capitalism. And because, for Marx (1977:932), "capital is not a theory, but a social relation between persons which is mediated through things," we can agree with Cleaver (1979:71-2) when he maintains that the commodity-form, understood as the embodiment of the social relation of the class struggle, "*is the fundamental form of capital*... In fact, *we can define capital as social system based on the imposition of work through the commodity-form.*" Furthermore, "capital's power to impose the commodity-form is the power to maintain the system itself" (Cleaver 1979:73). To this we may add the observation that to extend the imposition of the commodity-form to new areas is to expand the system.

Capitalism grows in two ways: through the processes of accumulation and primitive accumulation. The latter is necessarily the historically prior form, because the initial separation of the worker from the means of production is not in the first instance "the result of the capitalist mode of production but its point of departure" (Marx 1977:873). It is for this reason that Marx writes of primitive accumulation as the "prehistory of capital." But once primitive accumulation has sundered the worker from land or tools and has established labor power as a commodity, "Capitalist production stands on its own feet, it not only maintains this separation, but reproduces it on a constantly extending scale" (Marx 1977:874). Because the capitalist is able to force workers to produce more than is necessary for their reproduction, a surplus is available for reinvestment in the purchase of additional labor power or other means of production. And this in turn sets the stage for another round of self-expansion through the accumulation of additional surplus-value.

The growth of capital in this fashion is limited by the range of primitive accumulation on which it is based. The self-expansion of capital on the basis of surplus-value grows vertically but requires a constantly enlarging labor pool and market if it is to avoid top-heavy stagnation. The establishment of capitalism as the predominant mode of production depended on the extension of its foundation of capitalist commodity relations to new spheres. This was no simple task, for everywhere independent producers resisted the expropriation of their lands, their means of subsistence, and their instruments of production. Early methods of primitive accumulation differed, but all rested on coercion and, Marx (1977:915-6) noted,

> All employ the power of the state, the concentrated and organized force of society, to hasten, as in a hot-house, the process of transformation of the feudal mode of production into the capitalist mode, and to shorten the transition. Force is the midwife of every old society which is pregnant with a new one. It is itself an economic power.

The bloody discipline that characterized primitive accumulation in the European heartland was reenacted globally as the emergent bourgeoisie sought to create a world after its own image.[7]

Even so, the universal generalization of commodity relations has been incompletely achieved. At the time Marx wrote *Capital*, the process of primitive accumulation even in Europe had only been partially accomplished, and he noted the existence of "social layers, which, although they belong to the antiquated mode of production, still continue to exist side by side with [capital] in a state of decay" (Marx 1977:931). The penetration and imposition of capitalist commodity relations is, in time, space, and economic sector, exceedingly uneven. As Mandel (1978:46) observes, "Primitive accumulation of capital and capital accumulation through the production of surplus value are, in other words, not merely *successive* phases of economic history but also *concurrent* economic processes." This is particularly obvious in the Third World, where the most advanced forms of capitalist relations coexist with social forms characteristic of pre-capitalist modes of production. Analysis of the structural articulation of the different moments of accumulation in Third World nations has received much attention.[8]

The extent to which primitive accumulation has been and still is a phenomenon of advanced industrial societies is less well recognized. Perhaps this has something to do with Marx's formulation of primitive accumulation as the "pre-history" of capital, or with the very ubiquity of the commodity-form in countries such as the United States and Japan. But Marx meant "pre-history" in the sense that primitive accumulation is a necessary precursor to accumulation of capital through the extraction of surplus-value. And the apparent ubiquity of commodities should not blind us to the fact that capital constantly seeks to force *all* use-values to submit to the commodity-form and to convert simple commodity production to capitalist commodity production wherever and whenever it can. Indeed, primitive accumulation may be a *permanent* process, because capital systematically seeks not only to make a commodity of all use-values but also to create *new* needs whose satisfaction entails new use-values that in turn can be commodified.[9]

We have seen that, for Marx, primitive accumulation is the historical process of separating the independent producer from the land, tools, materials, and other inputs that constitute the objective means and conditions of production. This could, and frequently did, mean total and immediate expropriation. The draconian approach had the advantage of instantaneously establishing both labor power and means of subsistence as commodities, creating labor pool and market in one fundamental transformation. The application of direct extraeconomic force was initially necessary for capitalism to set down roots. But as it matured, mechanisms working through the "silent compulsion of economic relations" (Marx 1977:899) proved that they could bring the independent producer gradually but effectively into capitalist commodity production.

Marx states that where capitalism establishes itself alongside petty commodity production, it has a destroying and dissolving effect independent of extraeconomic compulsion. It does this subtly but effectively through the vehicle of the commodity-form. Capital

> makes the sale of the product the main interest, at first without apparently attacking the mode of production itself... Once it has taken root, however, it destroys all forms of commodity production that are based either on the producer's own labour, or simply on the sale of the excess product as a commodity. [Marx 1981:120]

Where the immediate and complete expropriation of the independent producer is constrained, capital seeks to establish the hegemony of exchange-value as opposed to use-value by binding the autonomous producers inextricably to the commodity-form, to bring them ultimately under capitalist relations of production.

The dynamic involved is illustrated by Marx in that section of the *Grundrisse* that treats primitive accumulation. He describes the gradual and piecemeal subjugation of independent weavers to capital. They begin by selling cloth to a merchant, and as the market expands, they restrict other activities and come to purchase raw materials instead of producing them themselves. Ultimately we find that whereas the capitalist

> bought their labor originally only by buying their product; as soon as they restrict themselves to the production of this exchange value and thus must directly produce *exchange values*, must exchange their labour entirely for money in order to survive, then they come under his command, and at the end even the illusion that they *sold* him products disappears. He buys their labor and takes their property first in the form of the product, and soon after that the instrument as well, or he leaves it to them as *sham property* in order to reduce his own production costs. [Marx 1973:510]

While Marx here analyzes one facet of the transformation of commodity production, he neglects to explain why independent producers should reduce the scope of their productive activities.

The answer is that capital views the independent producer as a potential market as well as a potential laborer. Capital must sell commodities to realize the value of labor. Because capitalist production represents a social concentration of both labor power and means of production, the products of capitalist enterprise tend to sell more cheaply than their equivalents produced under non-capitalist modes. The independent producer will be disposed to replace self-supplied means of production with purchased inputs insofar as they lower the costs of producing a commodity for sale. This tendency gains strength to the extent that the petty commodity producer competes with other producers in a similar situation. These two moments are mutually reinforcing, and insofar as the independent producer loses the ability to reproduce autonomously the means of production, it is a route that once pursued cannot be retraced. The petty commodity producer is bound ever more firmly and more completely to capital.

Moreover, as the dynamic described earlier progresses, class *differentiation* occurrs as individual producers find differential success in accumulation and become increasingly heterogeneous. Marx saw petty commodity producers as a transitional class that would, under the solvent effect of expanding capitalist commodity relations, decompose into bourgeoisie and proletariat.

Agriculture and social change

The goodness of fit of the agricultural sector with the classic Marxian analytical framework has long been a matter of considerable debate. By 1890 the rapid growth and centralizing tendency of industrial capital in the United States was so unambiguously predatory that the Sherman Anti-Trust Act was passed to restrain monopolistic consolidation and anti-competitive pricing by powerful cartels. The situation in the agricultural sector was quite different. In that year the number of owner-operated farms was at an all-time high and would not reach its apogee until 1935. Kautsky's classic *The Agrarian Question*, written near the turn of the century, was produced in part as a rejoinder to those who saw in the persistence of agrarian petty commodity production in Europe a refutation of the Marxian theory of capitalist development (Banaji 1980:39).

Similarly, the apparent vitality of the American "family farm" has been touted as evidence that Marxist interpretations of the agricultural sector are flawed (Soth 1957). The number of farms may have been decreasing, but they were not for the most part being replaced by capitalist farms based on the principle of wage labor. Scholars of the political economy of agriculture using Marxist concepts in their analyses were faced with an important anomaly: the persistence and coexistence of rural petty commodity production alongside a dominant capitalist mode of production. As Mann and Dickinson (1978:467) put it, "Capitalist development appears to stop, as it were, at the farm gate."

This issue has engendered in recent years a substantial body of literature that revolves around three central questions:

1. Is independent commodity production in agriculture[10] decomposing into a classically capitalist mode of production characterized by a distinct bourgeoisie and proletariat?
2. If not, what are the obstacles preventing the development of a fully capitalist agriculture?

3. Given the existence of obstacles, are they truly constraining features, or has capital nevertheless found ways to extract surplus from superficially "independent" petty commodity producers?

Classically oriented Marxist scholars do not admit of the theoretical possibility that petty commodity production could fail to decompose into the two opposed classes of capitalist society. Kautsky asserted that while agriculture might follow a complicated and contradictory course of development distinct from that of industry, the general tendency for proletarianization of the bulk of the peasantry was firm (Banaji 1980). Lenin (1967) held similar views, and he is echoed by de Janvry, the most prominent of contemporary decompositionists, who comments of peasants and family farmers: "However lengthy and painful the process may be, their future is full incorporation into one or the other of the two essential classes of capitalism" (de Janvry 1980:159).

This position has been criticized as unacceptably teleological. There has been a series of analysts who have argued that the historical persistence of agrarian petty commodity production is evidence of relative structural stability in the articulation of capitalism with the simple commodity mode of production. Mann and Dickinson (1978) argue that certain inherent characteristics of agricultural production effectively exclude capital from that sector. Others contend that petty commodity production is fully integrated with the dominant capitalist mode of production (e.g., Amin and Vergopoulos 1974; Friedmann 1980; Lewontin 1982) or is suspended and transfixed by the balancing vectors of "contradictory combinations of contradictory class locations" (P. H. Mooney 1983:576). What these approaches share is an understanding of petty commodity production not as a transitional form but as a potentially permanent element of advanced capitalism.

Arguments for the persistence of petty commodity production are frequently linked to the existence of what, following Mann and Dickinson's (1978) convenient formulation, I shall refer to as "obstacles" to the penetration of agriculture by capital; that is, features that limit or even preclude the generalization of production based on wage labor. Many of these obstacles are related to the unique conditions of farming as a production process. The role of land is particularly important. In the first place, its availability is fixed, so that amassing contiguous acreages for large-scale production can be accomplished only through cannibalization of smaller ownership units, a process that can be lengthy and difficult. Moreover, outright purchase of farm land is an expensive proposition that effectively freezes the mobility of large amounts of capital and ties it to a highly uncertain market. Absolute rent also becomes a constraint (Massey and Catalano 1978).

Mann and Dickinson (1978) focus their attention on certain natural factors that appear as barriers to extension of the specific social relations of production associated with capitalism. They note the disjuncture of production time and labor time in much farming, arguing that the excess of the former over the latter adversely affects the rate of profit and turnover rates, because value is created only as labor is applied to the production of commodities. Also, the seasonal and sequential nature of agricultural production presents difficulties with regard to securing and maintaining a labor force, a requirement that is particularly crucial at harvest. The physically extensive nature of farming makes labor especially difficult to manage and control effectively. The perishability of many agricultural commodities further underlines growers' vulnerability to labor unrest. And, of course, farming is a tremendously risky business, subject as it is to the unpredictable vagaries of the environment.

To this set of natural obstacles may be appended certain characteristics of the "family-labor farm" itself: a powerful subjective commitment to farming as an occupation, a consequent willingness to engage in self-exploitation (Lianos and Paris 1972), and the willingness to accept returns below the average rate of profit. Availability of off-farm work and the willingness of the state to (in some measure) underwrite the support of the independent commodity producer for purposes of legitimation have also had an impact.

These are the obstacles that may explain the apparent failure of capital to penetrate agriculture.[11] There is no agreement as to which of these proposed barriers are the most effective or exhibit the most explanatory power.[12] But the variety of explanations offered is persuasive testimony – especially with the parallel and corroborative statements of business analysts (e.g., Cordtz 1972) – to the existence of features that at least *slow* the penetration of agriculture by capital, or shape its character.

However, on empirical grounds it is difficult to substantiate the position, asserted or implied by many of those who identify obstacles to capitalist penetration, that simple commodity production in agriculture has in fact found stability in its articulation with the dominant capitalist mode of production in which it is embedded. Historical changes in farm structure reflect a progressive differentiation among producers that is currently resolving itself into a distinctive pattern of dualism (Buttel 1983). A small segment of extremely large operations (4.5 percent of all farms) relying heavily on wage labor now accounts for 47.5 percent of the value of U.S. farm production. At the other pole is an increasing number of small operations that, comprising some 71.9 percent of all farms, account for only 13.2 percent of sales but some 79 percent of off-farm income (USDA 1981:43; United States General Accounting Office 1985a). Sandwiched between the narrow capitalist apex and a broad proletarianizing base is a disappearing middle of family-labor farms that continue to approximate the circumstances of independent petty commodity production. If there are obstacles to the development of capitalist relations in agriculture, they would appear merely to slow the process of decomposition, not to preclude it entirely. Flinn and Buttel (1980) are thus led to propose a dialectic of the family farm to describe the interaction of the tendencies for transformation and the braking effect of the various obstacles.

There is, however, a question as to whether or not the alleged obstacles to the penetration of capital into agriculture have had as much of an insulating effect as has been ascribed to them. They may have slowed the formal decomposition of simple commodity production, but they may not have precluded the imposition by capital of various mechanisms for the extraction of surplus from nominally independent producers. Davis (1980) interprets such phenomena as contract farming, indebtedness, and integration into monopoly-controlled factor and product markets as capitalist relations of exploitation insofar as they are based on coercion, contract, and control. The farmer in such circumstances is little more than a "propertied laborer" despite ownership of the means of production. The family farm, far from being an institutional impediment to capital, becomes a basis for its further development. Working along similar lines, P. H. Mooney (1983) formulated a theoretical framework for analysis of the detours capital takes in establishing such relations. He complains that "much effort has been expended discovering 'obstacles' that explain capitalism's supposed inability to penetrate an agriculture that this model would suggest it has already penetrated" (P. H. Mooney 1983:578).

I am in substantial agreement with this position. But I believe that acceptance that capital has penetrated agriculture in ways that do not entail the wholesale destruction of simple commodity production is compatible with both the existence of a tendency to decomposition and the existence of barriers that slow that trend. That capital is able to find methods for the extraction of value from producers other than the wage labor relation is itself a kind of barrier to the establishment of capitalism in agriculture in its purest form. All these processes are interacting moments of capitalist development. But there is a fourth facet to this dynamic that is often touched upon but rarely developed. In an effort to contribute to the resolution of this debate by broadening its scope, I offer the following analysis of the technical transformation of the agricultural sector and the role of agricultural research in that process.

Primitive accumulation and agricultural research

Kautsky exhorts us to look for "*all* the changes" (Banaji 1980:40, emphasis added) that agriculture experiences as capitalism develops. In seeking to

Table 2.1. *Changes in percentage composition of agricultural inputs, United States, 1870–1976*

Year	Labor (%)	Real estate (%)	Capital (%)
1870	65	18	17
1900	57	19	24
1920	50	18	32
1940	41	18	41
1960	27	19	54
1970	19	23	58
1976	16	22	62

Source: Cochrane (1979:Table 10.2).

understand how and why capital has or has not penetrated agriculture, we have not sufficiently followed his advice. The problem has been a preoccupation with the *farm*-level production process and a tendency to conceive of "agriculture" as synonymous with "farming." The debate as to whether or not capital has penetrated agriculture has really been couched in terms of whether or not it has penetrated farming. If we cannot assume that capital stops at the farm gate, neither can we assume that agriculture does not extend beyond the farm gate. This is a point of great importance, for the most significant change in agricultural production experienced under the impact of capitalist development is the displacement of production activities *off-farm* and into circumstances in which fully developed capitalist relations of production can be imposed. The industrialization of agriculture (Danbom 1979) is a well-recognized phenomenon. But it means much more than the use of machinery and chemicals in farming. It also means the production of these inputs in an industrial setting.

Table 2.1 shows the changes that have occurred in the mix of inputs used in farming in the United States since 1870. The relative importance of land has remained more or less constant. But over the past hundred years the contributions of labor and capital have been almost completely reversed. Capital, largely in the form of new technologies, has displaced labor as the chief component of production in farming. Indeed, farming has become the most capital-intensive sector of the modern capitalist economy, and labor productivity there has outstripped that of industrial production.

Table 2.2 provides more detailed insight into this trend. Between 1935 and 1977 the total volume of productive resources used in farming changed little. But as the role of labor in the mix of factors of production declined by a factor of four, capital inputs in the form of machinery more than tripled, and capital use in the form of agrichemicals increased fifteenfold. Machinery and agrichemicals are not produced on the farm; they are purchased inputs. The reversal in the relative positions of labor and capital is therefore paralleled by a similarly dramatic transposition of the importance of purchased and non-purchased inputs. As Richard Lewontin (1982:13) has succinctly phrased it, "Farming has changed from a productive process that originated most of its own inputs and converted them into outputs, to a process that passes materials and energy through from an external supplier to an external buyer."

A corollary to the shifting mix of purchased and non-purchased inputs has been the historical rise of agribusiness. Farmers no longer produce their own seed corn; they buy it from Pioneer Hi-Bred or Northrup King. They no longer use mules, oxen, or horses for their motive power. None of these creatures can compete with the well-known mechanical ungulate now found on every farm; after all, "Nothing runs like a Deere," or a Ford, or an International Harvester. And those tractors and combines run not on home-produced hay but on petroleum products from Mobil and ARCO. Fields are spread not with manure from the farm's livestock but with ammonium nitrate from W. R. Grace or superphosphate from Occidental Petroleum. And these inputs are paid for with money that is itself a purchased input obtained from Bank of America or Continental Illinois. Produce does not go direct to the consumer after processing on the farm, but to Heinz, or General Foods, or Cargill, or Land O'Lakes.[13] Currently, on-farm production accounts for only 13 percent of the total value of finished agricultural products. Thirty-two percent of the value added derives from commercial inputs, and 55 percent is added in the post-farm stages of processing, transportation, and distribution (Manchester 1985:11). Farming is only one part of the agricultural sector, and it is not even the part that is most productive of value.

The rise of agribusiness has by no means gone unnoticed; quite the contrary. Goss et al. (1980:97) explicitly recognize that the "input and product market stages have bid traditional activities away from the farm enterprise" and note the need to view farming as but one of the components of agricultural production. But even when this is understood, attention tends to focus on the extension of relations of control from agribusiness to the petty commodity producer. This is crucial, but incomplete, because it focuses on the problematic concept of exploitation in the sphere of circulation (Friedmann 1980). An important addition is made by Friedland et al. (1981:15), who have proposed the concept of "differentiation" to describe the displacement of elements of production off the farm. But Friedland et al. consider that, having been moved into city and factory settings, these dis-

Table 2.2. *Farm input indexes, United States, 1935–1977 (1967 = 100)*

Year	Total inputs	Nonpurchased inputs	Purchased inputs	Farm labor	Machinery	Agrichemicals	Farm productivity
1935	91	158	46	299	32	8	57
1940	100	159	58	293	42	13	60
1945	103	161	62	271	58	20	68
1950	104	150	70	217	84	29	71
1955	105	143	76	185	97	39	78
1960	101	119	86	145	97	49	90
1965	98	103	93	110	94	75	100
1970	100	97	102	89	100	115	102
1975	100	92	107	76	113	127	115
1977	103	88	118	71	116	151	118

Sources: USDA (1978), Cochrane (1979).

placed pieces of the production process are no longer appropriate parts of a sociology of agriculture.

To the contrary, understanding this process of differentiation is a crucial key to the construction of a political economy of agriculture. The differentiation of functions off-farm provides capital with a way around obstacles to its penetration that does not necessitate exploitation of the farm producer in the sphere of circulation. John Deere produces tractors on the basis of wage labor on the assembly line. The relations of production are purely capitalist in nature, and the extraction of surplus value is unambiguous and uncomplicated. The labor of the farmer, from which surplus value can be extracted only indirectly through unequal trade relations in the sphere of exchange, is replaced with the directly exploitable labor of the roughneck on Exxon's drilling rig, or the labor of the fermenter technician at Monsanto. The fact that certain productive activities have been moved off-farm while others have not is persuasive evidence of the reality of particular barriers to production.

Moreover, the productive activities that are taken off-farm are not just any activities; they are those that reproduce the farmer's means of production. To the extent that provision of seed, motive power, etc., is undertaken by capital and not the farmer, the autonomy of the petty commodity producer is eroded. The means of production come to confront the farmer as commodities – they can be purchased but they cannot be autonomously reproduced. By binding the farmer firmly to off-farm capital, this process of stripping functions away from the farm not only allows for the extraction of surplus value in industrial settings but also sets the preconditions for the sort of indirect exploitation of the farmer described by Davis (1980) and P. H. Mooney (1983).

Recall now Marx's description of the subjugation of independent weavers given in the preceding section. Just as the weavers were restricted little by little to a limited range of activity and finally so dominated by capital that they could even be left ownership of the means of production as "sham property," so is the farmer rendered a "propertied laborer" through the replacement of non-purchased inputs by purchased ones and by increasing integration with factor markets. And just as the subordination of the weavers was, according to Marx, one moment of primitive accumulation, so is the subjugation of the farmer also a manifestation of that process. What is primitive accumulation, after all, but "the transformation of [the] means of production into capital" (Marx 1977:932). That the emasculation of the independent producer appears to proceed voluntarily makes no difference. The imperatives of the market are just as effective, if more subtle (and therefore more legitimate), as measures of primitive accumulation involving bloody discipline.

But whence came the commodities that are substituted for the farmer's self-sufficing provision of the means of production? They are the products of scientific and technological research performed in both public institutions and private laboratories. New knowledge produced by agricultural science has increasingly reached the farmer not as a public good supplied by the state but in the form of commodities supplied by private enterprises. Agricultural research has greatly facilitated the "differentiation" of activities off-farm and into industrial settings and can therefore be understood as an essential component of the contemporary dynamic of primitive accumulation in the agricultural sector.

The adoption by the farmer of new technologies that are made available as commodities is enforced by the operation of the "technological treadmill" (Cochrane 1979). In the United States, farming is an archetypally atomistic and competitive sector. No individual producer can influence selling price. The profitability of any operation is largely a function of unit costs of production. New technologies offer a means of reducing these costs. Early adopters of new technologies enjoy windfall innovators' rents, but these disappear as adoption spreads and the cost curves for all operations converge. Because the adoption of new technologies results in increased production, there is a tendency for prices to fall. This merely sets the stage for another round of innovation. Those who fail or are unable to adopt the new technologies suffer economic loss. Marginal producers are continually forced out of business, and their operations are absorbed by more successful operators. The treadmill fosters cannibalistic centralization in farming while simultaneously ensuring a secure and expanding market for the purveyors of new technologies.

The benefits of new technologies deployed in American agriculture have accrued principally to agribusiness and to the small group of farm operators in the technological vanguard. For the vast majority of farmers, admits economist Willard Cochrane (1979:352), "the agricultural development process based on rapid and widespread technological advance has been a nightmare." Agricultural research and technical innovation have been principal mechanisms by which the conditions for the elimination of many farm operations have been created, even as surviving farmers are bound to and dominated by the upstream suppliers of inputs and the downstream purchasers of an ever increasing agricultural product.

It is important to understand that this is a relatively recent phenomenon. Only in the 1930s did science become an important and transformative productive force in agriculture. Recall Engels' and Bernal's formulation of the historical sequence in which scientific disciplines have matured. Biology has indeed been a late-blooming field (Mayr 1982). Now, the parent discipline of the agricultural sciences (e.g., agronomy, plant breeding, horticulture, dairy science, poultry science, etc.) is biology. Inasmuch as biological knowledge has been relatively underdeveloped, the agricultural sciences did not provide the tools with which capital could overcome the natural barriers that constrained its direct, and even indirect, penetration of farming. Lack of biological understanding also precluded change in areas that were ready to serve as vectors of capitalist penetration, because the characteristics of living organisms materially condition the feasibility of achieving mechanical or chemical solutions to the problems posed by agricultural production. For example, mechanization of harvest depended upon the ability to manipulate plant architecture. And the development of pesticides and herbicides had to await the elucidation of hormonal processes and other pathways of biological action in insects and weeds. Lack of biological knowledge was itself a barrier to capital.

Marx (1973:511) himself noted the great irony that although capitalist development "in the countryside is the last to push on towards its ultimate consequences and its purest form, its beginnings there are among the earliest." That is, although primitive accumulation of the means of production in land by expropriation of the peasantry constituted the initial moment of capitalist development, agricultural petty commodity production has proved the most persistent form of pre-capitalist relations of production. Agriculture is clearly regarded as a special sector that poses a unique set of barriers to the extension of capitalism. But Marx did not regard these barriers as permanent, and he predicted that the development of the forces of production would ultimately provide the erosive power to eliminate them:

> Agriculture is claimed for capital and becomes industrial only retroactively. Requires a high development of competition on one side, on the other a great development of chemistry, mechanics, etc., i.e., of manufacturing industry. [Marx 1973:511]

That is, the development of fully capitalist relations of production in farming requires both science and the technological treadmill. In sum:

1. There is a wide variety of obstacles to the penetration of agriculture by capital.
2. The process of decomposition of petty commodity producers into the two fundamentally opposed classes of capitalist society is thereby slowed but not precluded.
3. In addition to the slow extension of the wage relation in farming proper, capital historically penetrates agriculture along two other interlinked paths:
   a. the extraction of value from the farm-level producer in the sphere of circulation through contract farming, terms of trade, etc., and
   b. the differentiation of activities (especially the reproduction of the

means of production) off-farm and the direct exploitation of wage labor in the production of agricultural commodities (e.g., inputs, processed food).

4. The differentiation of activities off-farm facilitates the extraction of value in the sphere of circulation.
5. The development of the forces of production, of agricultural science especially, has provided the technical basis for the differentiation of activities off-farm.
6. Agricultural research can therefore be considered an important moment of the contemporary and continuing process of primitive accumulation in agriculture.

Seeds and the circuits of capital

The dynamics outlined in the foregoing sections can be illustrated nicely with the example of plant breeding and seed production. For the independent petty commodity producer, the seed was the alpha and the omega of agricultural life. As seed, it is the beginning of the crop production process, and as grain it is its endpoint. But because the seed is a living and reproducible package of DNA, the endpoint of one cycle of production merely sets the stage for the next. Seed is grain is seed is grain; the option to produce or to consume is there in each seed. And even in consumption there is the element of reproduction. In growing food crops, the farmer may provide for the farm family's means of subsistence. Growing a forage or silage crop provides the means for reproducing livestock (and therefore motive power, food, and fertilizer). Growing a fallow or nitrogen-fixing crop provides the means for reproducing the fertility of the soil. And, of course, seed in the form of grain is a commodity that can be sold for the cash to purchase items of all sorts. Upon the seed depends ultimately the capacity to reproduce a large part of the farm operation. And in control over the seed is a measure of real independence. The seed is the biological nexus of farm-level production.

As such, it would appear that the seed might be a strategic point of interest in the development of a capitalist agriculture – and so it has been. Yet the seed has only grudgingly and incompletely assumed the commodity-form. Only, as we shall see, under the direct application of science and law has the seed submitted to this imposition. For the seed presents capital with a simple biological obstacle: Given appropriate conditions the seed will reproduce itself manyfold. This simple yet ineluctable biological fact poses significant difficulties for commercial interests that would engage in the development of new plant varieties for profit.

Capital is not a thing, but a set of social relations that, in their totality, must be continually reproduced if capital is to survive and grow in a particular sector. Figure 2.1 shows how money is used by a capitalist to purchase the labor power and means of production (as commodities) to be used in a production process for the creation of a second set of commodities that can be realized in the market for a profit. The new set of money may be set into action once again, and the process moves through another circuit. Should either exchange or production relations be broken at any point, the reproduction of capital is short-circuited. How this can happen in the case of seed production is formally specified in Figure 2.1.

Assume that the process outlined in Figure 2.1 represents seed production. Using money capital that they have amassed, the owners of a seed company set in motion labor power (plant breeders, laboratory technicians, field and warehouse workers, etc.) and means of production (land, equipment, laboratory apparatus, warehouse facilities, etc.) in order to develop a new crop variety. In order to realize the value of the labor employed in the production process, the seed must be sold. If the new variety is economically superior to existing cultivars (or if advertising can convince potential customers that it is), then farmers will purchase the seed. The seed company's commodity (C′) becomes part of the farmer's means of production as a result of this transaction. The seed company realizes a profit of M′ – M and uses its augmented capital to begin another round of seed production.

The farmer uses the purchased seed of the new variety to grow grain for sale. But grain, in its alter ego, is also seed. The farmer can save some of the grain produced from the new variety as "bin-run" seed (so called because it comes not from a seed company but from the bin of the farmer's harvester) to be used as planting stock in the next year. Should the farmer decide to do this, the exchange relation anticipated by the seed company between its C‴ and the farmer's MP′ is not consummated. Business success depends on recurring sales of seed. But a farmer need enter the market only once to obtain seed of a new cultivar and can then supply his own needs (and those of his neighbors) indefinitely. Moreover, competing seed firms can simply multiply and sell seed of a popular variety originated by another company. The reproducibility of the seed furnishes conditions in which the reproduction of capital is highly problematic.

The growth of capitalism necessarily entails the destruction of modes of production based on the personal labor of independent producers. The most elementary moment of this dialectical process is primitive accumulation, which involves the transformation of means of production into capital. That there now exists a vital and expansive capitalist seed industry in the United States is undeniably true. How the farmer has been separated from repro-

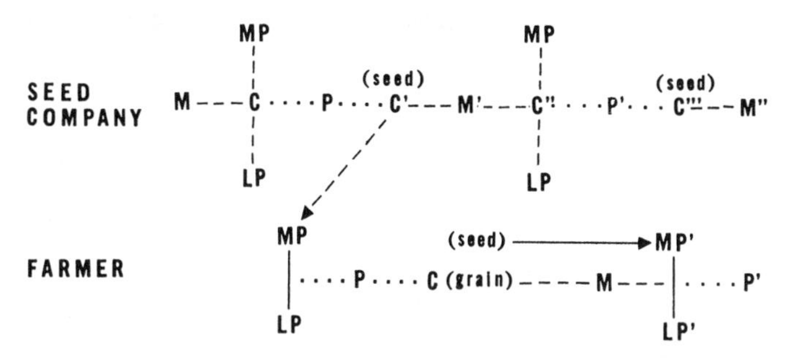

Figure 2.1. Seeds and the circuits of capital. Key: M = investable funds; C = commodities; MP = means of production; LP = labor power; P = production process; --- = exchange relations; ··· = production relations.

duction of the seed, which is perhaps the element of agricultural means of production most central to the entire farm production process, is the central question of this book.

Basic research, applied research, and the commodity-form

As we shall see, capital has historically penetrated plant breeding along two principal vectors, one technological and the other social. In both cases this penetration has been predicated on the active support of the state. That the state should seek to provide the conditions for profitable capital accumulation is not surprising. Most critical analyses of agricultural research share a perception of the land-grant complex as the handmaiden of industry (e.g., Hightower 1973; Perelman 1977; Lewontin 1982). And in fact, this history of plant breeding will show just how extensively public research agendas reflect the character of the capitalist mode of production in which they are embedded.

But to view the public research institutions as mere automatons blindly advancing the interests of capital is both to misread history and to fail to assess the contemporary political possibilities for enhancing the degree of public control over important and productive organs of the state apparatus. Capital has in fact had to struggle to move public research efforts in desired directions. The pivot of this struggle has been the question of the appropriate division of labor between public and private science. This debate has, in turn, historically turned upon a distinction made between "applied" and "basic" science. This section critically examines the empirical and conceptual content of these terms.

Ruth Hubbard (1976:iv) has aptly written, "There is no such entity as science. There are only the activities of scientists." Science is to be understood as a *social* process. Moreover, it is to be understood as a *labor* process, for scientific research as a human activity is in simplest terms the application of labor to the production of new knowledge of the natural world. Tables 2.3 and 2.4 display data taken from the National Science Foundation's (NSF) state-of-science report, *Science Indicators, 1985* (National Science Board 1986). These tables reveal what I regard to be fundamental features of the manner in which the social labor process we call science is organized in the United States and to what ends it is directed.

First, note that scientific effort is not conceived as undifferentiated in character but is disaggregated into "basic," "applied," and "development."[14] These categories bear some scrutiny. The question of how to distinguish between basic and applied work has been a continual theme of debate at least since the time of Pasteur. Definitional efforts have focused on formulations that classify scientific research either on the basis of the investigator's motives or on the character of the work itself. Approaches to the specification of the content of "basic" science typically emphasize the scientist's subjective orientation:

> The man working at the "pure science" end of the spectrum, whether in a university or in an industrial laboratory, pursues a problem because it is interesting or because it appears to have a certain relevance to fundamental knowledge [Kidd 1959:368-9]

> By *pure science* or *basic research* is meant a method of investigating nature by the experimental method in an attempt to satisfy the need to know. [Feibleman 1982:144]

> [Basic research] encompasses inquiries that grow out of healthy intellectual curiosity and represent the pursuit of knowledge for its own sake. [Ashford 1983:19]

"Pure" and "fundamental" have become synonyms for "basic." It is assumed that the scientist gravitates toward "pure" research, and the difference between basic and applied work is sought in the degree to which the scientist's institutional milieu permits him or her to realize this natural inclination. This distinction is phrased as a contrast between research that is "committed" or "uncommitted" (Kidd 1959), "free" or "oriented" (Salomon 1973), "unarticulated" or "articulated" (McElroy 1977).

Though attractive to sociologists interested in scientists' ethos, and ideologically useful in maintaining an image of an ethical "Republic of Science," such definitions were and are inadequate as a practical framework for the evaluation of the distribution of research funds in the post-World War II world of "big science." A second approach to classification has been to define scientific work according to its substance (Kidd 1959). This produces conceptions of a distinction between basic and applied research based generally on the notion of utility. Scientific research is "applied" to the extent that it has the capacity to generate immediately useful results. Definitions of basic and applied research incorporate such dichotomous attributes as "general significance/focus," "long-term/short-term," "fundamental/practical," "knowledge/application," and "unforeseeable results/expected results." Thus, for Jantsch (1972:98), "Fundamental research is simply research on the fundamentals." But, as is well documented, solutions to practical problems have often had implications of more general scientific import, and discoveries of fundamental laws of nature have revolutionized production. It is impossible to operationally define a distinction between basic and applied research because "findings are an end product and research is a process" (Kidd 1959:370).

Just so, research is indeed a process. It is a process that encompasses not only the discovery of new knowledge but its application to the material world as well. In *The German Ideology*, Marx and Engels observe that

> Feuerbach speaks in particular of the perception of natural science; he mentions secrets which are disclosed only to the eye of the physicist and chemist; but where would natural science be without industry and commerce? Even this "pure" natural science is provided with an aim, as with its material, only through trade and industry, through the sensuous activity of men. [Marx and Engels 1970:63]

And Pasteur (quoted in Salomon 1973:81) had in his day vociferously insisted, "No, a thousand times no, there is no category of sciences which can be called applied sciences. *There is science and the applications of science* bound together like the fruit and the tree which bears it." Any individual research undertaking is part of a generalized flow from the generation of new scientific understanding to its application in production. At one end of the research continuum may be "pure" knowledge, but at the other there is a new good or a new production process. And in capitalist society these new entities appear as products in commercial circulation.

The NSF, charged with the task of monitoring and guiding the allocation of research funds in the context of the real-world economy, necessarily augments conventional idealist notions of basic and applied research with language that goes to the material heart of the matter. Summaries of the NSF definitions on the basis of which the categories of Tables 2.3 and 2.4 were constructed are given below:

1. Basic research is directed toward "a fuller knowledge or understanding of the subject under study, rather than a practical application thereof." More specifically, basic research projects represent "original investigations for the advancement of knowledge . . . which do not have specific *commercial* objectives."
2. "Applied research is directed toward practical applications of knowledge" and covers "research projects which represent investigations directed to discovery of new scientific knowledge and which have *specific commercial objectives* with respect to either products or processes."
3. Development research is "the systematic use of the knowledge or understanding gained from research directed toward the *production* of useful materials, devices, systems or methods, including design and development of prototypes and processes" (National Science Board 1986, emphasis added).

At the core of the distinctions made between basic, applied, and development research is not the motivation of the researcher nor the technical character of the research, but the relationship of the research to the commercial product, to the *commodity-form*.

With this point in mind, let us examine the data displayed in Tables 2.3 and 2.4. First, it is clear that the state and industry are the principal sources of funding for scientific research in the United States. Together they account for 97 percent of expenditures, each providing roughly half of total disbursements. But this balance between the federal government and private enterprise is not reflected in research actually performed. The state's main function is to provide other sectors with the resources to engage in research. Industry, universities, and non-profit institutions all perform more dollar value of research than they themselves fund. It is the state that makes up the difference. And while the government supplies 47 percent of total research dollars, it actually spends only 15 percent in the performance of its own research activities. The principal beneficiary of federal largesse is industry, which receives nearly $27 billion in state contracts and support for research and development. Universities and non-profit institutions together receive about one-third of the level of aid provided to business.

Of particular interest is the distribution of expenditures across and within the different classes of research. The share of research funded and performed progressively decreases for the federal government, universities, and

Table 2.3. *Sources of national research and development funding by institution and character of work, 1986*

Institution	Basic $mil	Basic %	Applied $mil	Applied %	Development $mil	Development %	Total $mil	Total %
Federal government[a]	9,240	64	10,250	41	35,760	46	55,250	47
University	1,530	10	820	3	150	—	2,500	2
Nonprofit	685	5	480	2	210	—	1,375	1
Industry	2,995	21	13,750	54	42,730	54	59,475	50
Total	14,450	100	15,300	100	78,850	100	118,600	100

[a]Includes federally funded research and development centers.
Source: National Science Board (1986).

Table 2.4. *Performers of national research and development by institution and character of work, 1986*

Institution	Basic $mil	Basic %	Applied $mil	Applied %	Development $mil	Development %	Total $mil	Total %
Federal government[a]	3,150	22	4,150	16	10,300	13	17,600	15
University	7,100	49	2,900	12	600	1	10,600	9
Nonprofit	1,100	8	950	4	1,350	2	3,400	3
Industry	3,100	21	17,300	68	66,600	84	87,000	73
Total	14,450	100	25,300	100	78,850	100	118,600	100

[a]Includes federally funded research and development centers.
Source: National Science Board (1986).

non-profit institutions as one moves across the continuum of the research process from basic investigation to development. The pattern is directly the converse for industry. Thus, universities, on the basis of some $7 billion in federal subsidy, accounted for about half of basic research performed, but only 12 percent of applied work and a mere 1 percent of development. Industry, on the other hand, undertook 21 percent of basic research, but 68 percent of applied effort and, with $24 billion in federal aid and contracts, fully 84 percent of development work.

A number of simple but important points follow from the straightforward accounting in Tables 2.3 and 2.4:

1. Public monies are used by the state to undergird scientific research and development.
2. The state is the source of 64 percent of university research funding. Public funding underwrites nearly two-thirds of all scientific endeavor in American universities.
3. In absolute terms, state support of universities and non-profit institutions is dwarfed by massive subsidization of private research and development capabilities. Fully 43 percent of state science funding goes to the support of private development research alone.
4. This pattern of public expenditures in support of scientific research reinforces an existing social division of labor in which the state – in large measure through the vehicle of the university – undertakes responsibility for providing the fundamental research effort on which continued technological growth depends. Capital concentrates its efforts on applied research and especially on development.
5. As is clearly implied by the definitions used by NSF in organizing the data, this social division of labor is structured around the commodity-form. Private enterprise invests principally in those areas most proximate to a finished commercial product. In general, the greater the distance of a particular research project from the product, the greater will be the tendency for capital to permit the state or its proxies to fund and to perform the research. This distance is defined by social and economic considerations relating to control over the production of specific real or potential products and will vary across and within industrial sectors and scientific disciplines. The terms "basic," "applied," and "development" are not only crude divisions of the complex continuum of the scientific research process, they are equally crude reflections of the variability of the circumstances in which the commodity-form can be successfully imposed.[15]
6. Tables 2.3 and 2.4 report data for the year 1986 alone. Examination of equivalent statistics for the years since 1953 showed that though the overall shape of the division of labor between state and capital has been stable, its internal composition is constantly shifting (National Science Board 1986). Science and technology are, after all, in constant flux. Kettering is reported to have observed that "the only real difference between basic and applied research is a matter of about 10 years" (U.S. House of Representatives 1984:499). There must be a continual readjustment of the particulars of the public/private division of labor as new

scientific, technological, and social developments alter the distances to the commodity.

Most conventional conceptions of basic and applied research are idealizations. They persist because of their ideological utility in focusing attention on the search for knowledge rather than on the search for the commodity. One of the most powerful ideological notions held by scientists is that whatever the uses to which technology is put, science remains "pure." In *The Logic of Liberty*, Michael Polanyi (1945:6) argued

> that the essence of science is love of knowledge and that the utility of knowledge does not concern us primarily. We should demand once more for science that public respect and support which is due it as a pursuit of knowledge and knowledge alone. For we scientists are pledged to values more precious than material welfare and to a service more urgent than that of material welfare.

Science might be unambiguously underwritten by both state and capital, but as long as the scientist remained true to the Mertonian ethos – universality, communalism, disinterestedness, organized skepticism – the autonomy and integrity of scientific practice were automatically assured (Merton 1970; Polanyi 1945, 1962). Faced with any attempt to exert lay or democratic control over their activities, scientists have time and again retreated into the protective arcanity of their own expertise. They have argued that any interference in management of the republic of science would kill the goose that lays the golden eggs so coveted in modern technological society.

Contrary to Polanyi, science must be understood as a process linking knowledge and application. Marx's observation, quoted earlier, that under capitalism, invention "becomes a business" is quite correct. "The point," observes Landes (1969:538) in *The Unbound Prometheus*, "is that man can now order technological and scientific advance as one orders a commodity." Quite so: According to Tables 2.3 and 2.4, industry now performs some 73 percent of scientific research and development in the United States. We cannot understand science without reference to the commodity. When we see the words "basic" and "applied" alongside the word "science," in this book and elsewhere, we must think first not of values or of technics, but of social relations as expressed in the commodity-form.

Plant genetic geography

Earth, wind, and fire – soil, air, and water – have been regarded by many peoples as the earth's fundamental natural resources. But, of course, what really differentiates our planet from the other bodies of the cosmos is the existence of life. And germplasm, the hereditary material contained in every cell, must be counted as a fourth resource of prime importance. The term "plant genetic resources" (PGRs) encompasses the total range of plant germplasm available in the global gene pool. And it is plant germplasm that is the raw material of plant biotechnologists of all historical eras. As such, an awareness of the sources and nature of this raw material is necessary for a proper understanding of how it has been appropriated and used.

The vagaries of natural history have meant that PGRs are not distributed evenly across the face of the globe. Biotic diversity is concentrated in what is now the Third World. Moreover, it is in the Third World that the domestication of plants first occurred and systematic crop production was first initiated (Hawkes 1983). In domesticating plants, humans added intense artificial selection pressures to the ongoing process of natural selection. They also carried crops away from their centers of origin, thus facilitating recombination with other plant populations and forcing adaptation to the exigencies of new environmental parameters. The result of these complex interactions was the development within any one species of thousands of "land races."

Land races are genetically variable populations that exhibit different responses to pests, diseases, and fluctuations in environmental conditions (Harlan 1975b). The genetic diversity in these land races was, and remains, a form of insurance for peasant cultivators. By planting polycultures comprising genetically diverse varieties, peasant farmers made certain that, whatever the year might bring in the way of weather or pests, some of the seed sown would grow to maturity and provide a crop. The objective of these early breeders was not high yield but consistency of production. And the result of their efforts was the development of great inter- and intra-specific genetic variability in particular and relatively confined geographic regions.

The existence of such areas was first recognized in the 1920s by the Soviet botanist N. I. Vavilov. He identified a variety of these areas that he considered to be the centers of origin of most of the world's economically important crops. The "Vavilov centers of genetic diversity," as they have come to be called, are identified in Figure 2.2. Table 2.5 lists the crops for which each of these regions is a center of diversity.

It should be clear from Figure 2.2 that the Vavilov centers are situated predominantly in what is now known as the Third World. Ironically, the gene-rich are currently the world's least developed nations, while the gene-poor are the advanced industrial nations. Of crops of economic importance, only sunflowers, blueberries, cranberries, pecans, and the Jerusalem artichoke originated in what is now the United States and Canada. An all-American meal would be somewhat limited. Northern Europe's original genetic poverty is only slightly less striking; oats, rye, currants, and raspberries constitute

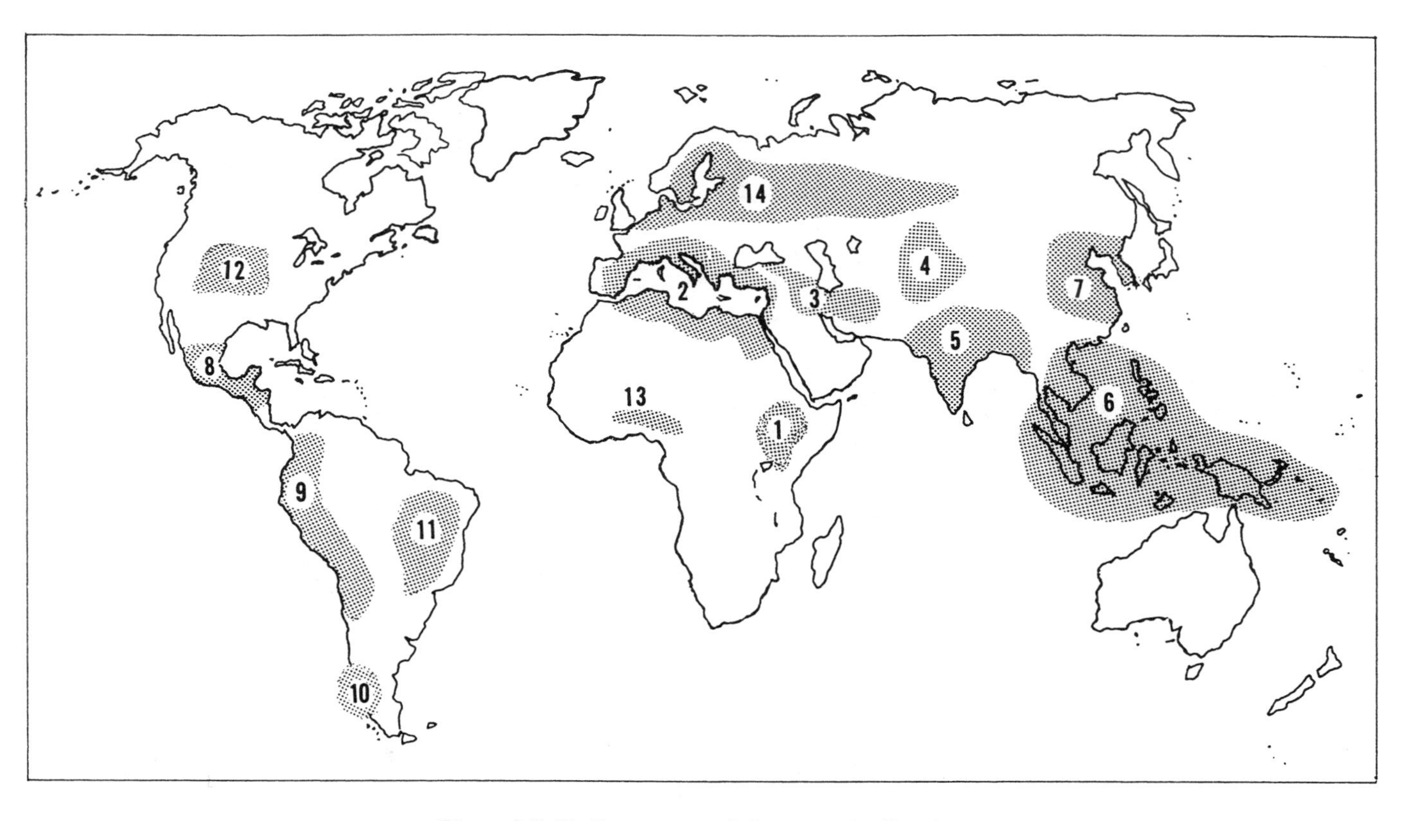

Figure 2.2. Vavilov centers of plant genetic diversity.

Table 2.5. *Crop species associated with Vavilov centers of plant genetic diversity*

1. *Ethiopia*	4. *Central Asia*	6. *Indo-Malaya*	9. *Peru-Ecuador-Bolivia*
Barley	Almond	Banana	Bean (*P. vulgaris*)
Castor bean	Apple	Betel palm	Bean (*P. lunatus*)
Coffee	Apricot	Breadfruit	Cacao
Flax	Broadbean	Coconut	Corn (maize)
Okra	Cantaloupe	Ginger	Cotton
Onion	Carrot	Grapefruit	Guava
Sesame	Chickpea	Sugar cane	Papaya
Sorghum	Cotton	Tung	Pepper (*Capsicum*)
Wheat	Flax	Yam	Potato
2. *Mediterranean*	Grape	7. *China*	Quinine
Asparagus	Hemp	Adzuki bean	Quinoa
Beet	Lentil	Apricot	Squash
Cabbage	Mustard	Buckwheat	Tobacco
Carob	Onion	Chinese cabbage	Tomato
Chicory	Pea	Cowpea	10. *Southern Chile*
Hop	Pear	Sorghum	Potato
Lettuce	Sesame	Millet	Strawberry
Oat	Spinach	Oat	11. *Brazil-Paraguay*
Olive	Turnip	Orange	Brazil nut
Parsnip	Wheat	Mulberry	Cacao
Rhubarb	5. *Indo-Burma*	Peach	Cashew
Wheat	Amaranth	Radish	Cassava
3. *Asia Minor*	Betel nut	Rhubarb	Mate
Alfalfa	Betel pepper	Soybean	Rubber
Almond	Chickpea	Sugar cane	Peanut
Apricot	Apricot	Tea	Pineapple
Barley	Cowpea	8. *Central America*	12. *North America*
Beet	Cucumber	Amaranth	Blueberry
Cabbage	Eggplant	Bean (*P. vulgaris*)	Cranberry
Cherry	Hemp	Bean (*P. mult.*)	Jerusalem artichoke
Date palm	Jute	Bean (*P. lunatus*)	Sunflower
Carrot	Lemon	Bean (*P. acut.*)	13. *West Africa*
Fig	Mango	Corn (maize)	Kola nut
Flax	Millet	Cacao	Millet
Grape	Orange	Cashew	Oil palm
Lentil	Pepper	Cotton	Sorghum
Oat	Rice	Guava	14. *Northern Europe*
Onion	Sugar cane	Papaya	Currant
Opium poppy	Taro	Pepper (*Capsicum*)	Oat
Pea	Yam	Sisal	Raspberry
Pistachio		Squash	Rye
Rye		Sweet potato	
Pomegranate		Tobacco	
Wheat		Tomato	

Sources: Grigg (1974), Wilkes (1977).

the complement of major crops indigenous to that region. Australia has contributed nothing at all to the global larder.

But in spite of the poverty of their original genetic endowments, North America, northern Europe, and Australia can today hardly be considered genetically underdeveloped. Indeed, they enjoy the world's most productive agricultures and reputations as the globe's breadbaskets. But, as Figure 2.2 and Table 2.5 show, the crops that now dominate the agricultural economies of the advanced industrial nations are not, for the most part, indigenous species. They have been introduced from elsewhere, principally from what is now the Third World. The development of the advanced capitalist nations has been predicated on transfers of plant germplasm from the periphery. If the United States now has a food weapon, as former Secretary of Agriculture Earl Butz so bluntly put it, it is because nations such as Nicaragua, Ethiopia, Iran, and China have supplied, respectively, the corn, wheat, alfalfa, and soybean germplasm for its arsenal. The global distribution and transfer of PGRs have been and still remain crucial elements of the political economy of plant biotechnology.

3

The genetic foundation of American agriculture

> The greatest service which can be rendered to any country is to add a useful plant to its culture.
>
> Thomas Jefferson

It was with some reluctance that I settled on the preceding quotation as an opening for this chapter. These words of Jefferson's have been repeated so often that they have been rendered time-worn and hackneyed through overuse. But buried in the cliche, wrapped in onion-like layers of associations with agrarianism and the nobility of the yeoman-farmer, is a core of profound truth.

Of crops of economic importance, only sunflower, blueberry, cranberry, and Jerusalem artichoke originated in North America. This simple fact of natural history has had important ramifications for the economic, political, and social development of the United States. The introduction of plants into America has been much more than a great service; it has been an absolute imperative, a biological *sine qua non* upon which rests the whole complex edifice of American industrial society.

The North American continent appeared as a plant genetic *tabula rasa* on which the Indians had inscribed maize, beans, and squash, but which was otherwise devoid of plants that would support either settlement or commerce. Here was a situation in which, *pace* Jane Jacobs (1969), the development of an agricultural base was a precondition for urban productivity. Without a substantial infusion of exotic germplasm and its adaptation to American conditions, European colonization could not be sustained. In an even more fundamental way than Hacker (1940:397) has suggested, American agriculture has indeed been the "cat's-paw for our industrial capitalism."[1]

Early plant introduction in North America

It is said that an army travels on its stomach. Ultimately we all travel – and live or die – on our stomachs, and this is, of course, no less true of the first

European settlers in what was to become the United States than it is of us today. An adequate food supply was the material prerequisite for the establishment of a permanent European presence in the new land of America. The Jamestown settlers brought with them seed of the English crops they were accustomed to grow in the expectation of self-sufficiency. William Bradford records: "Some English seed they sew, as wheat & pease, but it came not to good" (Rasmussen 1975a:94). The result that winter of 1609-1610 was the "starving time" in which two-thirds of the colony died.

A variety of factors, including the late planting and the questionable quality of the seed, may have contributed to this initial crop failure. But it is clear that English plant varieties were by no means well adapted to the different growing conditions of Virginia. Of necessity the settlers turned to the Indian crops, which were not necessarily to their taste but which meant life. The colonists themselves recognized the need to identify varieties that would thrive in the American environment. An early history of the Massachusetts Horticultural Society states that in 1621 the governor of Plymouth Colony requested the Indian Massasoit "to exchange some of their corn, for seed, with ours, that we might judge which best agreed with the soil where we lived . . . They possessed varieties adapted to the warmer or colder parts of the country" (quoted in Webber 1900:468). In the spring of that year the colonists planted 20 acres of maize and 5 acres of English grains, and Edward Winslow wrote: "We had a good increase of Indian corn, and our barley indifferent good, but our pease not worth gathering" (Rasmussen 1975b:10). Though the Indians themselves were to be driven to near-extinction, the squash, bean, and maize varieties they had developed sustained the colonists while European and other exotic crops made the slow adjustment to new ecological niches. The names of the maize varieties grown by the early settlers are testimony to their origins: Tuscarora, Golden Sioux, King Philip.

Each wave of new settlers brought with it a new set of crops and cultivars. The 1628 Endicott expedition to Massachusetts Bay colony, for example, brought with it seeds of wheat, rye, barley, oats, beans, peas, peaches, plums, cherries, filberts, pears, apples, quince, pomegranate, woad, saffron, licorice, madder, potatoes, hops, hemp, flax, currants, cabbage, turnips, lettuce, spinach, radishes, onions, and peas (Klose 1950:5). The complement of introduced species grew rapidly and with extraordinary variety. The English brought cowpeas, the French what is now known as Kentucky bluegrass, the Dutch clover, and the black slaves who worked the tobacco fields millet and sorghum.

Many varietal introductions failed; some succeeded. Of necessity early American farmers were continually experimenting with their crops – "sometimes desperately" (Zirkle 1969:26). Annually, absent a total crop failure, they selected the best individual plants to be saved as seed for the subsequent year's planting, a breeding process now called "simple mass selection" by plant scientists. Insofar as they augmented their planting stock with seed from other farmers or newly arrived immigrants, they were providing the conditions for natural crossing of plants and the generation of additional variability for successive cycles of selection. Though unaware that they were "breeding," farmers were engaged in building an adapted base of germplasm for American agriculture.

In 1699 this informal experimentation was given its first institutional expression with the establishment of an experimental farm by the Lords Proprietors of South Carolina. The purpose was to test the adaptability of mulberry trees, indigo, tobacco, hemp, flax, and cotton to local conditions (Klose 1950:10). It is clear that the Lords Proprietors, sponsors of South Carolina's imperial venture in colonization, were interested in cash rather than subsistence crops. The mulberry tree had been introduced as early as 1621 in Virginia in an effort to encourage the production of silk in the American colonies. Rice was brought to South Carolina in 1688, and sugar cane to Louisiana by 1718.

But exotic plantation crops were not necessarily any better suited to American conditions than were European crops. It was not until 1794 that a commercially useful sugar cane cultivar was developed and brought into production in Louisiana. Rice cultivation became profitable only at the turn of the century, when a variety from Madagascar proved suitable to the South Carolina environment. In 1733 James Oglethorpe established the Trustees Garden of Georgia, which, like its Lords Proprietors' predecessor, was designed to facilitate the introduction of plants of commercial promise.

The interest of propertied elites in plant introduction carried through into the period of independence. George Washington, like many other large landowners, imported large quantities of seed from Britain and other European countries. His papers record orders of a wide variety of species from an English supplier and, in 1794, thirty-nine kinds of tropical plants, including the breadfruit tree then popular in the West Indies as a food for slaves (Klose 1950:15). As might be expected, Jefferson was extremely active in procuring exotic seed. He arranged to receive an annual shipment of seeds from the superintendent of the Jardin des Plantes at Paris and searched diligently if unsuccessfully for a wheat variety resistant to the Hessian fly. He was particularly interested in rice and obtained varieties from China, Italy, Egypt, Palestine, and equatorial Africa.

Many of Jefferson's acquisitions were distributed to friends in the agricultural societies that were springing up in the early nineteenth century. Indeed, organizations such as the South Carolina Society for the Promotion of Agriculture (established 1785) and the Berkshire Agricultural Society (established 1811) were, through mutual exchange, instrumental in dispersing introductions over a wide area and into numerous ecological niches. The members of such societies were by no means common farmers. Jefferson's own Albemarle Agricultural Society of Virginia (established 1817) included among its thirty organizers not only Jefferson and James Madison, but two later governors, a future senator, a justice of the U.S. Supreme Court, and sundry other statesmen and professionals (Klose 1950:20). Such men were generally more interested in indigo, rice, cotton, sugar cane, and olives – crops whose production could be organized as a plantation operation – than they were in staple food crops. Their elite positions enabled them to obtain germplasm through channels closed to the average farmer. Thus, Jefferson was able to obtain rice from the young prince of Cochin China. And Elkanah Watson, merchant, banker, farmer, and founder of the Berkshire Agricultural Society, used his substantial personal means in 1818 to systematically request seeds from American consuls all over the world (Baker et al. 1963:4).

The U.S. Patent Office and germplasm

It may well have been Watson's action that motivated Secretary of the Treasury William L. Crawford to make a similar request of the young nation's foreign consuls and naval officers in 1819. The agricultural societies had, since Washington's presidency, solicited federal aid for the promotion of agriculture and for the introduction and trial of new crops and varieties. Jefferson had noted that

> In an infant country, as ours is, these experiments are important. We are probably far from possessing, as yet, all the articles of culture for which nature has fitted our country ... *to find these out will require abundance of unsuccessful experiments*. But if, in a multitude of these, we make one useful acquisition, it repays our trouble. [quoted in Klose 1950:17, emphasis added]

Secretary Crawford's directive calling for the assistance of consular and naval personnel made a similar argument, but added an additional important rationale for state support of germplasm collection:

> The introduction of useful plants, not before cultivated, or of such as are of superior quality to those which have been previously introduced, is an object of great importance to every civilized state, but more particularly to one recently organized, in which the progress of improvements of every kind has not to contend with ancient and deep rooted prejudices. The introduction of such inventions, the results of the labour and science of other nations, is still more important, especially to the

> United States, *whose institutions secure to the importer no exclusive advantage from their introduction.* Your attention is respectfully solicited to these important subjects. [quoted in Klose 1950:26, emphasis added]

Both Jefferson and Crawford recognized the fundamental import of plant collection and evaluation for the future of American agriculture and the nation. The object of both yeoman-farmer and planter was to permanently establish new crops of subsistence and commerce on the American landscape. This was an enormous task. There would be the occasional brilliant success – the rice from Madagascar that thrived in South Carolina – but, as Jefferson noted, the rule would be failure.[2] Many crops and most varieties would be unsuited to the particularities of the American climate. Yet the successes would ultimately repay those failures. And success would depend on the capacity to draw for introductions on as large and diverse a gene pool as possible. Plant introduction in the nineteenth century was a numbers game; the more one played the better your chances of winning.

But, as Crawford understood, it was not a game that could be played successfully by the individual entrepreneur. True, Jefferson might have access to Chinese and Middle East rice, and Elkanah Watson might use his wealth to funnel some of the world's germplasm back to his farm in Massachusetts, but neither individual altruism nor individual wealth could sustain plant collection over the time and at the scale needed to provide the country with the adapted base of germplasm it required for rapid agricultural development. Moreover, as Crawford notes, while the United States had by 1819 already had a Patent Act for thirty years, it occurred to no one that the legislation might cover plants, and American institutions therefore secured to the importers of germplasm "no exclusive advantage from their introduction." Being indefinitely propagable, it was unlikely that plant material would return much profit to an entrepreneurial collector even if he or she could beat the odds and identify a superior variety.[3]

The United States in 1819 was an agricultural society. Yet the vagaries of natural history had not provided it with a foundation of plant genetic resources that would permit expansive growth of population and commerce. There was a clear and crucial social need for the introduction and adaptation of exotic crop species and varieties. The unique characteristics of the material in question (seed), the magnitude of the undertaking, and the inability of individuals to recoup investment in plant exploration militated against systematic private efforts in this field. Essential services that are not privately profitable in capitalist society fall to the public. The result was one of the first significant interventions by the American federal state to provide the conditions for accumulation and growth. Crawford's directive mandating global collection of germplasm was the first formal institutional step toward the massive public commitment to agricultural research that the USDA today represents.

No expenditure of monies was initially authorized for the collection activities of consuls and naval officers, and acquisition of materials went slowly at first. A second Treasury circular in 1827 encouraged more attention to collection work and provided detailed instructions for the preservation and shipping of seed. The Navy proved particularly cooperative,[4] and exotic germplasm began to flow to the United States from its far-flung diplomatic representatives and military officers. Collection and evaluation activities were greatly stimulated by Henry Ellsworth, who, as Commissioner of Patents between 1836 and 1849, considered that provision of novel plant varieties was as much the concern of the Patent Office as was the encouragement of new mechanical inventions. Ellsworth himself had purchased large tracts of land in Indiana and other prairie states (Baker et al. 1963:5) and was an enthusiastic supporter of the extension of settlement to those areas. Whatever his motives, he succeeded in 1839 in obtaining congressional funding for the collection and distribution of seeds, plants, and agricultural statistics.

The year before, the Navy had authorized the first official plant exploration expedition. Between 1838 and 1842, Commander Charles Wilkes' ship cruised the Pacific under orders to secure new agricultural plants (Klose 1950:29). By 1848, ships of the East India Squadron were regularly collecting plants. The Perry naval expedition of 1853 is best known for forcing open the harbors of Japan to American commerce. Perry's gunboats also brought home a tremendous variety of seeds and plant materials obtained from Japan, China, Java, Mauritius, and South Africa. The genetic fruits of this imperial adventure included seeds or cuttings of vegetables, barley, rice, beans, cotton, persimmon, tangerine, roses, and "three barrels of the best wheat of Cape Town" (Klose 1950:33). Other expeditions sent plants from South America, the Mediterranean, and the Caribbean. With funds available to support collection work, consuls also began sending seed in quantity: wheat from Poland, Turkey, and Algeria, rye from France, sorghum from China, cotton from Calcutta and Mexico City, peppers and maize from Peru, rice from Tokyo. In 1842 a greenhouse was established in Washington for the preservation of botanical collections, and in 1857 a propagating garden was established for the multiplication of introduced varieties.

Crucial to the success of the introduction program established by the Patent Office was distribution of the seed of exotic varieties. Even before he had obtained a congressional mandate, Commissioner Ellsworth took it upon himself to ensure that foreign germplasm enjoyed wide dissemination. Novel varieties were sent to farmers under the postal frank of sympathetic members of Congress. Sending plants and seeds in the mail was a significant innovation, and packages of plant material were shipped through the postal service some five years before parcel post arrangements were established for other items (Powell 1927:3). By the time he left the commissionership in 1849, Ellsworth was sending out 60,000 packages of seed each year.

With Ellsworth's retirement, the Patent Office was moved from Treasury to the Department of the Interior, but the work of the agricultural division remained largely unaltered. Ellsworth's successors enjoyed a steadily growing appropriation for the seed collection and distribution activities that continued to be one of their chief concerns. Commissioner Charles Mason (1853-1857) was especially committed to plant introduction, and he made a point of sending small packages to a large number of farmers in an effort to ensure the dispersion of seed to as many locations as possible. By 1855, over a million packages had been distributed.

There is no question that the Patent Office program of plant introduction resulted in substantial infusions of foreign germplasm into the American gene pool prior to the Civil War. But if the federal effort was important in providing variable plant material, it was the farmers of the nation who molded it into useful form. This was necessarily the case. The Patent Office had neither the personnel nor the facilities for extensive trials and evaluation of the varieties they so assiduously collected. Moreover, given the botanical knowledge of the time, only the most general predictions could be made regarding the likelihood that a particular variety would be appropriate for a particular area. Under these circumstances, wide distribution of exotic seed to farmers was the most efficient means of developing adapted and improved crop varieties.

The development of the adapted base of germplasm on which American agriculture was raised is the product of thousands of experiments by thousands of farmers committing millions of hours of labor in thousands of diverse ecological niches over a period of many decades. Introductions might or might not be successful, but in any case they had an opportunity to cross naturally with established land races, so that, even where they failed, they might leave a useful legacy of genetic variability. The spread of crops to new areas was made possible only through the adaptation of a few varieties without which the move would not have been possible. Between 1839 and 1859 the center of wheat production moved from western Pennsylvania to western Ohio, and only then was the crop "beginning to look like a native in America" (Reitz 1962:108). Individual farmers were responsible for developing improved cultivars, among the most famous being Red Fyfe wheat, Grimm alfalfa, and Rough Purple Chili potato, the germplasm sources being, respectively, Poland, Germany, and Panama.

The breeding method used by farmers was essentially no different from that employed by their neolithic forbears. Camerarius had demonstrated the existence of sexual reproduction in plants in 1694. By 1800 it was recognized among an elite of naturalists that cross-fertilization could be used to produce a new plant variety that would combine the characteristics of both parents, and in the first half of that century a few horticulturalists attempted to apply the method practically. More influential, however, was the work of the Frenchman Jean Baptiste van Mons, who emphasized the principle of selection: "To sow, to resow, to sow again, to sow perpetually, in short to do nothing but sow, is the practice to be pursued, and which can not be departed from" (quoted in Webber 1900:469). Van Mons' words graphically describe the manner in which the nation's farmers employed simple mass selection to improve the land races of the crops they grew by screening out poorly adapted types and saving superior individuals and populations for seed. Such improvement occurred with native species as well. Yellow dent maize, which now dominates the United States and much of Europe, did not exist in precolonial times. It is the result of crosses between dent corns of Mexican origin and the flint corns of the American east coast (Pioneer Hi-Bred 1984:28).

The institutionalization of agricultural research

And the farmer-breeders were eminently effective. By 1860, a host of crops was firmly established and formed the base for a variety of regional agricultural economies: a commercial feedgrain/livestock economy north of the Ohio River, with a wheat belt farther north, specialized dairy and vegetable production in the Northeast, tobacco, rice, cotton, and sugar cane in the South (Cochrane 1979:76). Agricultural production supported a population of 31.5 million in twenty-eight states, with enough surplus to export some twenty million bushels of wheat and ten million bushels of corn. Foreign plant genes had been successfully domesticated and a firm agricultural foundation prepared for the rise of industrial capitalism.

There were, however, those who wanted for agriculture something more than the trial-and-error methods of the individual farmer. Jefferson and Washington were curious men, it is true, but they were also plantation owners with all the commercial concerns and financial acumen that implies. Lewis Mumford (1963:25) has argued that capitalist rationality "preceded the abstraction of modern science and reinforced at every point its typical lessons and its typical methods of procedure."[5] In the first half of the nineteenth century there were few Americans active in business, politics, or the professions who did not have some direct dealings in farm property as proprietors or landlords (True 1900:159). For such men, the backbone of the agricultural societies, "venerable tradition no longer served when it was a question of getting the greatest returns from the land" (Bernal 1965:653). These gentlemen farmers advocated new agricultural practices, endorsed state geological surveys, and experimented with new breeds and varieties, manures, and crop rotations (Gates 1960:315; Rossiter 1975:8).

They also supported the institutionalization of agricultural education and research, not always disinterestedly. In 1824, Stephen van Rensselaer founded an agricultural school that would ultimately become Rensselaer Polytechnic Institute. One wonders if the prospect of increased production from his 3,000 New York tenant farmers had anything to do with his generosity. The American publication of Justus Liebig's book *Organic Chemistry and Its Applications in Agriculture and Physiology* in 1841 stimulated a tremendous amount of interest in the application of science to agricultural production, especially in areas of the east suffering from declining soil fertility.[6] By 1850, agricultural societies and journals were agitating vigorously for elevation of the Patent Office Division of Agriculture to department status and for the establishment of agricultural colleges. Over the next decade, several agricultural schools were founded in various states: Michigan, 1855; Iowa, 1858; Pennsylvania, 1859; New York, 1860.

In 1862, a banner year for agriculture, the USDA was established, the Morrill Act authorizing creation of the LGUs was passed, and the Homestead Act was approved. These events are often regarded as signal victories for the farmer (Paarlberg 1978:137) and, even among contemporary critics of the LGU complex, as evidence of the democratic origins of agricultural research institutions (Hightower 1973:8). Actually, there appears to have been no consensus in the agricultural community regarding the desirability of any one of these "victories." Creation of the USDA was due principally to the lobbying of the agricultural societies and journals (Gaus et al. 1940:5). Common farmers resented the influence of the wealthy gentlemen-farmers and the editors and, as the nation's "most self-conscious taxpayers" (Danbom 1979:17), were suspicious of an enlarged government presence in agriculture (Gates 1960:313; Simon 1963:103).

Most farmers were even less enthusiastic about the prospect of agricultural colleges. Their uneasiness about the hybridization of these two words centered on a fear that education would teach them nothing about farming they did not already know and yet cost them much. And indeed, they had good reason for the apprehensions that led them, in the words of the farm paper *American Farmer*, to the "warmest opposition" (quoted in Gates 1960:360) to the proposals for supporting agricultural education, The proposed "land-grant" institutions were, after all, to be financed through the sale of *land grants*. Farmers would pay in soil for the privilege of educating their children. The roots of the movement for educational reform were not even agricultural in nature. The impetus for "practical" training found its origins and firmest commitment in the industrial sector, which saw an opportunity to use the sale of public land to finance the training of a skilled manufacturing work force (Danbom 1979:17).

Little wonder, then, that an eastern senator, Justin Morrill of Vermont, introduced the land-grant legislation and that it was in the states where agriculture was least important that the bill found its principal support.[7] Opposition from a speculation-leery West and a South fearful of federal power[8] led President Buchanan to veto the first attempt at passage of the Morrill Act in 1857. But in 1862, with secession removing southern resistance and a Homestead Act assuaging western worries, the Land Grant College Act was approved. In practice, what looked like a contradiction – the juxtaposition of free homestead land and a policy of land-grant sales – was more apparent than real. The abuses that had characterized American land distribution programs up to 1860 were soon realized in homestead dispositions. Similarly, disposal of land-grant tracts soon degenerated into a pattern of "neglect, carelessness, incapacity, and something akin to corruption" (Gates 1943:295).

The upheaval of the war years delayed the establishment of many of the new colleges. Even when they did get under way they immediately encountered significant difficulties. The mass of common farmers had viewed the creation of the colleges with little enthusiasm and felt that the institutions that had been foisted upon them should at least address their immediate needs. Preoccupation with practical results was no less characteristic of the members of agricultural societies and the journal editors whose support for the colleges had been wooed by the promises of a handful of evangelistic, German-educated scientists that agriculture could be made rational and scientific (Rosenberg 1976; Rossiter 1975). The problem was that the colleges at first had little to offer. As Regent J .M. Gregory of Illinois Industrial University admitted in 1869, "Looking at the crude and disjointed facts which agricultural writers give us, we come to the conclusion that we have no *science of agriculture.* It is simply a mass of empiricism" (quoted in Busch and Lacy 1983:9).

The newly created Department of Agriculture was established with fewer scientific pretensions. Congress declared in the Organic Act that

> There is hereby established at the seat of Government of the United States a Department of Agriculture, the general designs and duties of which shall be to acquire and to diffuse among the people of the United States useful information in subjects connected with agriculture in the most general and comprehensive sense of that word, and to procure, propagate, and distribute among the people new and valuable seeds and plants. [Baker et al. 1963:13]

The collection and dissemination of germplasm would be the center of the department's activities, just as those functions had been the core of the work of the Patent Office Division of Agriculture. The first commissioner,[9] Isaac Newton, authorized expansion of the propagation garden and initiated formal bilateral exchanges of seeds with foreign governments. Newton's successors followed his lead, and wide arrays of species and varieties were introduced before 1900. With population moving west, the identification of crops and varieties suited to the arid regions of the High Plains and the Southwest assumed particular importance. Between 1860 and 1900, the center of wheat production moved west and north from Ohio into Minnesota as the Dakotas and Nebraska became major producing areas on the basis of newly introduced varieties (Reitz 1954:253). Chinese and African germplasm brought extensive sorghum cultivation to Kansas and Texas, and forages such as Japan's lespedeza and Johnson grass from Africa facilitated livestock production. The navel orange, sent by the Brazilian consul in 1871, established the backbone of California's citrus industry. In a self-conscious effort to gain for the United States a diversified and fully self-sufficient agriculture, the department encouraged trials of virtually every world crop of any economic importance. Between 1860 and 1900, Klose (1950) records trials of tea, coffee, opium, vanilla, ginger, castor bean, gum arabic, camphor, yam, almond, walnut, cork, arrowroot, licorice, fig, date, pomegranate, olive, guava, nectarine, pineapple, pistachio, madder, rubber, frankincense, balsam, and senna.

Though the department continued to expand the scope of its work, it remained substantially identified with plant introduction and seed distribution late into the nineteenth century. In 1881, for example, J. W. Covert, chairman of the House Committee on Agriculture, remarked, "The controlling idea involved in the creation of the Department is that our wide domain should be tested, to ascertain what can be most successfully grown in its various sections" (quoted in Baker et al. 1963:28). As late as 1878, fully a third of the department's annual budget was being spent on germplasm collection and distribution (Klose 1950:62).

With the passage of the Hatch Act establishing the state agricultural experiment stations (SAESs) in 1887,[10] a regular program of exotic plant and seed distribution to these new research institutions was arranged. Norman Colman, commissioner of Agriculture in 1887, welcomed the SAESs and expressed the hope that they might one day "do the testing and experimental work for the whole body of agriculturalists" (USDA 1888:35). He nevertheless noted the success of the distribution program and observed that "the increased production of wheat, oats, and other cereals and grasses, has, by reason of the wide distribution of improved varieties, paid tenfold the entire amount expended by the Department of Agriculture since it was established." Why then, in 1893, did Secretary of Agriculture J. Sterling Morton recommend in his annual report the "retirement of the Department from the seed business" (USDA 1894:391) and the reduction of the budget for germplasm distribution by 75 percent?

Seed distribution: public duty or private prerogative?

When Henry Ellsworth began disseminating exotic germplasm in 1836, it was his intention to ascertain its suitability to diverse American growing conditions. Because farmers were doing the actual experimentation, and because there was no assurance that the new varieties would be of any value, seed was distributed free of charge. Initially, this practice elicited no opposition from private purveyors of seed. The specialized commercial seed trade was embryonic and was limited largely to merchandising small lots of European vegetable varieties to home gardeners (Pieters 1900). Most farmers produced their own seed, and what trade existed was dominated by farmers themselves.[11] Objections to the free seed program were raised first by agricultural journals. By 1850, the farm papers had begun to solicit and maintain subscriptions by offering packets of seed of new varieties,[12] and they resented the Patent Office largesse.

Their opposition had little effect, however. The popularity of the government-supplied seed grew rapidly, and the congressmen and senators under whose postal frank the packages were sent out responded by increasing the appropriations for germplasm dissemination. Between 1850 and 1856 the Patent Office agricultural division budget increased from $4,500 to $75,000. Plant exploration and seed distribution accounted for much of the increase. By 1854, demand outstripped the Patent Office supply of new and exotic introductions, and in an effort to at least maintain the genetic diversity of the seed disbursed, a representative was sent to Europe to purchase substantial quantities of grains, grasses, and legumes as a supplement to the exotic stock (Klose 1950:43). By 1861, a total of 2,474,380 packages of seed, the bulk of which contained common vegetable and flower varieties, were being distributed through congressmen to their constituents.

In the mid-nineteenth century, seed firms had as yet been unable to penetrate the market for field-crop seed to any significant degree. That area was still almost exclusively characterized by on-farm production or inter-farmer commerce. But a growing landless urban population stimulated the rise of the specialized market-garden business, and because vegetables usually are not grown to full maturity, there was a small but significant market for vegetable seed among the growing class of commercial growers of fresh produce. Home gardeners, too, augmented demand and formed the basis of an expanding mail-order market. By 1860, many companies had been established that specialized in seed production and served these markets. A few of these firms were substantial operations, with trial fields and even a breeder or two (R. F. Becker 1984).

This nascent capitalist seed sector was thoroughly alarmed by the explosive growth of what they regarded as the government "seed business." The Patent Office action in countenancing the inclusion of common vegetable and flower seed in its distribution program was especially unwelcome because it tended to undercut the seed trade position in the one market where it had been able to establish a presence. The Civil War forced something of a retrenchment of the free seed program, but its opponents did not succeed in eliminating it, though they did succeed in having supply contracts awarded exclusively to American firms.

The Commissioners of agriculture in the post-Civil War period were not unmindful that much of the seed they sent out was of common varieties that needed no regional testing and that the recipient might not even be a farmer. They made periodic attempts to limit what they perceived as abuses, but were unwilling to risk the negative effects on their appropriations that a confrontation with Congress might have had. Besides, as the second commissioner, Horace Capron, noted, "If nine-tenths of the seed distributed are sheer waste, and the rest judiciously used, the advantage to the country may be tenfold greater than the annual appropriation for agriculture" (Klose 1950:59). The commissioners continued to encourage plant exploration and introduction even as Congress mandated progressive increases in free seed distribution and required in annual appropriations acts that 75 to 90 percent of the seed should be sent out under its members' auspices.

Most of the increase in volume after 1875 was accounted for by vegetable and flower seed. Again, the resurgence of free dissemination of germplasm directly threatened the seed trade at an extremely sensitive point. The year 1875 had seen the invention of the refrigerated railway car, and large-scale commercial vegetable production was beginning to appear. California was becoming a center for this new industry and for specialized seed production as well. In 1883 the representatives of thirty-four seed companies met in New York City to found the American Seed Trade Association (ASTA) as a vehicle to promote their interests before the government.

The seed industry found a champion in J. Sterling Morton, a conservative journal editor who became Grover Cleveland's Secretary of Agriculture in 1893. In his first annual report, Morton observed that the seed trade was no longer an infant industry and that private enterprise could put new plant varieties into the hands of farmers and consumers two to three years more quickly than could the government. He contended that the USDA's "Seed division has outlived its usefulness, and that its further continuance is an infringement of the rights of citizens engaged in legitimate trade pursuits" (USDA 1894:391). This initial sally was ignored in Congress, so he again called for cessation of free seed distribution in his report for 1894, asking, "Is it a function of government to make gratuitous distribution of any material thing... If, in a sort of paternal way, it is the duty of the government to distribute anything gratuitously, are not new ideas of more permanent value than old seeds?" (USDA 1895a:69-70).[13] There was no response this time either. He resolved to take the matter into his own hands and, justifying himself on the grounds that the seed offered was not "rare and uncommon" as required by law, purchased no seed for distribution in 1896. In a lengthy explanation of his action in the 1895 annual report he concluded:

> The work of distributing new, rare, valuable, or other seed should be left entirely in the hands of the branch of industry to which it lawfully belongs, leaving the right of selection to the individual consumers, and to their individual efforts or the associated work of their class the determination of its value. [USDA 1896:211]

Congressional reaction to this effrontery was swift, and a joint resolution was passed ordering the secretary to resume distribution. In 1895 a record number of seed packages was distributed, and Morton was reduced to caviling in his 1896 report that

> Briefly, the seed gratuitously sent about the country would have planted ... a strip of ground 1 rod in width and 36,817 miles in length. Such a strip would reach one and a half times around the globe, and a passenger train going at the rate of sixty miles an hour would require 51 days 3 hours and 14 minutes to travel from one end of this gratuitously seeded truck patch to the other . [USDA 1896:xxxix]

This quotation graphically illustrates the tremendous scale of the federal program of seed distribution. And, as Morton so often reminds us, the seed was being disseminated "gratuitously," that is, without charge, *in a manner antagonistic to the seed as a commodity-form and in direct competition with the private seed trade.*

It would be convenient to interpret Congress' overriding of Morton as a commitment to the right of the state to provide for its citizenry those goods or services it believes they should have regardless of private interests in the matter. This is not the case, however. There is no evidence that, for most representatives and senators, support for the seed distribution program was motivated by anything other than a desire to maintain a convenient means of ingratiating themselves with their constituents. But it was the demand of farmers and other consumers for the seed that was ultimately the determining factor in congressional support, and there seems to have been ample justification for farmer and consumer preference for government seed.

In 1897 the volume of seed distributed reached an all-time record of 22,195,381 packages. Because each package contained five packets of different varieties, the government actually sent out over 1.1 *billion* seed packets. There is little cause for wonder as to why the seed trade objected to government competition. The seed companies believed that the government was catering to a demand that they could be profitably satisfying in the open market. Moreover, the competitive bidding process under which the USDA purchased its supply of seed for free distribution kept profit margins narrow. Supplying the government with seed might be bread and butter, but it was not particularly lucrative.

Also, the department insisted on purchasing seed of good quality. Realizing that there was little he could do about the congressional fondness for the program, Commissioner of Agriculture Frederic Watts resolved in 1874 that if the department had to send out common as well as exotic seed, it would at least be of good quality (Klose 1950:61). The department became increasingly discriminating in its purchase of seed, and by 1886 it had established testing procedures to assess germination rates and cleanliness and was requiring that the seed it bought meet particular standards of quality (Galloway 1912:8). Thus, the seed sent out in the congressional distribution may have been old varieties, but certainly was not old seed as Morton charged.

Indeed, it was very likely to be fresh seed of top quality, and it achieved a popular reputation for being such. The integrity of the seed distributed by the USDA contrasted sharply with that for much of the seed purveyed by the commercial seed trade. Seed in commerce frequently was old and low in germination capacity and contained substantial amounts of weed seed and grit. Moreover, in an effort to distinguish their products and enhance their market, seedsmen in their catalogs and journal advertisements publicized similar or identical varieties under a wide range of synonyms and made extravagant claims for their novelties. The chief of the USDA's Seed Division refers in his annual report to the commercial "... tendency to place all sorts of vegetable seeds upon the market under new names... without [some] check new names for old varieties unworthy of dissemination will continue to be increased" (USDA 1888:655). It is no accident that the principal item of business for participants at the ASTA's inaugural meeting was adoption of a common disclaimer to be printed on all seed packages repudiating responsibility for the performance of their product and intended to protect them from the damage claims with which they were continuously plagued (*Seed World* 1983:32). Rather substantial swindles involving field seeds were relatively commonplace, and it is notable that the come-on for many of these was the claim that the purportedly superior new variety had its origin in a special germplasm imported by the USDA from some exotic land (Hayter 1968:188).[14]

Thus, there appears to have been good reason for the popular enthusiasm for USDA seed, which spurred legislators to sustain the free distribution program in the face of the seed-trade objections and a Secretary of Agriculture bent on defending the rights of private enterprise. As one citizen supporter of the program observed,

> Scores of varieties of most excellent seeds have been put within the reach of the masses of people, who would not otherwise have obtained them because of the exorbitant prices charged for them by unscrupulous dealers who have been among the first to condemn the Agricultural Department. [USDA 1888:653]

Conclusion

By the end of the nineteenth century a state presence was firmly established in the plant sciences. This presence resulted from the need for the state to undertake the germplasm collection and research that were not privately profitable but were essential to both agricultural and industrial progress. If, as Louis Hacker (1940) suggests, American agriculture undergirded the rise of industrial capitalism, foreign germplasm and the labor of the farmer-breeder undergirded agricultural development.

Government seed distribution activities, which had originally been an organic component of germplasm collection, were transformed by political pressures and an ideological commitment to the welfare of the farmer/consumer into an institutional impediment to the expansion of private enterprise in the seed business. The state, in distributing large volumes of quality seed without charge, put itself in the potentially contradictory position of constraining private capital accumulation.

4

Public science ascendant: plant breeding comes of age

There is a great deal of art to plant breeding, but more science.

H. Nilsson-Ehle

The eighteenth and nineteenth centuries saw the development of a plant genetic foundation on which American agriculture could successfully expand. This was accomplished principally through the appropriation of plant germplasm from other parts of the globe, a process that was underwritten and performed almost exclusively by the government. Varietal development consisted primarily of simple selection procedures at which farmers were no less adept than state experiment station or land-grant university personnel. But the turn of this century saw developments in science that would catalyze the transformation of plant breeding and establish the hegemony of the scientist, rather than the farmer, as the principal producer of new crop varieties. The rediscovery of Mendel's work in 1900 promised to put plant improvement on a much more sophisticated basis and make a "science" of what was until that time recognized as an "art."

Yet it would be a quarter of a century before the promise of Mendelian genetics was substantially realized. The immediate demands made by farmers on the experiment stations created an environment in which opportunities to pursue basic scientific research were limited. Concerned by stagnating agricultural productivity between 1900 and 1930, non-farm business interests championed the cause of agricultural science and the rationalization of farm production. These efforts on the part of the business community resulted in the passage of a series of legislative acts creating financial and institutional space for basic agricultural research. Analysts of the agricultural features of the New Deal era have concentrated principally on the state-sponsored social programs of the period (e.g., Saloutos 1982; Kirkendall 1982). Less well recognized is the extent to which the enhanced state capacity for intervention enjoyed by the federal government in the 1930s (Skocpol and Finegold 1982) was used to greatly strengthen agricultural science. This ultimately confounded the objectives of the social programs designed to stabilize the farm sector and provided private enterprise with a technical solution to the creation of space for capital accumulation in plant breeding.

The promise of Mendel

James "Tama Jim" Wilson succeeded Morton as Secretary of Agriculture in 1897, and the annual report for his initial year in office avoided any mention of the controversy that had so preoccupied his embattled predecessor. Morton's attempts to eliminate the free seed program were not without impact, however. He had argued that

> The reason and necessity for such distribution was removed when the experiment stations were established in the several States and Territories. Those stations are in charge of scientific men. They are, therefore, particularly well equipped for the trial, testing, and approval or condemnation of such new varieties as may be introduced from time to time. [USDA 1895b:70]

This was in large measure true. The plant introduction activities of the Patent Office and the USDA had successfully established a multiplicity of the world's botanical species as American crops. But continuous infusions of germplasm were needed as crops spread to new areas or as disease or pest problems rendered other varieties obsolete. The free seed program was no longer principally serving the purpose for which it had originally been initiated. Wilson chose to uncouple the political congressional distribution from the scientific collection, evaluation, and dissemination of exotic germplasm.

He did this by establishing in 1898 a Section of Seed and Plant Introduction within the USDA whose function was to coordinate the department's plant exploration and introduction activities. A staff of professional botanists was employed who were able to recognize plant diseases and pests and assess the agronomic value of varieties. With N. E. Hansen's 1898 journey to Russia in search of hardy alfalfas and forage crops was launched the "golden age of plant hunting" that over the next quarter century would see some 48 expeditions scour the world for useful germplasm. Accessions were, and still are, recorded in the *Inventory of Foreign Seeds and Plants*. The first entry is a cabbage variety, "Bronka," collected near Moscow. In the inaugural issue of the inventory, O. F. Cook, Special Agent in Charge of Seed and Plant Introduction, observed:

> It should be repeated here that our efforts are in a line quite distinct from that of the Congressional seed distribution... Importations are

> accordingly made, in the great majority of cases, in experimental quantities only, for the use of the experiment stations and private parties having special knowledge and experience in the cultivation of particular crops. [USDA 1899:4]

The greater part of the flow of exotic germplasm was thus directed to the state experiment stations, as Morton had advised.

However, a substantial portion was also sent out to individual farmers for trial. In 1916, 337,442 packages of what the USDA called "new and rare" seed were sent to private parties. Farmers continued to be regarded as an important component of the plant breeding community, for both experiment station personnel and farmers continued to employ the same principal breeding method: simple mass selection. While the station scientists might take a more intensive and systematic approach to the selection process, the difference was one of degree, not of kind. Farmers were no less competent selectors than the scientists, and because there were far more of them, they produced a great many varieties via simple selection from exotic introductions.

Beginning about 1890, workers in the land-grant universities and later the SAESs became increasingly interested in hybridization as an adjunct to selection. It is important to understand that during this period hybridization simply meant the cross-breeding or sexual combination of two varieties of plant a or animal. A hybrid was simply the product of such a union. After 1935, the term "hybridization" assumed a much narrower meaning in reference to a combination of two inbred lines, as in hybrid corn. In general, "hybrid" now carries this more restricted meaning. However, the reader must always be alert to the particular context in which the term appears in order to avoid misunderstanding.

In 1899 the Royal Horticultural Society organized an International Conference on Hybridisation and on the Cross-Breeding of Varieties. Presenting papers were several prominent Americans, including H. J. Webber of the USDA's Plant Breeding Laboratory, Liberty Hyde Bailey of Cornell University, and W. M. Hays, who represented the American Association of Agricultural Colleges and Experiment Stations. Hays (1900:257-8) admitted that farmers "select in a crude, yet sometimes very effective manner," but looked to the professional scientist for what he anticipated would be rapid progress: "One does not need to be a prophet to see in the handwriting upon the wall, that science is soon to make in plant breeding, and in animal breeding as well, greater achievements than heretofore, because more people with scientific training are devoting themselves to it."

Less than a year later, in the spring of 1900, the European botanists Hugo de Vries, Carl Correns, and Erich Tschermak all published papers detailing independent discoveries of rules of heredity that they subsequently found had been proposed by the Austrian monk Gregor Mendel in 1865; see Mayr (1982) for a detailed account of these events and their position in the history of biology and biological thought. Washington State Experiment Station wheat breeder W. J. Spillman was also very close to an independent rediscovery of Mendelian inheritance, and his 1901 paper establishing the existence of predictable recombinations of parental traits in hybrid progeny helped to ensure the rapid acceptance of the new theories in the United States. In fact, scientific opinion proved tremendously receptive to the elegant simplicity of Mendel's work, which "appeared with almost the power of revelation" (Rosenberg 1976:90-1). At the Second International Conference on Plant Breeding and Hybridization held in New York in 1902, William Bateson, who was shortly to confer the name "genetics" on the new science, explained why the practical plant breeder found hybridization so promising in the light of Mendelism:

> *He will be able to do what he wants to do instead of merely what happens to turn up.* Hitherto I think it is not too much to say that the results of hybridization had given a hopeless entanglement of contradictory results. We crossed two things; we saw the incomprehensible diversity that comes in the second generation; we did not know how to reason about it, how to appreciate it, or what it meant... The period of confusion is passing away, and we have at length a basis from which to attack that mystery such as we could scarcely have hoped two years ago would be discovered in our time. [Bateson 1902:3,8]

While such optimistic hopes were not to be fully realized, the emergence of Mendelian theory certainly gave to plant breeding a great stimulus and engendered among breeders a climate of confidence in the future of their work that helped to define and shape the discipline. W. A. Orton, a USDA cotton breeder who had just succeeded in developing a wilt-resistant cotton variety via hybridization, echoed Bateson and called for a commitment to a whole new approach to the work of plant improvement:

> The plant breeder's new conception of varieties as *plastic groups* must replace the old idea of fixed forms of chance origin which has long been a bar to progress... Since the science of plant breeding has shown that definite qualities may be produced and intensified as required, it is no longer necessary to wait for nature to supply the deficiency by some chance seedling. [American Breeders Association 1905:204, emphasis added]

Orton was writing of a radical shift in attitude, of the development of an active relation to the plant rather than a passive one. Instead of selecting from the diversity in nature, he saw germplasm as something to be molded in a predictable fashion. Wilkes (1983:141) sums up the impact of the rediscovery of Mendelian genetics: "for the first time the plant breeder had a clear idea of how to proceed with crop improvement."

The new vision and excitement were given institutional expression in 1903 with the creation of the American Breeders Association (ABA). W. M. Hays, L. H. Bailey, and W. J. Webber, all of whom had attended the London conference on hybridization in 1899, were members of the organizing committee. They acted ostensibly at the suggestion of Secretary of Agriculture Wilson, who was strengthening the scientific work of the department and who in 1901 had centralized all USDA plant-related work under a Bureau of Plant Industry. Hays was to be selected by Wilson as his Assistant Secretary of Agriculture in 1904, and the two men shared similar concerns and outlooks (Baker et al. 1963:40).

Hays was elected the ABA's first chairman, and his address at the inaugural annual meeting was a profoundly revealing statement of his intentions for the ABA in the brave new world Mendel had opened. Hays would certainly have concurred with Marx's last thesis on Feuerbach: The point is not simply to interpret the world, but to change it. So far as he was concerned, scientists had been concentrating far too much on interpretation, and he complained that they had "hardly grasped the vast economic interests which are at stake, nor have they seen the open doors of opportunity which might be entered by cooperation with the men who control the breeding herds and the plant-breeding nurseries" (ABA 1905:11). The ABA was to be no cloistered scientific society, but a purposive juxtaposition of science and business:

> The producers of new values through breeding are here brought together as an appreciative constituency of their *servants*, the scientists... All that these two classes of men lack to bring them together in a grand cooperative effort to improve those great staple crops and those magnificent species of animals – which combine sun-power and soil and air into the useful products upon which the human family live – is a plan of working together. [ABA 1905:10, emphasis added]

In Hays' view the ABA was created as an institutional mechanism for the determination of a division of labor between state and private entities engaged in the creation of novel forms of plant and animal life. There was no doubt in his mind as to which group was to serve the other.

Hays explicitly articulated the substance of what has become a pivotal issue of agricultural research policy. But if in 1903 he was prescient, he was also some three decades premature, and the ABA was never to become the institutional means for fostering the transfer of knowledge from the scientific community to private industry that he hoped. There were several reasons for this. The association's membership was originally drawn from the ranks of agricultural scientists and businesses, and its early proceedings read much like the yearbooks of the USDA. Founded on the heels of the rediscovery of Mendelian inheritance, it soon attracted a large set of members with increasingly diverse scientific and commercial interests. Ultimately the ABA proved too narrow an institution to encompass the rapidly growing and differentiating fields associated with the development and elaboration of genetics. As agricultural disciplines coalesced around departments in the land-grant universities and founded their own professional societies (e.g., the American Society of Agronomy, 1907), agriculturally oriented scientists tended to transfer their principal allegiance elsewhere. This process was accelerated by the growing eugenic tendencies within the ABA and by certain internal tensions among prominent officers (Kimmelman 1983). In 1912, Hays resigned from the association he had worked so hard to shape, and following a substantial reorganization, the ABA changed its name to the American Genetic Association and effectively ceased to play an influential role in agricultural matters.

The position of the seed industry

Another reason for the failure of the ABA to galvanize Hays' hoped-for "grand cooperative effort" between public science and private business was, at least in plant breeding, the relative weakness of the private sector. Hays' disingenuous rhetoric gave the seed companies more credibility than they actually merited. Before embarking on the pursuit of new avenues opened by Mendelian genetics, the seed trade had first to contend with the historical residue of the previous era. Seed companies' first priority was simply to establish a market, and they continued to view the congressional distribution as a principal constraint.

In his 1899 annual report, Secretary of Agriculture Wilson reported a continuous flow of letters from seedsmen urging discontinuance of this work. The seriousness with which the seed trade viewed the program is evident in the vitriolic language with which free seed distribution is excoriated in ASTA proceedings[1] and by the association's constant lobbying efforts against it. It was not until 1924 that the seed trade was finally able to persuade Congress to eliminate the free distribution of seed to the public. At the time it was abolished, the program was the third largest line item in the USDA budget, with only salaries and the expenses of the Bureau of Animal Industry receiving larger appropriations.

The seed industry also had to contend with the constraints and problems posed by the biological nature of the seed and its natural reproducibility. It is no accident that private companies had established a presence not in the grain crops – in which the end product can also be used as seed – but in vegetables and forage grasses, which are harvested not for grain but for use of other plant parts such as leafy growth or immature fruits. The farmer's ability to provide for his or her own seed requirements effectively excluded private industry from the most widely grown and therefore most potentially lucrative crops. In wheat, for example, 97 percent of the seed used in 1915 was sown on the farm where it was produced. The remaining 3 percent was almost entirely accounted for by sales between farmers. A further complication arose from the social relations characteristic of commercial seed production. Seed companies do not grow the seed crop themselves, but contract this activity to independent farmers. The seed companies had no wish to purchase more than what they expected to sell. But failure to acquire the entire seed crop left a surplus of seed to the grower, who was likely to put it on the market himself, frequently at prices that undercut those established by the commercial retailer. New varieties had a tendency to "leak" in this way. By contracting to farmers, seed companies were continually in danger of reconstituting their own competition and creating a depressing effect on retail seed prices.

Market enlargement was further inhibited by the very uneven quality of the seed produced in the private sector and by the often inflated claims made in an effort to increase sales. Despite his enthusiasm for the future of the private breeder, Willet Hays (1900:263) had to admit that "The public has so little confidence in new things, because they have been asked to pay long prices for the privilege of experimenting with so many newly originated varieties sent out before their values were fully determined experimentally." J. D. Funk of Funk Seeds, proprietor of one of a very few seed corn firms, noted at the ABA's first annual meeting in 1903 that the farmer had come to view commercial seed corn as "a huge 'gold brick', and he is afraid that it is just another scheme to get him to bite" (Funk 1905:30). In 1905 the USDA was given authority to purchase and test samples of seed in the commercial market. Between 1912 and 1919, 20 percent of the 15,000 samples tested were found to be adulterated or mislabeled (Copeland 1976:330).

The agrarian struggles of the Populist movement in the 1890s were not completely dissipated by the electoral defeats of 1896. Farmers spearheaded the drive for much of the regulatory and reformist legislation of what has become known as the "Progressive Era."[2] It was they who saw most clearly the discrepancy between the raw materials they produced and the quality of final product offered in the market (Kane 1964:161). Popular struggles for regulation of industry extended to concern over the quality of seed as well. Between 1899 and 1908 eleven states passed seed laws, and in 1909 fifteen more states had legislation pending that stipulated such provisions as "seed true to name," "guarantee of purity and germination," and "cleanliness."

Speaking at the 1909 ASTA convention, E. H. Jenkins of the recently founded Association of Official Seed Analysts told the seedsmen that the farmer was demanding protection and described the situation in the seed trade as "caveat emptor with a vengeance... The farmer's crops are his livelihood, but for his seed no one will be in any degree responsible" (ASTA 1909:58). Watson S. Woodruff (ASTA 1909:23), president of the ASTA, denounced the subject of seed legislation as "one of the most serious matters for consideration that our Association has ever had to face. I might go so far as to call it a crisis." Efforts to eliminate the congressional seed distribution were added to ASTA's efforts to "shape state legislation in a safe and sane direction."[3]

In the first quarter of this century the private seed industry faced a series of difficulties that placed constraints on its flexibility and the possibilities of accumulation and expansion. It was not in a position to pursue the promise that the rediscovery of Mendelian theory held for crop improvement. The state, on the other hand, was unfettered by the need to turn a profit. Indeed, it was scientifically and institutionally positioned to move with alacrity further into plant breeding. Moreover, this move was assured because important segments of capital supported and encouraged the strengthening and expansion of public agricultural research.

Capital and country life

The farmer took a certain pleasure in viewing his occupation as the fundamental material base of society. Danbom (1979:21) quotes Grange Worthy Master N. J. Bachelder in a 1908 address as follows:

> The prosperity of other industries is not the basis of prosperity in agriculture, but the prosperity of agriculture is the basis of prosperity in other industries.... Immense manufacturing plants and great transportation companies are dependent upon agriculture for business and prosperity.

The relation is in general far more reciprocal in nature than Bachelder admits, but his statement contains a good deal of truth with reference to those industries that articulate most closely with agriculture. The railroads and a significant part of the banking community depended heavily on farm production for their revenues. Both the productivity and total output of American agriculture had risen substantially between 1870 and 1900, largely as the result of increases in the amount of land under cultivation and the application of machine technology. But for three decades after 1900, production grew quite slowly, and productivity held constant as land was added less rapidly and few important new technologies were developed.

The apparent stagnation of agriculture attracted considerable attention throughout the nation. Rising food prices evoked Malthusian concerns, and a deteriorating agricultural trade balance raised fears as to the capacity of the United States to sustain its position in international trade. The Country Life Movement is probably the most widely known and studied response to what was perceived as a problem of national significance. Organized around prominent agricultural scientists such as Cornell's Liberty Hyde Bailey and Gifford Pinchot of the USDA, the movement was funded and supported by urban-based interests and industries that depended on the flow of agricultural commodities for their revenues. What these groups shared was an interest in the rationalization of agriculture through science. If the agricultural scientists were motivated by a vision of a transformed and improved rural society, the business interests were unambiguously interested in restoring productivity advance as a necessary condition for their own continued capital accumulation.

If increased productivity was the answer to an expanding flow of commodities, it was widely assumed that what was needed to improve productivity was quicker and more widespread adoption of improved farming practices and new technologies. The policy expression of the movement that gave birth to the Country Life Commission focused on educating the farmer. The Rockefeller-endowed General Education Board encouraged the land-grant universities to adopt agricultural demonstrations and a systematic approach to extension (McConnell 1953:25).

Impatient with the USDA's dilatory approach, private enterprises such as Sears Roebuck & Co. pushed ahead with their own extension programs. It was the railroads that were most active. By 1910 most of the major railroad companies had established agricultural departments to, in the words of Frisco Line executive B. W. Redfearn, "promote better agricultural methods among our farmers, interest them in the scientific side of the work, and . . . prevail on them to adopt it" (quoted in Scott 1962:14). Agricultural demonstration trains were detailed for traveling programs that encouraged the use of more productive plant varieties. Between 1904 and 1911 the Burlington and Rock Island Railroad's sixty-two "Seed Corn Specials" covered 35,705 miles carrying Iowa State agronomists to 740 lectures attended by 939,120 persons (Scott 1962:4). R. B. White of the Baltimore and Ohio's Agricultural Development Department explained the motivation behind the similar efforts of his company:

> The Railroad's interest in problems of the farm is not prompted by any philanthropic motive, but purely because we believe it good business to take an active interest in what the territory we serve produces. The wisdom of such a policy is indicated in the greatly increased traffic of farm supplies and farm products. [quoted in Jones 1957:67]

With passage in 1914 of the Smith-Lever Act, which institutionalized the county agent system, private industry was able to reduce its own efforts in favor of federally subsidized extension activities.

The Smith-Lever Act was largely the product of lobbying by such groups as the American Bankers Association, the Council of North American Grain Exchanges, and the National Soil Fertility League. Members of this latter organization were leading transportation companies, banks, and manufacturing concerns. Their purpose was frankly not to improve soil fertility but to achieve passage of extension legislation (McConnell 1953:32). Farmers had little to do with the Smith-Lever Act, and North Dakota senator Asle J. Gronna declared on the Senate floor that he had "yet to find the first farmer who has asked for this appropriation" (quoted in Danbom 1979:73). Passage of the act put in place the institutional mechanism for systematic transfer of practical scientific knowledge from college and experiment station to the farm. Danbom (1979:74) notes that in 1914 "the potential means of revolutionizing agriculture were complete."

But the revolution would not come for another twenty years or so. Not until 1935 would productivity move off the horizontal and begin its precipitous climb to current levels. Danbom was therefore quite correct in using the qualifier "potential" with "means of revolutionizing agriculture," but he did so for the wrong reason. He asserted that "The problem faced by scientists was less one of finding ways of increasing production than of getting farmers to adopt the innovations they had developed" (Danbom 1979:39). He thus located the constraints on improved productivity in farmer resistance to productive innovation and their highly "traditional orientation."[4] Farmers were, however, much more receptive to innovation than Danbom gave them credit for. Indeed, given the wide range of seed swindles and snake oil schemes to which so many fell prey (Hayter 1968), we may wonder at the credulity of many farmers. Most were, by 1900, tied to product markets. This market integration was enough to facilitate the operation of the technological treadmill and thus to ensure adoption of those innovations that *actually did* improve productivity. The failure of productivity to rise before 1935 had as much to do with the constraints on scientific output as it did with a lack of extension programs and farmer recalcitrance.

The rapid acceptance of new plant varieties and such innovations as bordeaux mixture and the Babcock butter test before 1900 testify to the willingness of farmers to employ effective new technologies. But none of these technologies were transformative; they tended to provide immediate solutions to discrete problems, but did not move productivity to qualitatively different levels. Experiment station research focused on the "putting out of fires" rather than on more theoretical work that might have led to a radical reconstitution of the agricultural production process. This applied tendency of public research was directly a function of farmer demands. Rosenberg

(1976:156) gives an excellent account of the scientists' struggle to pursue basic research in the face of a farm constituency "aggressive in its rectitude and casual in its assumption of the right to enforce demands upon the performance of station scientists." Farmers expected experiment station personnel to test seed, fertilizer, and soil samples and to answer the practical questions they had regarding farming operations. They saw the public institutions as regulatory and advisory agencies and had little use for the abstract and apparently impractical pursuits of "research."

The scientists, who wanted to be something more than fertilizer analysts, complained bitterly of the farmers' insistent demands and the way in which such demands limited their potential. In 1906, scientifically minded experiment station directors and land-grant university deans were able to use the influence of the Association of American Agricultural Colleges and Experiment Stations (AAACES) to achieve passage of the Adams Act. This bill appropriated funds only for payment of the necessary expenses of conducting "original researches or experiments bearing directly on the agricultural industry of the United States" (34 Stat 63, Sec. 1) and empowered the secretary of agriculture to ascertain that the money was so used. The 1908 report of the AAACES' Commission on Agricultural Research looked hopefully to the end of "an era of the diffusion, rather than the acquisition of knowledge" and to some insulation from the farmers' "coercing influence in the direction of superficial inquiry and immature conclusions" (AAACES 1908:7). The USDA enforced the Adams Act provisions through the Office of Experiment Stations, but change was slow in coming: "Lots of people think they are doing investigational work if they are conducting variety tests," complained the office's assistant director A. W. Allen (quoted in Rosenberg 1976:183).

Popular pressure for practical results was not so easily escaped, and while assisting the farmer might be duty, scientists felt called to a higher task. Kansas researcher Waters told the National Association of State Universities and Land-Grant Colleges (NASULGC, successor to AAACES) that

> It has been a fundamental mistake to assume that the duty of the experiment station is solely or even principally to benefit the farmer directly. A larger responsibility rests upon it – that of making an exact science of agriculture. [Waters 1910:81]

This was a motto to which business as well as the scientist could subscribe. And when in the midst of the farm depression of 1922 H. L. Russell, dean of the College of Agriculture at the University of Wisconsin, testified before Congress in support of the Purnell Act raising appropriations for agricultural research at the experiment stations, it was to business rather than the farmer that he appealed:

> The business interests of this nation are ready to support this [research] effort in a better way than they have ever before, because business recognizes now more than it ever did the absolute fundamental necessity for having prosperous agriculture in order that business may continue. [U.S. Congress 1922:36-7]

It was only after some twenty-five years into the twentieth century that a really significant institutional and financial space was opened for agricultural research that was relatively autonomous of the direct farmer demand that constrained acquisition of basic understanding that could lead to the development of transformative technologies.

Public breeding ascendant

Agricultural science is heavily grounded in biology, and that field was, in 1900, still immature. The truth is that agricultural science did not have much to offer the farmer that would greatly increase productivity. The Country-Lifers and their corporate supporters may have envisioned a rationalized agriculture, but agricultural science was not yet ready to deliver on its promises.

The rediscovery of Mendelian theory may have generated a euphoric sense of anticipation among those involved in plant improvement, but in fact it was only anticipation. Mendel's work was less a Rosetta Stone, providing the key to the mysteries of heredity, than an agenda for further research. An understanding of the mechanisms of inheritance was to be a crucial tool for the control of transmitted characters, but before the new science of genetics could really begin to contribute to breeding practice, a host of inconsistencies had to be clarified, interpreted in a Mendelian framework, and unified in a coherent corpus of theory. Theoretical work that was ultimately to have tremendous practical impact was begun in the early 1900s by such men as E. M. East at the Connecticut Experiment Station and George Shull at Cold Spring Harbor. East produced a discussion of the work of such biologists as Lamarck, Weismann, Darwin, DeVries, Johannsen, and Mendel in *The Relation of Certain Biological Principles to Plant Breeding* (1907), and in 1905 Shull began studies of quantitative inheritance in maize. But for every East with freedom to pursue a scientific problem *per se* there were many more station scientists who lacked the imagination or, more important, the autonomy to investigate matters that did not relate directly to a particular practical problem of farm production. If Mendel was necessary for rapid progress he was not sufficient, and despite the hopes of some, there was to be no swift outpouring of markedly superior new plant varieties.

Indeed, the most prominent successes of the early years of the twentieth century had nothing to do with the new science of genetics, but were products of that old workhorse of plant improvement, varietal introduction. In 1895 and 1900 the newly established Section of Seed and Plant Introduction sent M. A. Carleton on germplasm collection expeditions to Russia. Carleton brought back varieties of wheat that were to transform the production of that crop in this country. With the introduction of these varieties, the United States went from the production of only soft wheats to production of three classes of hard wheats for milling and export, and by 1921 the variety Kharkov was grown on over 21 million acres (Klose 1950:116).

Other important introductions included Acala (introduced 1907) and Yuma cottons (introduced 1900) from southern Mexico and Egypt, respectively. Victoria oats were brought from Uruguay in 1927 to save U.S. growers from crown rust. Most impressive, however, was the introduction program for soybeans. Between 1900 and 1930 over four thousand varieties were obtained from Japan, Korea, and China. After initial screenings on USDA and experiment station plots, the most promising varieties were distributed to farmers for localized adaptation and testing. By 1914 the soybean had begun its meteoric rise to prominence, and in 1924 it was a crop grown on 2.5 million acres and worth some $24 million.

If the farmer was still a significant factor in the breeding process, the USDA and station plant breeders were gradually developing more sophisticated techniques. Farmers were effective because of their sheer numbers, but the scientists began to make up in systematic intensity of effort what they lacked in scale of operation. At the turn of the century the method of "single-line selection" was well established. This method consisted of segregating and reproducing the seed from single plants, applying continuous selection to subsequent generations, and paying attention to the value of the variance revealed in the populations. The specific protocol often followed the "centgener" approach: a plot of one hundred plants spaced a certain distance apart (Clark 1936:219).

Gradually, the Darwinian attention to selecting better-adapted individuals was joined to the Mendelian analysis of hereditary differences (Simmonds 1979:13). Hybridization came increasingly to be used in conjunction with selection. Two varieties would be cross-bred, and new genetic variability generated by the combination of their hereditary characters. Single line selection was then applied to the progeny of the cross. It became apparent that individual characters could be transferred from one variety to another via a modification of this approach called backcrossing. An elite variety susceptible to a disease could be crossed with an exotic variety that contained genes for resistance to the disease. The progeny of the cross would be selected for possession of the resistance character and mated again to the elite variety. This process of backcrossing to the elite parent could be recurrently performed until there was a new variety with all the characteristics of the elite precursor *plus* the disease resistance of the exotic.

The coupling of Darwinian and Mendelian thought, and the practical advantages of the new techniques, had a profound effect on plant breeding. Such methods as backcrossing involved more time-consuming and elaborate operations than most working farmers could afford. Moreover, as the complex nature of the chromosome basis of heredity was progressively revealed by research, an understanding of such genetic features as linkage, multiple and modifying factors, and factor interactions became increasingly necessary for the breeder. The development of increasingly sophisticated statistical methods made biometry a vital tool for the analysis and interpretation of experiment results.

In short, plant breeding was becoming less of an art and more of a science, with a corresponding change in the character of its practitioners. The individual farmer practicing simple mass selection in his fields was no longer the equal of the experiment station researcher. The noted corn breeder H. K. Hayes recalled in his 1957 presidential address to the American Society of Agronomy his collegiate introduction to plant breeding in 1907 at Massachusetts Agriculture College:

> I took a course in thremmatology, a dressed up name for plant breeding. It was taught by a distinguished professor of horticulture. He taught us to say, "similar begets similar" for the well-known precept that "like begets like." He taught us Mendel's Laws but I did not understand them at the time. At graduation I did not know that such a thing as a chromosome existed. [Hayes 1957:626]

By 1920 the breeding of rust-resistant wheats involved knowledge not only of the chromosomes' existence but also of their number and genomic relationships.

New knowledge affected the way breeders viewed their plants. The shift from selection to hybridization was paralleled by a reductionistic shift in focus from the whole organism to its constituent genetic components. Attention was focused on the gene rather than on the plant. Recall W. A. Orton's realization that plant varieties were "*plastic groups*," a genetic vocabulary capable of being recombined with specific objectives in mind. The breeder's awareness was broadened to include not only the limited set of varieties that under selection might be adapted to commercial use but also a whole new set of strains and land races that previously had seemed too "primitive" or "wild" for consideration. Breeders no longer sought in plant introductions new varieties that might be superior to current ones; they looked at exotic germplasm for *specific traits* that could be transferred to established varieties. This was a change of the greatest importance. It marked a watershed in the development of plant breeding in the United States. No longer was the breeder's task to *adapt* elite germplasm from other countries to American conditions, it was now to *improve* established varieties by incorporating particular exotic characters.

This change in emphasis occurred gradually but was substantially complete by 1925. It naturally implied a different strategy for germplasm collection activities. K. A. Ryerson (1933:124), plant explorer with the USDA, summarized the new perspective: "Species and varieties which in themselves have little or no intrinsic value become of first importance if they possess certain desirable characters which may be transmitted through breeding." Though the number of expeditions remained relatively constant over time, the actual number of accessions per five-year period nearly doubled between 1925 and 1930 and has continued to grow, save for a hiatus in collection during World War II. Plant explorers looked not so much for new introductions as for breeding material, not so much for a superior variety that might be adapted to American conditions but for a plant with perhaps only one superior characteristic. Hence, they collected a much broader range of germplasm. Since 1925 the contribution of exotic plant introductions to American agriculture tends to be described not in terms of the new species and varieties established but in terms of the particular characters that exotic germplasm has furnished for incorporation into elite American breeding lines.[5] Plant exploration became a search not for useful *plants* but for useful *genes*.

From the turn of the century, public plant breeders produced a growing stream of new crop varieties. Many of these were selections out of introduced strains, but there was an increasing proportion that were the result of hybridization. As the complement of new public varieties grew, the experiment stations confronted the problem of getting the new plant types to the farmer. As early as 1897 the Minnesota station began its own distribution program. In a survey of the state experiment stations Ten Eyck (1910) found that twenty-four SAESs had some arrangement for the distribution of seed of new varieties directly to the farmer. Such arrangements proved more or less impractical as demand grew and farmers accustomed themselves to what would soon become, as Richey phrased it, the "time-honored method of variety replacement" (Richey 1937:973).

The Wisconsin station set up the Wisconsin Cooperative Experiment Association, a group of some 1,500 University of Wisconsin graduates who agreed to multiply and disseminate newly released varieties (Ten Eyck 1910:72). In a variety of permutations this basic arrangement was the model for the crop improvement associations established between 1915 and 1930 in virtually every state. These associations were closely allied to the colleges and experiment stations, and the public breeders were assured that their labor would not be counterfeited by the unscrupulousness or poor management of growers through imposition of a legal framework of seed certification providing for inspection and regulation of seed production. The "college-bred" varieties thus became available as certified seed.

The seed industry was notably absent from these arrangements, and this was deliberate policy on the part of station administrators concerned with both quality and equity. Seed companies were welcome to purchase certified seed from crop improvement association growers, but public breeders had no intention of permitting the seed trade to become an exclusive conduit for dissemination of their work to the farming community. The certification program was built around commitment to quality; certification became almost synonymous with superior varieties, genetic purity, and high seed quality standards (Copeland 1976:313). Seed companies were not very happy with such arrangements. Public breeders were in effect setting benchmarks of quality, and the association of certification with quality leveled prices among different varieties. The "blue tag" that identified certified seed greatly reduced the possibilities for product differentiation.

The ASTA had established a Committee on Experiment Stations to monitor public activities and invited written submissions from station personnel. A sampling of these comments published in the 1923 ASTA annual meeting proceedings is indicative of the character of relations between public breeders and the seed industry. New York observed that, "Unfortunately, seed houses generally have not been willing to serve as a medium for the introduction or distribution of certified pedigreed seed." Kansas bluntly asserted that in the seed trade there was "a need for somewhat more of old-fashioned honesty," and Pennsylvania reported, "We believe that the interests of the [farmers] would not be best served if the distribution [of new varieties] were carried on only by seed firms" (ASTA 1923:36-8).

The seed industry might not like its situation, but there was little it could do. The crop improvement associations established the International Crop Improvement Association (ICIA) in 1919 to protect the interests of the seed growers, and in the ICIA the seed industry faced an organization with as much political influence as the ASTA. The new techniques of breeding had begun to show results and were products of the public institutions. If private companies were to pursue these new methods for the production of improved varieties, they would have to obtain their breeders from the public sector.[6] Even then, in the absence of any kind of legal protection for newly developed varieties there would be difficulty in obtaining adequate returns on research investment. The seed industry was locked into a subordinate position to a public sector aggressive in its approach to applied science and ideologically committed to a mission of serving the farmer.

The position of the seed industry was further undermined by deteriorating economic conditions after 1925. By 1933 the annual volume of seed sales had fallen to *half* of the 1925 level, as farmers responded to financial difficulty by reducing their use of purchased inputs and reproducing more of their own seed. Moreover, in 1933 the Red Cross began distributing *free* seed to families on relief, and other assistance agencies followed suit. These organizations contracted for large volumes of seed direct from growers and then distributed the seed themselves. The seed trade saw this new threat as even worse than the congressional distribution they had scuttled less than a decade earlier, after nearly sixty years of battle. A hastily appointed Committee on Free Seed Distribution reported at the 1934 ASTA meeting that the activities of the relief agencies, if continued, could "result in an entirely new type of seed growing, one definitely for relief purposes, not for commercial purposes" (ASTA 1934:56). The hard-won status of seed as *commodity* was once again under assault, and the committee resolved "to go to the top at Washington, to bring all the influences to bear, that we can, from the seedsman, from the dealers throughout the country, Senators, Congressmen, every avenue of approach, where we have influence" in order to "throw this business back into normal channels" (ASTA 1934:56). By "normal channels" was meant, of course, the seed trade. The ASTA ultimately settled on a "seed stamp" system in that coupons would be issued to the needy by the government for redemption at stores and companies *selling* seed. It was this plan which was adopted by the government and the relief agencies. Unlike the congressional distribution, subsidized seed distribution during the Great Depression retained a commodity character and was accomplished within a normal commercial framework.

Once again, the seed trade was concerned less with moving into new areas than with defending the space it had so laboriously carved out for its activities. Thus, there developed an unambiguous hegemony of public science in the field of plant breeding. Table 4.1 provides some empirical contours to the relative position of public and private breeding and also illustrates the shift in techniques occurring in plant breeding methodology. Of the 128 principal wheat varieties grown in 1934, 100 (78 percent) were of public origin. Among private breeders, individual farmer-growers dominated, and commercial seed company presence was negligible. Also, note that while selection was the method used to produce most varieties, hybridization established a strong position.[7] This contrasts sharply with private efforts, which were almost wholly characterized by introduction or selection of new varieties rather than by use of the more advanced techniques of cross-breeding. An examination of varieties registered with the American Society of Agronomy between 1926 and 1936 reveals a similar pattern of public dominance in other important crops.

This preeminence of public plant breeding over the seed industry should

Table 4.1. *Principal wheat varieties grown in 1934 by breeder and breeding method*

Method	Public[a]	Private individual	Seed company	Total
Introduction	11	9	4	24
Selection	59	12	1	72
Hybridization	30	2	—	32
Total	100	23	5	128

[a]USDA, experiment station, or land-grant university.
Source: Reitz (1962).

not be taken to imply that public agricultural research was independent of the influence of business interests. We have seen the important role played by large-scale non-farm capital in the Country Life Movement. The climate of support for steps taken to increase productivity thus generated was instrumental in facilitating passage of the Adams and Purnell acts, which financially undergirded the development of a competent and sophisticated scientific approach to agricultural research. Rosenberg (1976) has persuasively detailed the circumstances in which station scientists found their natural allies in progressive, highly capitalized farmers and in what we would now term agribusiness.[8] Station administrators and college deans, now apotheosized as "research entrepreneurs," turned early to these sources for financial as well as political support. Charged with establishing a plant pathology department at Cornell, H. H. Whetzel found himself short of funds and successfully sought industrial fellowships from chemical companies in 1909. By 1944 he had attracted $265,920 in support (Whetzel 1945). H. L. Russell began accepting industrial fellowships at Wisconsin by 1920 and explained his approach to industry support to his colleagues at the NASULGC 1931 annual meeting in these terms:

> Personally, I have always felt a good deal like President Hadley of Yale did, when someone accused him of taking money for his institution that reeked of taint. He fully admitted the allegation by saying the only taint he found was that it "taint" enough. [Russell 1931:226]

The seed industry, however, had little influence. It was a school of rather small fish when compared with other capitalist enterprises. The Burlington Northern and the Rock Island Line were interested in hauling grain, and the more productive the farmer the better they liked it. Large-scale capital pushed public agricultural research, to the detriment of the seed industry, because only public science was ready to contribute to productivity. The seed industry, as a small and weak fraction of capital, was sacrificed to the interests of capital as a whole.

This does not mean that capital got the productivity it was looking for; as we have seen, productivity was static between 1900 and 1934. Despite advances made in plant breeding, yields were also static during this period, as indeed they had been since the 1860s. The reason for this apparent stagnation lies partly in the fact that crops were still moving into new niches on what was often more marginal land. Between 1900 and 1935 the center of wheat production moved from southern Minnesota to central Kansas. The new varieties that flowed in an ever greater stream from public researchers were not raising average yields, but were permitting extension of production into new areas: wheat into Kansas, Colorado, and Oklahoma; cotton into the High Plains; rice into Louisiana; soybeans into the Midwest. Disease problems were also appearing, and there was a continuous search for resistant strains. Also, nearly 40 million new acres were brought into production shortly after 1917 in response to rising prices and slogans like "plow to the fence for national defense" (Saloutos 1982:3). Advances in plant breeding served to *maintain* levels of yield that might otherwise have declined. Also, it should be recognized that not all research made progress. Corn was a noteworthy example. The methods of seed-corn selection recommended to farmers by public breeders were based on a faulty understanding of hereditary mechanisms in corn and actually tended to reduce yields.

At least as regards the fruits of plant breeding, the first quarter of this century does not seem to have witnessed resistance to the revolution, to use Danbom's formula. For example, the rapid adoption of the Russian wheat introductions is testimony to the receptivity that most farmers exhibited toward new public plant varieties. Danbom is ultimately correct. There was a resisted revolution, but it occurred not between 1900 and 1930 but between 1930 and 1940. And resistance was expressed not in recalcitrant adoption of innovations but in political terms. The Country Life Movement's Liberty Hyde Bailey was, as we shall see, but John the Baptist to Henry A. Wallace's messiah.

New genetics, New Deal, new agriculture

As the agricultural economy's slide from the high prices of its Golden Age began to accelerate after 1920, farmers began to question the benefits of research. In an article entitled "The Responsibility of the Agricultural Experiment Station in the Present Agricultural Situation," the director of

Ohio's facility complained that some "have gone so far as to hold the experiment stations responsible *for* the present situation." He admitted that "This is very likely true," but asserted that the solution was *more* research, not less, for research meant more "economical production" (Williams 1928:519-21). It was this point that would become the principal rallying cry for embattled supporters of public research.

With the Depression in full swing by 1932, opposition to agricultural research had reached Capitol Hill. During appropriations hearings in that year, Senator Kenneth McKellar (D-Tennessee) bitterly attacked all items in the USDA budget that appeared to fund scientific work (*Science* 1932:10). In the House, Congressman Allgood, who had been Alabama's Commissioner of Agriculture for four years, suggested that in view of the problem of overproduction it might serve the farmer better if research was undertaken to *propagate* rather than eliminate plant diseases.[9] The Roosevelt administration inherited a budget in which funding for agricultural research was substantially reduced over historic levels, and its first inclination was to continue this trend by virtually eliminating extension and research activities. Threats of political retaliation from such organizations as the American Engineering Council and firms like Armour and Company and the Champion Fibre Company prevented the demise of such programs, but failed to halt the decline in research appropriations (Pursell 1968:233).

Secretary of Agriculture Henry A. Wallace, in his 1934 report to the president, admitted that there was an apparent paradox in the relation of Agricultural Adjustment Act (AAA) policies and scientific research:

> In these efforts to balance production with demand, and to prevent useless farm expansion, it may seem that the farmer has a quarrel with science; for science increases his productivity, and this tends to increase the burden of the surplus. Some farmers take this view. They believe they got into the present economic jam partly as a result of technical efficiency. They ask why the Government agencies help farmers to grow two blades of grass where one grew before, and simultaneously urge them to cut down their production. They declare it is almost criminally negligent for a Government to promote an increase of production, without facing the results of that increase. These ideas lead to something of a revolt against science, and to demands for a halt in technical progress until consumption catches up with production. [Wallace 1934b:25]

Wallace recognized the central paradox clearly, and he came to the same solution that Williams had in 1928: "It is undeniable that science creates problems; but the remedy is not less but more of the disturbing ferment. What one needs is not less science in production, but more science in distribution" (Wallace 1934b:25). The AAA and the much ballyhooed social research and planning programs of the Bureau of Agricultural Economics had put science in distribution in place.[10] It remained for science in production to be reconstituted, and it was to this task that Wallace turned by the time he wrote his report for the yearbook.

Plant breeder, founder of the first hybrid seed-corn company, editor of the influential farm journal *Wallace's Farmer*, son of a former Secretary of Agriculture, Henry A. Wallace was a complex man with a visionary conception of the beneficent power of science. He also understood, perhaps better than any American of his generation, the process by which agricultural production was being integrated into modern industrial capitalism.

In an article entitled "Give Research a Chance," written for the *Country Gentleman*, Wallace warned of the consequences of neglecting agricultural science (Wallace 1934a). When shown the article, Congressman James P. Buchanan, chair of the House Appropriations Committee and member of the Agriculture Committee, responded, "Tell Wallace that I will give him all the money he wants for fundamental research and I will give it to him in a lump sum so that he can formulate his own program" (Rose 1935:20). The result was the Bankhead-Jones Act of 1935, which authorized the expenditure of $20 million between 1935 and 1940 for research into "laws and principles underlying basic problems of agriculture" *in addition to* research otherwise provided for (Wallace 1936:84). The act also appropriated funds for nine regional research centers and strengthened the extension service.

There is no evidence that farmers' organizations played any significant role in achieving passage of the act, or even that they participated at all in the process. Like the Hatch, Smith-Lever, Adams, and Purnell acts before it, the Bankhead-Jones legislation was principally the product of an articulate scientific elite allied with private interests and represented by agricultural journals and corporations. Certainly, decision makers were worried about farmers' lack of appreciation, if not outright hostility, to agricultural research. Wallace's comments, quoted earlier, on the "quarrel between farmers and science" are paralleled by the presidential addresses of H. K. Hayes and F. D. Richey to the American Society of Agronomy in the years 1935 and 1937. Hayes (1935:957) noted the fact that "some have maintained that one of the causes of overproduction has been the development of high yielding varieties." By 1937 Richey was complaining that breeders "have been confronted recently with something of an obsession about the responsibility of plant research for the crop surpluses" (Richey 1937:969).

One New Hampshire farmer wrote that research "can be divided into two groups, that which increases consumption which is beneficial, and that which increases production which is harmful" (quoted in Pursell 1968:233). Along the same lines, the Grange's 1938 national convention was marked by the

National Master's demand that research appropriations be shifted from production research to utilization investigations: "Long years ago the National Grange took exception to the slogan so widely popular in that period, about making two blades of grass grow where only one grew before, and vigorously warned against the danger of overproduction of food" (Gardner 1949:131).

The National Master's plea was a bit late. The horse had already left the barn, and its name was Bankhead-Jones. Farmers naturally gravitate toward concern with the prices they receive rather than toward what research might or might not do. The farmer, like any producer, has as his principal preoccupation the valorization of his product. Political struggle in the agricultural sector during the Depression focused not on research but on product markets and on such issues as price support, loans, land tenure, and parity. The political actions taken by numbers of diverse farm organizations were, on the face of it, quite successful in forcing concessions from the state. A vast and intricate set of social programs was established by the federal government in order to address the problems of agriculture and aid the beleaguered farmers and their families.

But ultimately, these programs failed to materially slow the process of differentiation under way in the American countryside. They failed to do so in no small part because Wallace had been able to realize his goal of releasing more of the disturbing ferment of scientific research. Senators Bankhead and Jones collaborated more than once, and they are much better known for the Bankhead-Jones Farm Tenant Act of 1937 than they are for the Bankhead-Jones Act of 1935. Ironically, the former piece of legislation has the most historical visibility despite its failure in what it was intended to do (Kloppenburg and Geisler 1985). On the other hand, the latter bill was an outstanding, if little recognized, success. Their respective records of failure and success are not unconnected.

Social programs were unable to contain the forces unleashed by the development of science. Historians and other social scientists have been preoccupied with New Deal policies in the sphere of circulation; they have failed to look closely at the production of new scientific knowledge. The New Deal saw a tremendous increase in support for agricultural research, and it was that feature that gave to postwar agriculture its distinctive character and galvanized the migration from the land that Goodwyn (1980:31) has described as "a frantic sojourn of a defeated peasantry that had lost all hope of economic justice in its homeland."

The financial and infrastructural strengthening of public agricultural research came at a moment when plant breeding was coming of age. One of Wallace's first acts as Secretary of Agriculture was to appoint a Secretary's Committee on Genetics. Fifteen prominent plant and animal breeders were given the task of surveying the theory and practice of germplasm utilization in the improvement of species of agricultural importance as it had developed over the previous three decades. The 1936 and 1937 yearbooks of agriculture report the results of this survey, and they mark the coming of age of agricultural genetics. These volumes dispensed with the statistical summaries that had always been included in yearbooks and concentrated on the subjects of plant and animal breeding. The 1936 yearbook focused on 18 major species, and the 1937 issue followed on minor crops.

Despite the diversity of species discussed, the yearbook articles exhibit a remarkable thematic continuity, and there are two principal themes upon which nearly every author touches. The first of these is the debt to Mendelism. All authors pay tribute to the theoretical work that illuminated the practical problems of plant improvement. In his article on wheat, Clark notes that "The importance of genetics in solving the underlying principles of inheritance, thus facilitating more efficient breeding operations, cannot be overemphasized" (Clark 1936:240). And Jenkins goes so far as to say that, in corn, "Genetics forms the basis on which practical breeding methods are formulated and is solely responsible for the development of present corn-breeding methods" (Jenkins 1936:493). By 1936, breeders understood the inheritance of some 350 genes in corn and could map the location of about a hundred on the chromosome. Every article contains a section on the genetics of the particular species in question and the manner in which basic research has informed and shaped breeding practice.

The second theme is connected to the first. If the rediscovery of Mendelism in 1900 generated a euphoric but premature anticipation of a qualitatively different capacity to produce directed alteration in plants, by 1935 this anticipation was being realized. The shift from selection to hybridization as the chief tool of the breeder was virtually complete, and authors wrote not of whole plants but of the importance of "gene content" and the salience of exotic germplasm as a source of new "characters." In the 1936 yearbook's summary essay, Hambidge and Bressman (1936:130-1) observe:

> Now the breeder tends rather to formulate an ideal in his mind and actually create something that meets it as nearly as possible by combining the genes from two or more organisms... In this connection, he has a new confidence... he has a vision of creating organisms different from any now in existence, and perhaps with some remarkably valuable characters.

Whereas in 1900 the confidence of breeders was based on promise, in 1935 it was based on achievement and thirty years of theoretical and practical advances. This time the breeders' anticipation was not premature.

At the time the 1936 yearbook was published, a dramatic secular upturn in yields was already under way in all crops. Between 1935 and 1970, yields of wheat, cotton, soybeans, and corn would more than double. The trends

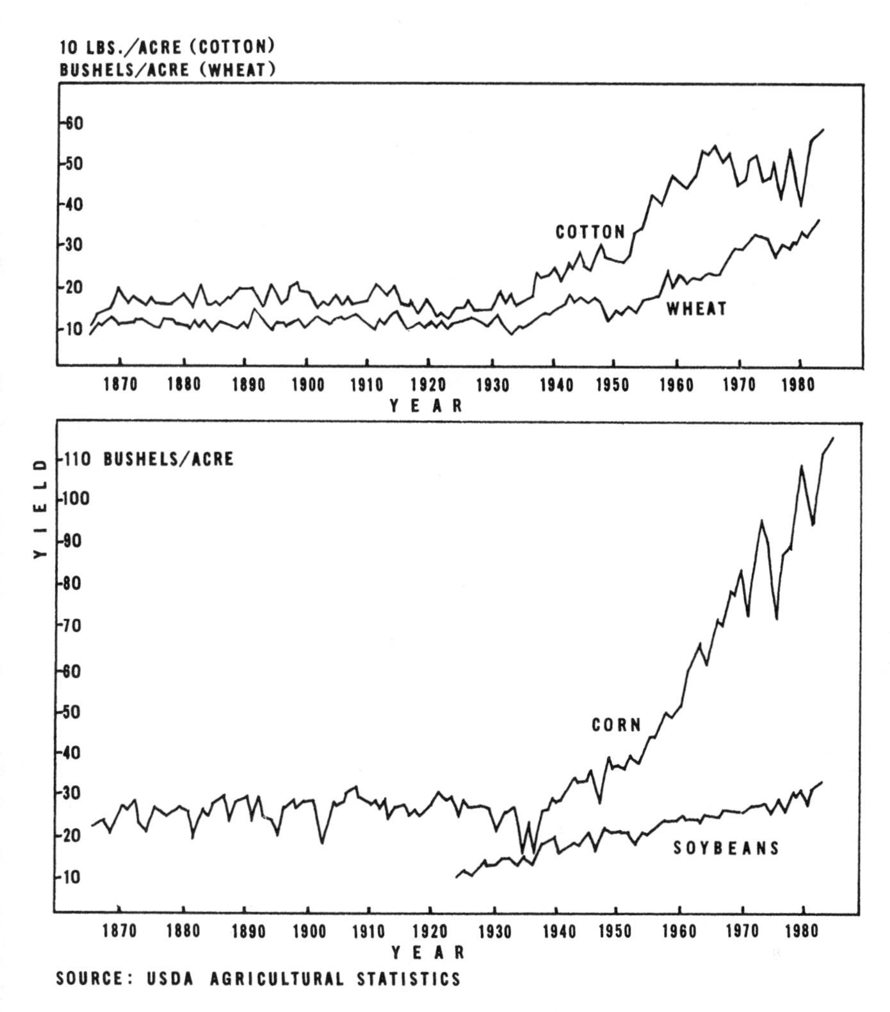

Figure 4.1. Yields of selected crops, United States, 1866–1982.

graphically and dramatically depicted in Figure 4.1 speaks volumes about the subsequent development of the structure of agriculture. In the face of such yield vectors, reformist New Deal social programs would necessarily prove inadequate to the task of slowing differentiation among farmers. As the inadequacy of production controls from the AAA of the 1930s to the Payment-in-Kind (PIK) program of 1983 has clearly shown, no amount of policy tinkering in the sphere of circulation can cope with a problem that originates not in the field but in the lab.

Conclusion

Notably absent from the pivotal yearbooks of 1936 and 1937 is any substantial consideration of the activities of private breeders. In the 1936 *Yearbook of Agriculture*, Harlan and Martini (1936:325) succinctly observed that "the accomplishments of private breeders are not extensive," and this appears to be a general consensus. According to Hambidge and Bressman's article in the *Yearbook*,

> The field of breeding and genetics has become so large, it is so dependent on progress in basic research, and it requires such continuous effort on projects running over many years or even more than one generation that it obviously becomes a function of governmental institutions capable of devoting the necessary money and time to the work and doing it with a sufficiently disinterested attitude. This is especially true because the results are for the benefit of all people rather than one group.

In their article, Hambidge and Bressman cite the development of commercially acceptable hybrid corn as perhaps the most recent striking success of public breeding efforts.

Ironically, it was this very success that ensured that plant breeding would not long remain the more or less exclusive province of the public sector. Nor would public breeders be able to maintain a strictly disinterested attitude toward their work. For reasons that will become clear in the following chapter, hybridization made corn breeding privately profitable. In 1926, Henry A. Wallace founded the first company devoted specifically to the commercialization of hybrid corn. Private investment in the plant sciences was to galvanize a whole set of changes in social relations of production on the farm and in the test plots of plant breeders.

5

Heterosis and the social division of labor

> We hear a great deal these days about atomic energy. Yet I am convinced that historians will rank the harnessing of hybrid power as equally significant.
>
> Henry A. Wallace

The development of hybrid corn has long been regarded as the supreme achievement of public agricultural science. Given Soviet premier Nikita Khrushchev's 1958 visit to hybrid seed-corn producer Roswell Garst's Iowa farm, and his plans for a Soviet Corn Belt based on American genetic technology, we may forgive Henry Wallace the hyperbole evident in the passage quoted. Without doubt the statistics commonly associated with the introduction of hybrid corn and the phenomenon of heterosis – or hybrid vigor – are impressive. Corn yields, which had actually been declining in the United States, began to climb sharply after hybrid seed became commercially available in the mid-1930s. The shift from open-pollinated to hybrid varieties was completed by Corn Belt farmers in a single decade, and by 1965 over 95 percent of U.S. corn acreage was planted with the new seed. The remarkable rapidity with which the innovation spread and the classic S shape and dramatic slope of the adoption curve for hybrid corn attracted the attention of sociologists. With the deployment of hybrid corn was born a genre of sociological inquiry known as diffusion-adoption research (Ryan and Gross 1943; Ryan 1948; Rogers 1962; Fliegel and Van Es 1983).

The effect of hybrid corn on physical output was no less dramatic. Despite a reduction of 30 million acres on which grain corn was harvested between 1930 and 1965, the volume of production increased by over 2.3 billion bushels. In the early 1950s, economist Zvi Griliches set out to assess empirically the magnitude of the social benefits accruing to hybrid corn. He compared the value of the increased corn output attributable to hybrids with the public and private research expenditures used in developing the new varieties and concluded that "*At least* 700 percent per year was being earned, as of 1955, on the average dollar invested in hybrid corn research" (Griliches 1958:419). Given this tremendous social rate of return, Griliches might well have felt justified in calling hybrid corn "one of the outstanding technological successes of the century."

The series of articles Griliches wrote on hybrid corn (1957, 1958, 1960) provided a theoretical and methodological point of departure for a substantial body of "returns-to-research" literature, much of which has focused on the performance of agricultural science. This line of analysis now constitutes a major sub-area within agricultural economics and has engaged most of the luminaries of that discipline.[1] The generally high social rates of return that economists have found to be associated with agricultural research expenditures have been extremely useful in defending the LGUs against the attacks to which they were subjected in the 1970s by populist, environmentalist, and labor groups and in providing a rationale for research appropriations at the state and federal levels. And hybrid corn's 700 percent annual return on investment remains the much cited and archetypal example of the substantial returns society enjoys from agricultural research.[2]

Hybrid corn: fabulous or fable?

So hybrid corn has occupied a preeminent position in the annals of American agricultural science. Interestingly, it was this very success that engendered Griliches' second thoughts on his analysis:

> One troublesome problem, however, remains to haunt us. Does it really make sense to calculate the rate of return on a successful "oil well"? . . . What we would like to have is an estimate that would also include the cost of all the "dry holes" that were drilled before hybrid corn was struck. [Griliches 1958:426]

This is a point of the greatest importance, but Griliches did not push it to its logical conclusion. Even more significant would be inclusion of the cost of abandoning holes that might also have struck oil, and possibly at less cost. Though Griliches expresses the intention in his first article to learn something "about the ways in which technological change is generated and propagated in U.S. agriculture" (Griliches 1957:501), he does not critically examine the genesis of hybrid corn itself. Thus, he states, "As everyone knows, hybrid corn increased corn yields." That is true, but hybridization is but one of a variety of breeding techniques, and a full cost accounting should reflect the abandonment of productive wells as well as dry ones.

That alternatives to hybridization were available is not idle speculation. In a highly regarded plant breeding text, the British geneticist Norman Simmonds states:

> The realization . . . that refined modern methods of population improvement are very powerful indeed prompts the question: are hybrid varieties really necessary, even in outbreeders? In trying to answer this question, the undoubted practical success of hybrid maize is irrelevant; a huge effort has gone into it over a period of nearly sixty years; population improvement is recent in origin and small in scale by comparison, yet it is evidently capable of rates of advance which are at least comparable and may well be achieved more cheaply. [Simmonds 1979:161-2]

What Simmonds implies has been made explicit by Harvard geneticist Richard Lewontin (1982:16):

> Since the 1930's, immense effort has been put into getting better and better hybrids. Virtually no one has tried to improve the open-pollinated varieties, although scientific evidence shows that if the same effort had been put into such varieties, they would be as good or better than hybrids by now.

In other words, the tremendous "success" evidenced by hybrid corn might have been achieved just as well through population improvement techniques in open-pollinated varieties.

If this was true, why was only the hybrid course taken? Does it matter which breeding method one chooses if ultimately one obtains the same yields? Certainly any economist would be interested in the relative efficiency of the procedures and in the opportunity costs of selecting one breeding strategy over another. But there is an even more compelling reason to examine closely the historical choice of breeding methods in corn, for the use of hybridization galvanized radical changes in the political economy of plant breeding and seed production. There is a crucial difference between open-pollinated and hybrid corn varieties: Seed from a crop of the latter, when saved and replanted, exhibits a considerable reduction in yield. Hybridization thus uncouples seed as "seed" from seed as "grain" and thereby facilitates the transformation of seed from a use-value to an exchange-value. The farmer choosing to use hybrid varieties must purchase a fresh supply of seed each year.

Hybridization is thus a mechanism for circumventing the biological barrier that the seed had presented to the penetration of plant breeding and seed production by private enterprise. And industry was not slow to take advantage of the space created for it by scientific advance. Between 1934 and 1944, hybrid seed-corn sales went from virtually nothing to over $70 million as a wide variety of new and established companies entered production (Steele 1978:29). Seed-corn became the very lifeblood of the seed industry and now accounts for about half of the $6.4 billion in annual seed sales generated by American companies (McDonnell 1986:43).

The development of hybrid corn has been absolutely fundamental to the rapid growth the American seed industry has enjoyed since 1935. As Griliches (1958:421) noted, "If there were no hybrid corn, farmers would use mainly home-produced open-pollinated seed." In this era of "big science" and "hard tomatoes, hard times," it is natural to ask if the evident advantages hybridization offered for private enterprise influenced the development of that breeding technique, and if so if it was at the expense of other potentially productive methods such as population improvement. Reviewing the history of maize improvement, Simmonds (1979:159) has argued that the development of hybridization in corn was the result of "the historical accident that ideas on hybrids developed at just the time at which progress by other methods seemed poor." Other breeders express similar views (Frankel 1983). The pursuit of hybridization is seen to have been determined by individual breeders' objective scientific assessments that hybrids held the most promise for productivity gains. From this perspective, technical considerations internal to science dictated the choice of breeding method.

On the other hand, Berlan and Lewontin (1983:23) state flatly that "Hybrids opened up enormous profit opportunities for private enterprises and for this reason all efforts were shifted to the new technique." In their view, the development of hybrid corn was simply an artifact of the expression of particular class interests (Berlan and Lewontin 1986b). With these opposed positions in mind, let us examine the historical development of hybrid corn and attempt to understand the dynamic interaction of scientific possibility and social forces.

Corn breeding at an impasse

Before 1935, neither public nor commercial breeding efforts had much effect on the types and varieties grown in the cornfields of the nation. Farmers were the principal breeders. Early on, they mixed the Northern Flint and Southern Dent land-race complexes that were the genetic heritage of the American Indian. Crossing these two distinct reservoirs of maize germplasm provided a tremendous pool of genetic diversity from which a great variety of plant types could be extracted. Individual farmers had particular ideas regarding the traits each thought important (e.g., color, prolificacy, maturity, tillering, etc.), and rigorous mass selection was sufficient to produce a great number of varieties adapted to many different areas and exhibiting distinctive characteristics.

Moreover, genetic variability in corn varieties was continually enhanced as a result of the sexual morphology of the corn plant itself. Unlike most other principal crops, corn is an outbreeder (allogamous) rather than an inbreeder (autogamous). It is open- or cross-pollinated rather than self-fertilized, and each kernel on an ear of corn may be fertilized with pollen from a different plant. Thus, unlike a wheat variety, which mostly breeds true, a corn variety is in a constant state of genetic flux. Whereas in autogamous crops there is a small percentage of natural crossing, in corn nearly every plant is the product of a unique cross. Although distinct improved varieties could be developed, they were not very stable unless isolated from other varieties. C. P. Hartley (1905:34) complained, "The originator of a superior strain of wheat or other self-fertilized plant can feel assured that the results of his work will endure for years, but a valuable strain of corn can be lost in one year's time by being planted near another kind."

Between 1866 and 1900 the total number of acres planted to corn tripled, and production quadrupled. Chicago replaced Cincinnati as, in Sandburg's words, "Hog Butcher for the World... Player with Railroads and the Nation's Freight Handler." Now, a hog is little more than embodied corn, and agricultural products and inputs of all sorts constituted an important source of railway revenues. Corn became a mainstay of the midwestern economy. The rapidly growing economic importance of the crop generated an increasing amount of interest among both land-grant personnel and commercial interests in the efficiency with which it was produced.

The first decade of the 1900s saw not only the inauguration of "Seed Corn Specials" by railroads but also the rise to prominence of the "corn show." Corn shows were contests at which the season's finest maize ears would be selected by judges drawn from the experiment stations and college departments of agronomy. By 1908 there was a complex hierarchy of shows running from the local to the national level. They were sponsored by growers' associations and companies, such as Armour and International Harvester, that provided expensive pieces of farm equipment as prizes. At an annual National Corn Exposition a single ear would be declared champion of the world.[3]

The idea behind the corn show was that it would be an effective medium for encouraging farmers to adopt better varieties and would provide tangible examples of what were thought to be desirable ear types for seed corn. The corn shows succeeded in these ends beyond all expectations. In choosing ears for seed, farmers intensively selected for those characteristics weighted heavily on the "show card," which was the standard of judgment. The acceptance and distribution of a variety came to be largely dependent upon its performance in the corn shows (Sprague 1955:62), and winning samples could command high prices as seed stock. Commercial seed-corn companies received a boost as farmers sought to replace their own varieties with more prestigious strains.[4]

Ironically, the very success of this massive extension program ultimately had a *depressing* effect on corn yields. The corn show standards developed by the pre-Mendelian breeders attached to the agricultural colleges largely emphasized aesthetic factors. Such traits are not, however, necessarily correlated with yield or other economically important characteristics. What an ear looked like had little to do with the way plants grown from its kernels would perform in the field. Moreover, continued selection for ear type constituted a form of inbreeding that exerted a depressing effect on yield (Hayes 1963:26). The corn shows thus encouraged genetic uniformity and reduction in vigor, both through the replacement of a multiplicity of varieties with economically inferior but aesthetically pleasing show-derived strains and through their influence on the selection criteria employed by farmers (Wallace and Brown 1956:97; Hallauer and Miranda 1981:4-5). For three and a half decades after 1900, corn yields experienced a gradual decline. After 1910, plant scientists began to recognize the deficiencies of the corn shows, but by then the principles on which they were based were so well established among farmers, extension agents, and even many academics that more than a few scientific articles were necessary to effect a change of direction.

And those who were critical of the shows had nothing better to offer. Writing in a 1929 issue of *Country Gentleman*, the well-known agricultural writer Sidney Cates commented:

> Corn breeding for the greater part of the past generation has been a jogging-pace race between the always self-confident college group on the one hand and the say-little-but-do-much practical farmer group on the other. And no matter how confident the professional group has been, whenever it came to the scratch of competitive trials, some horny handed son of toil, who year by year rogues his big fields for exceptionally high-yielding stalks, has always won by several laps. It began to look as if the scientist had nothing worth while to offer in this important field. [Cates 1929:20]

This was not for want of effort. Corn improvement had been approached by scientific breeders in a variety of ways, but none had proved practical. Despite some promising results, corn's sexual promiscuity was a factor that continually confounded efforts to develop and fix superior varieties by varietal crossing. The USDA's C. P. Hartley was forced to conclude that "the ease with which strains can be crossed is more of a detriment than a help to corn improvement" (1905:33). Inbreeding did not appear to be the answer either. Enough self-fertilization studies had been undertaken by 1900 to convince most researchers that this technique was, as P. G. Holden put it, "disastrous – the enemy of vigor and yield" (quoted in Shull 1952:16).[5] The ear-to-row method of breeding pioneered and popularized by C.G. Hopkins at Illinois in 1897 also foundered. We now know that while many of the breeding procedures used were sound in principle, they were flawed in practice as a result of the inadequacies of test-plot technique and the lack of appropriate statistical methods (Jugenheimer 1976; Simmonds 1979; Hallauer and Miranda 1981:15). Scientific corn breeding seemed to be at an impasse, and, as Simmonds (1979:152) observes, "that the impasse was more apparent than real does not affect the fact that it was an impasse."

In his paper "The Composition of a Field of Maize," presented at the 1908 annual meeting of the American Breeders Association, George H. Shull (1908) offered a way out. Located at the newly established, Carnegie-funded Station for Experimental Evolution at Cold Spring Harbor on Long Island, Shull had in 1905 begun a theoretical investigation of Mendelian genetics using corn as a subject. He was studying the inheritance of numbers of rows of kernels as it is influenced by self- and cross-pollination. He noticed, as had many practical plant breeders before him, that inbreeding reduced vigor and yield. But he also noticed that crosses of inbred plants were in some cases more vigorous and higher-yielding than the open-pollinated variety with which he had begun the experiment. Inbred plants were quite variable, and inbreeding depression tended to level off in successive generations of self-fertilization.

From these observations he drew a most original conclusion: A field of maize is a population of very complex crosses, and inbreeding is not indefinitely deleterious, but serves to segregate the variety into the various distinct and homozygous genotypes (Shull used "biotypes") of which it is composed. It is only these inbreds that are pure lines. Certain combinations of these inbreds could outyield the original source population. However, seed of these hybrid combinations could not be saved and replanted as was customary with open-pollinated varieties. While a hybrid is completely uniform in the first (F_1) cross of its homozygous inbred parents, Mendelian segregation decrees that subsequent generations (F_2, F_3, F_4, . . . ,F_n) will be increasingly heterozygous and uneven in yield. Thus, although hybrid seed is not biologically sterile like the mule, it is in effect "economically sterile" (Berlan and Lewontin 1983). Shull concluded that

> The problem of getting the seed corn that shall produce the record crop of corn . . . may possibly find solution, at least in certain cases . . . by the combination of two strains which are only at the highest quality in the first generation, *thus making it necessary to go back each year to the original combination.* [Shull 1908:301, emphasis added]

At the ABA's 1909 winter meeting, Shull elaborated upon the practical implications of his ideas. In his paper "A Pure-Line Method in Corn Breeding," he outlined a breeding plan that is precisely that for the single-cross hybrids largely used today. He called upon the agricultural experiment stations to test the value of his model and asserted that "the object of the corn breeder should not be to find the best pure-line, but to find and maintain the best hybrid combination" (Shull 1909:52). Just under a year later, the ABA's annual meeting was held in conjunction with the National Corn Exposition in Omaha, and Shull had the opportunity to present his views to agricultural scientists for the third time in less than two years. In 1914 he coined the term "heterosis" to refer to the increased vigor exhibited by the first-generation progeny of a cross between two distinct varieties.

Shull's work evoked both enthusiasm and criticism. Edward M. East of the Connecticut Agricultural Experiment Station had also been pursuing inbreeding and crossing studies very similar to those at Cold Spring Harbor.[6] East had been in the audience when Shull presented "The Composition of a Field of Maize," and while he found that the theoretical points agreed with his own conclusions, he thought the breeding method proposed by Shull to be impractical even if theoretically correct. Inbreeding was too complex a process for the farmer, and East suggested dispensing with inbreds and simply encouraging the use of a "varietal cross" in order to utilize heterosis. The farmer would "*purchase* from the line breeder two strains of seed *each year*, and grow the F_1 generation of the cross between them" (East 1908:181, emphasis added). While it is unclear whether East means a "line breeder" to have a public or private character, he is unmistakably in accord with Shull in anticipating that the farmer would return to some supplier *yearly* for seed of the parent varieties, though the farmer would perform the actual crossing.

By "strains of seed," East meant standard varieties rather than inbreds. The principal practical problem associated with Shull's pure-line method was the weakness of the inbred lines. While they might exhibit considerable heterotic vigor when crossed, their individual unproductiveness meant that so little hybrid seed was produced on the female parent that it was not cost-competitive with open-pollinated seed even given its greater yield. Seed production was a bottleneck. Though Eugene Funk of Funk Farms, the nation's largest seed-corn producer, was intrigued by Shull's presentation in Omaha, he did not believe that the single-cross hybrid had any practical significance (Crabb 1947:108). G. N. Collins (1910), a USDA breeder who had undertaken inbreeding studies in corn and abandoned the technique as doing "violence to the nature of the plant," publicly declared Shull's method to be "dangerous." While a number of inbreeding projects were initiated in experiment stations, colleges, and the USDA, none produced useful lines. Even Henry A. Wallace, who had also been moved to begin inbreeding after reading an account of Shull's Omaha presentation,[7] abandoned his experiments as unproductive.

A decade and a half into the twentieth century there was no consensus as to the direction corn breeding should take. Nor was there any unambiguously persuasive empirical evidence to point the direction forward. Shull's scheme was practically and, in the opinion of many, theoretically flawed. Though he had moved to Harvard, East continued to direct corn breeding efforts at the Connecticut Experiment Station. In 1917, his student Donald F. Jones (1917) proposed a Mendelian interpretation of heterosis that did much to improve the theoretical plausibility of the phenomenon.[8] Jones followed this the next year with a solution to the main practical barrier to adoption of Shull's method: He proposed a "double-cross" hybrid (Jones 1918). A double cross is simply the product of crossing two single crosses. The crucial difference is that the seed to be used for farm planting is borne on one of the more productive single-cross parents rather than on the weak inbred grandparent, and seed yields per acre are therefore sufficiently large to make the double cross cost-competitive (Figure 5.1).

Three features of this double-crossing system are notable. First, it is too complex to be used by the average farmer. Second, seed from the double cross cannot be saved and replanted without substantial yield reduction; so, like the single cross, it is economically sterile. Third, proprietary control over the unique inbred lines means proprietary control over the hybrid. The implications of these characteristics were not lost on East and Jones. In 1919 they published the book *Inbreeding and Outbreeding: Their Genetic and Sociological Significance*. In it, they commented:

> The first impression probably gained from the outline of this method of crossing corn is that it is a rather complex proposition. It is somewhat involved, but it is more simple than it seems at first sight. It is not a method that will interest most farmers, but it is something that may easily be taken up by seedsmen; in fact, it is the first time in agricultural history that a seedsman is enabled to gain the full benefit from a desirable origination of his own or something that he has purchased. The man who originates devices to open our boxes of shoe polish or to autograph our camera negatives, is able to patent his product and gain the full reward for his inventiveness. The man who originates a new plant which may be of incalculable benefit to the whole country gets nothing – not even fame – for his pains, as the plants can be propagated by anyone. There is correspondingly less incentive for the production of improved types. The utilization of first generation hybrids enables the originator to keep the parental types and give out only the crossed seeds, which are less valuable for continued propagation. [East and Jones 1919:224]

East acted on his perception of the commercial potential of the double cross by approaching the tenanted 15,000-acre Sibley Estate of Ford County, Illinois, with an offer to develop hybrid corn for use in the Midwest. His offer was rejected (Crabb 1947:90).[9] Instead, George S. Carter of New

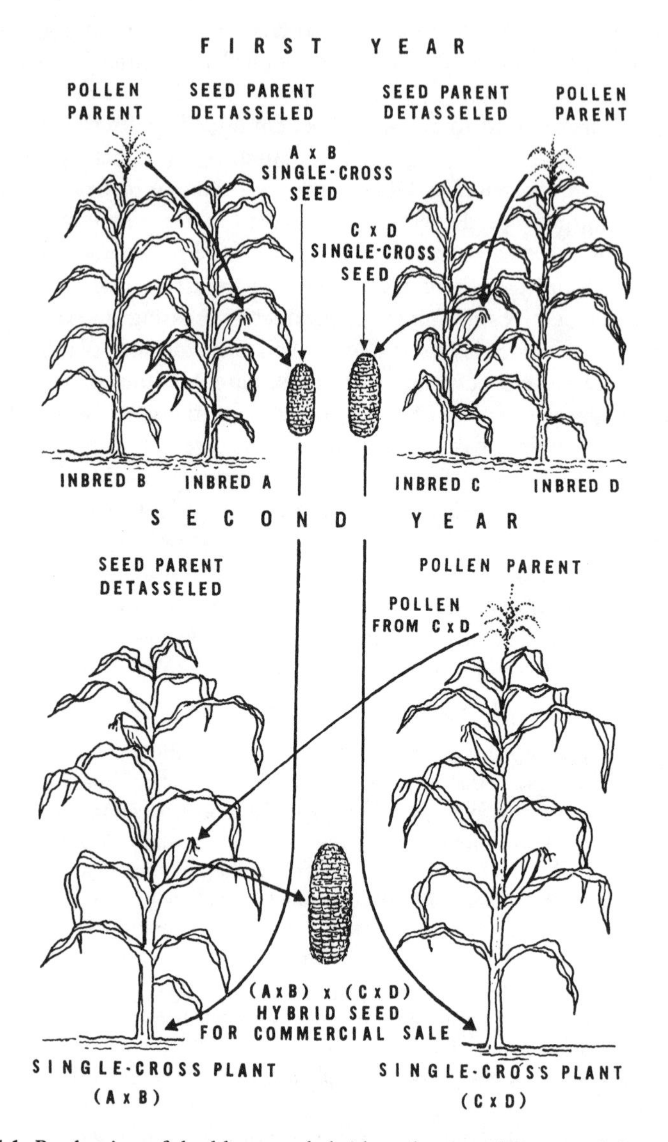

Figure 5.1. Production of double-cross hybrid seed corn using manual detasseling. The process begins with two pairs of homozygous inbred lines (A, B and C, D). Each pair is crossed (A × B, C × D) by planting the two lines in alternating rows and emasculating the female parent by manual removal of the pollen-shedding tassel (this process is known as detasseling). Only seed from the female parents is collected to ensure that no selfed seed is obtained. Plants grown from this single-cross seed are themselves crossed following the same procedure: (A × B) × (C × D). Seed is again collected from the female parent, and it is this germplasm that is the double-cross hybrid seed sold for farm production.

Haven, Connecticut, became the first commercial producer of hybrid seed in 1920. Jones supplied Carter with the single-cross parents, and Carter made the double crosses and sold the seed.

The work in Connecticut gave a modest impetus to what inbreeding studies were under way in public institutions, but it had greater impact on private breeders Eugene Funk and Henry A. Wallace, both of whom clearly recognized the commercial promise of the double cross. While on wartime special assignment to the USDA, Funk had in 1917 engineered a special appropriation for the establishment of six federal field stations for research into corn diseases (Crabb 1947:120). The first of these facilities was conveniently located on Funk Farms, and in 1919 Funk set his principal breeder Jim Holbert to the task of intensively developing inbred lines for use in what he hoped would be proprietary hybrids.[10] Jones' work reignited Wallace's interest in hybrids as well. He obtained inbred lines from nearly all public breeders then doing selfing studies – including East and Jones' material from Connecticut – and in 1919 he produced his first single crosses (Crabb 1947:147).

The value of inbreeding for the isolation of component genotypes and the heterotic effect of crossing were increasingly recognized by the more progressive public breeders. But Jones' double cross was not the only method that was suggested for utilizing these effects. In 1919, Hayes and Garber (1919:313) noted that while Jones' procedure merited further trials, an equally promising method was "the synthetic production of an improved variety by inbreeding and crossbreeding." Essentially, the proposed process involved selfing plants once or twice, selecting those lines for desired characters, intercrossing these superior lines in bulk, and continually repeating the process. This is what is now known as "recurrent selection," and it forms the basis for contemporary methods of "population improvement" (Simmonds 1979; Hallauer and Miranda 1981). There is an important difference between the procedure outlined by Hayes and Garber and the double cross advocated by Jones. Instead of an economically sterile hybrid, the outcome of breeding is a superior population that can be indefinitely propagated by random crossing within itself; that is, the farmer can save and replant the seed. Though both methods utilize inbreeding and heterosis, they have quite different socioeconomic results.

Jones (1920:87-8) understood the potential of population improvement, but regarded it as a method for the extraction of superior inbred lines rather than as a technique to be used as an alternative procedure for producing new commercial varieties. That population improvement did in fact become a simple adjunct to the dominant procedure of hybridization says less about biology than about political economy.

In 1920, Jones published in the *Journal of the American Society of Agronomy* an eloquent plea to his fellow breeders to adopt the "radical change in method" (Jones 1920:80) represented by the double cross. Among the advantages accruing to his procedure he includes "the fact that it gives the originator of valuable strains of corn the same commercial right that an inventor receives from a patented article" (Jones 1920:87). The extent to which his clear preoccupation with property rights influenced Jones' scientific work is difficult to assess. But in a letter written in 1925 to Henry Wallace, George S. Carter (Jones' commercial associate) notes that synthetic varieties resulting from population improvement might yield well but "would spoil the prospects of any one thinking of producing the seed commercially" (Carter 1925). Jones cannot but have been keenly aware of such points. And he concludes his 1920 paper:

> When, therefore, a method which is both commercially remunerative and scientifically exact is available, are the agronomists of this country going to be slow in applying it? [Jones 1920:98]

The answer was, yes, left to themselves they would be slow, if indeed they would apply it at all. Despite Jones' assertion of "astonishing yields" in his test plots (1919:2514), the breeding community had seen no persuasive evidence of the value of hybrids. Indeed, a Nebraska program begun in 1913 as a result of Shull's Omaha address had ended in failure. But the breeders were not left to themselves. By 1918, Henry Wallace was wholly committed to hybrid corn, and he threw the weight of the influential *Wallace's Farmer* behind the double cross, publicizing and praising those researchers who were pursuing inbreeding work. Addressing the annual meeting of the American Society of Agronomy in 1921, he unequivocally declared that as far as corn improvement was concerned, "The most certain and rapid progress can be made by developing pure strains by inbreeding and then combining inbred strains into . . . hybrids" (quoted in Crabb 1947:156).

But more than media pressure was required to make hybridization the central thrust of public corn work. According to Henry A. Wallace himself (1956:109), the USDA's Principal Agronomist in Charge of Corn Investigation, C. P. Hartley,

> was definitely opposed to hybrid corn. So were most of the men at the various experiment stations. The extension agencies were almost universally in favor of corn shows as the road to progress.

In 1920, Warren Harding was elected president, and he selected as his Secretary of Agriculture Henry C. Wallace, Henry A.'s father. Crabb (1947) reports that the elder Wallace asked his son to advise him of the quality of the corn breeding being done by the USDA. He was told that F. D. Richey, from whom Henry A. had been obtaining inbred material, was doing the best work. In February of 1922, Richey replaced Hartley as Principal Agronomist in Charge of Corn Investigations. Though in a 1920 address to the American Society of Agronomy Richey had cautioned that the possibilities for progress in the "pure line breeding" advocated by Jones were "largely theoretical so far," he was aware of what had brought him swift promotion. Less than two months after replacing Hartley, he issued a memorandum that left no doubt as to the path he expected his researchers to take:

> All experimental evidence indicated that the older methods of open-fertilized breeding of corn varieties are of little value . . . the fundamental problems of corn improvement can be solved *only through investigations based on self-fertilized lines*. [quoted in Crabb 1947:247-8, emphasis added]

The *coup d'etat* that brought Richey to power was followed by a rapid expansion in the number of inbreeding projects, especially in the Corn Belt (Jenkins 1936:472). However, progress did not come easily. It had been recognized that not all combinations of inbred lines were significantly more productive than open-pollinated varieties, but advocates of the hybrid had not anticipated just how few truly superior combinations there were.[11] Jones (1920) had used as one argument for taking the hybrid road the statistical impossibility of finding an open-pollinated plant with the arrangement of dominants that could be engineered via hybridization. Yet the magnitude of the task facing proponents of the double-cross hybrid was scarcely less daunting; given only 100 inbred lines, there are 11,765,675 possible double-cross combinations (Jenkins 1978:21). How could all those crosses be made and evaluated?

If the prospect of Augean labor was presented as a legitimate reason for the abandonment of open-pollinated varietal improvement, it was a spur to social and institutional innovation with regard to hybrid research. The Purnell Act of 1925 provided Richey with both the institutional authority and the financial clout to organize an unprecedented venture in directed scientific investigation. He appointed a committee to formulate a national corn breeding program in 1925, and the Maize Genetic Cooperative Group was established formally in 1928. Ten research leaders in a variety of locations were each charged with the responsibility of investigating one of the ten gene linkage groups then known in corn.[12] In addition, research in the Corn Belt stations was placed on a cooperative basis. Annual meetings among these groups provided a framework for the systematic exchange of ideas and germplasm. Moreover, the centralized source of Purnell funding allowed Richey to support hybrid development while isolating or bypassing those recalcitrant departments that resisted the new direction taken by research.[13] A sense of shared purpose, the financial support of the Purnell monies, and the leadership of the USDA were the components of a powerful new set of social relations of scientific production. Augmenting the social organization of research were important new tools. Powerful new statistical techniques supplied effective methods of predicting the combining ability and performance of the many new inbred lines being produced. The constraining factors of test crossing and evaluation were, if not eliminated, greatly eased (Sprague 1983:61; NRC 1985:30).

And the new arrangements were eminently successful. By 1935, public agencies had developed hybrid varieties that were some 10-15 percent better-yielding than their open-pollinated counterparts. We are now in a position to see that the "miracle" of hybrid corn is certainly impressive, but hardly miraculous. It was the product of political machination, a solid decade of intensive research effort, and the application of human and financial resources that, as breeder Norman Simmonds (1979:153) writes, "must have been enormous by any ordinary plant breeding standards." It also entailed the abandonment of the potentially productive well of population improvement. Two decades before the Manhattan Project, the agricultural sector had already witnessed the birth of "big science." Indeed, the development of hybrid corn can usefully be understood as agriculture's Manhattan Project, and as Henry Wallace noted in the quotation at the outset of this chapter, it may have had similarly important ramifications.

The triumph of hybrid corn required a reorientation of the farmer as well as of the scientific breeder. Corn growers were accustomed to selecting their own seed-corn ears on the basis of the corn-show standard, a technique that, as we have seen, owed as much to art as to science.[14] If farmers were to grow hybrid corn, they would have to be taught to accept the hybrid seed, which was far from the show ideal in form. Even more important, they would have to be shown that purchasing seed every year could be profitable to someone besides the seed dealer. Many scientific breeders had criticized the shows without much affecting the faith of the farming community in their time-honored practices.

Henry Wallace was wiser. He realized that the corn show, if nothing else, was superb popular theater. The corn show focused farmers' attention on corn varieties in a highly dramatic and concentrated fashion. Wallace recognized that if he could preserve the theater, but change the script, he could use the venerable shows to publicize hybrids. In the Christmas 1919 issue of *Wallace's Farmer*, he wrote: "If it is impossible to tell much about the yield of corn by looking at it, why shouldn't the corn show branch out into a contest of a new kind... a yield test under controlled conditions." He went on to publicly challenge the farmer who had won the grand champion 20-ear award at the 1919 International Livestock Exhibition to a yield contest between the prize-winning ears and 20 of Wallace's misshapen single-cross "nubbins." Wallace followed up his idea with the authorities at Iowa State University, and thus was born the Iowa corn yield test (Crabb 1947:148). From 1920 on, the yield test became increasingly popular as farmers gravitated naturally to interest in a show that emphasized actual potential for the production of a marketable commodity. The yield contest was an excellent vehicle for confronting farmers with a direct comparison between hybrids and open-pollinated varieties.

Of course, in the early years the hybrids did no better than the farmers' varieties. But in 1924 Wallace won a gold medal with Copper Cross, which, appropriately enough, was the progeny of two inbred lines, one of which had been supplied to him from Connecticut by Jones, and the other of which had come from Richey at USDA. Wallace contracted with the Iowa Seed Company for sale of the variety, and in 1924 Copper Cross was advertised in the company's catalog as "A novelty never before offered by a seedsman ... The yield is trebled, quadrupled, and in some cases increased by seven or eightfold" (quoted in Crabb, 1947:153). It was not the last time that publicly developed lines would bring private profit. In 1926 Wallace went one logical step further: He founded the Hi-Bred Corn Company (the forerunner of Pioneer Hi-Bred).

Dividing the labor: public and private research

Accepting the John Scott Medal in 1946, George Shull (1946:548) commented that Wallace's business venture had been "absolutely essential to the success which has come to hybrid corn." And Crabb (1947:265) declared that

> The long years of work done carefully, brilliantly, and patiently by East, Hayes, Jones, and all the other hybrid-corn makers would have come to little or nothing had not the often discussed system of private enterprise stepped in and matched, in a sense, the genius of corn breeders by raising up a new industry to convert the fruits of their research into something farmers could use.

Actually, the case is quite the opposite. Just as there were historical options for the directions breeding could have taken, there were options for the institutional mechanisms that could have been used to put hybrid varieties into farmers' fields. The commercialization of hybrid corn on a capitalist basis was no more an ineluctable necessity than was hybridization itself. Scientific research had been channeled toward elimination of the biological barrier that the seed posed to capital, but whether or not the opportunity for systematic penetration of the seed-corn market could be realized was a not a foregone conclusion.

Many experiment stations and land-grant college administrators assumed that farmers would produce their own hybrid seed with the support of public agencies. In the mid-1920s, a number of stations, including Iowa and Wisconsin, provided short courses for farmers interested in undertaking the production of hybrid seed corn (Sprague 1980:2). Wisconsin, in particular, was explicitly committed to provision of a technical and institutional environment that would encourage decentralized "*farmer* enterprise" (Neal 1983:2). The University of Wisconsin developed an administrative system and specialized drying and grading equipment specifically designed to facilitate small-scale production. The Agronomy Department was charged with the responsibility of providing farmers with parental foundation seed stocks and with controlling quality via certification procedures. Many other states implemented similar support systems. The result was that when publicly developed hybrids competitive with open-pollinated varieties became available in the late 1920s, many farmers were in a position to, and did, begin producing hybrid seed for themselves and for their neighbors (Steele 1978; Sprague 1980). By 1939, Wisconsin alone had 436 farmers engaged in the new enterprise (Neal 1983:3).

So the hybrid seed-corn market was not by any means automatically the domain of capital. It was something that had to be struggled over. Especially well positioned for this struggle were Pioneer, DeKalb Agricultural Association, and Funk Seed. The unique relationships these companies were able to establish with public research agencies gave them what was effectively preferential access to the techniques and breeding lines developed in many of the experiment stations and colleges. After 1935, as the market for hybrid corn was established and began to grow with unexpected rapidity, a great number of other seed companies entered hybrid production. In 1938, the American Seed Trade Association noted that the new type of seed accounted for a substantial proportion of industry profit, and they established a Hybrid Corn Group (ASTA 1938). Indeed, hybrid corn was a major factor in revitalizing a stagnant commercial seed trade, and sales increased rapidly as more and more acres were planted with seed that had to be purchased each year. Between 1940 and 1950, seed company revenues tripled as the proportion of corn acres planted to hybrid varieties jumped from 15 to 80 percent. As companies sought to gain expertise in the new technology, public breeders were hired away by the private sector, and some set up their own commercial seed operations (Crabb 1947; Hayes 1963). Funk closed the USDA research station on his farm and rehired its director as his vice-president for research.

After 1935 there was a very diverse group of enterprises producing and marketing hybrid seed corn. They ranged from a mass of individual farmers to a few well-established companies such as Funk Seed. But in these early years all producers shared one very important characteristic: They were producing publicly developed varieties (Jenkins 1936; Steele 1978; Sprague 1980). Exceptional double crosses were relatively rare, and superior varieties such as U.S. Hybrid 13 (a USDA line) enjoyed a wide use. This constituted a problem for private industry, however. As long as it relied upon "open-pedigree" public lines, its margins were determined by the farmer marketing that variety most cheaply. The principle of product differentiation and its benefits were recognized by seed firms. The obvious solution was research and the development of truly proprietary lines that could be marketed with "closed pedigrees" as unique and superior products.[15] But research was expensive, and any private research program would have to compete directly with what F. D. Richey and the Purnell Act had made a formidable public corn breeding program.

A second solution was simply to take public lines and slap a proprietary designation on them. In the 1936 *Yearbook of Agriculture*, USDA corn breeder Merle T. Jenkins (1936:479) complained that information and germplasm that had flowed so freely for so long now tended to go in only one direction:

> Among the private corn breeders and producers of hybrid corn, a tendency seems to be developing to regard the information they have on their lines and the pedigrees of their hybrids as trade secrets which they are reluctant to divulge... It would seem to be an extremely short-sighted policy, and one that probably will have to be modified in the future when the purchaser of hybrid seed corn demands full information on the nature of the seed he is buying.

By 1941 public breeders at the North Central Corn Improvement Conference were complaining that private companies were arrogating to themselves the achievements of the experiment stations by putting out publicly developed hybrids under all manner of "aliases" (Brink 1941). In 1948 over six hundred varieties of hybrid corn were available to farmers in Minnesota alone, and W. C. Coffey, president emeritus of the University of Minnesota, told the ASTA that

> farmers purchasing seed have no way of knowing whether different hybrids as advertised are practically identical or really different... I am informed, reliably I think, that in the administration of the Federal Seed Act, the situation relative to hybrid corn is not satisfactory. [Coffey 1949:126]

For their part, the seed companies worried about what might happen if the Federal Seed Act, which prohibited multiple names for a single variety, was ever satisfactorily enforced (ASTA 1944:137).

Relations between the seed companies and public agencies became increasingly strained and, in some cases, even antagonistic. Germplasm exchange was curtailed, experimental fields closed to private breeders, and charges of "theft" of inbred lines were traded (Brink 1941; Everett 1984). In an effort to protect farmers at least partially from the aggressive marketing practices of companies with closed-pedigree hybrids, a number of states adopted "delayed-release" policies for their varieties or instituted quality testing and registration requirements (Hayes 1963).

Although hybridization had removed the biological barrier to penetration of corn breeding by private enterprise, the state itself yet constituted a social constraint on capital accumulation. The state was in a contradictory position insofar as, in producing finished hybrid varieties, public agencies were performing an activity that private industry was now willing to undertake. Both state and capital were occupying similar positions in the social division of labor characteristic of agricultural research.

Private industry recognized this crucial contradiction early on and understood that its resolution depended upon the restructuring of that division of labor. In 1942 a seedsman identified only as "one of the deans of the industry" told the North Central Corn Improvement Conference:

> If our experiment stations will devote their energies to the advancement of *fundamental* research, they have it within their power to provide a basis for future progress in technical and practical corn breeding... around which active interest and cooperation of the entire hybrid industry, big and small, can rally and develop. [Ford 1942:13, emphasis added]

"Fundamental" meant moving away from the commodity-form, that is, away from the finished variety. Private-sector spokesmen urged public breeders to concentrate on the development of inbred lines and to leave the decision as to particular combinations of inbreds to be marketed as commercial hybrids to private breeders. The willingness of the seed companies to undertake the applied research of hybrid testing was largely a function of the development of efficient prediction and testing methods between 1930 and 1946 that reduced screening costs by several orders of magnitude (Hallauer and Miranda 1981:7).

But if private industry could see the contours of a division of labor it found desirable, there was no mechanism by which such a restructuring could be immediately achieved. Individual companies had had some success in influencing the directions taken by individual stations (ASTA 1938:66), but a more systematic vehicle of transformation was needed. Thus, in 1946, ASTA initiated the annual Industry-Research Conference as a means of bringing public and private breeders together for an exchange of views. In 1952 the Agricultural Research Institute (ARI) was established as a component of the National Science Foundation. Its purpose was explicitly to provide a "basis for integration" of public and private research activities (ARI 1952:11).[16] Representatives of private industry also began attending the meetings of the International Crop Improvement Association, the American Society of Agronomy, and the American Society for Horticultural Science. These bodies have provided the institutional forums in which the developing division of labor between public and private breeders has been effectively negotiated. This is clearly apparent in an examination of the historical record as represented in their published proceedings. They have, in fact, functioned as W. M. Hays had hoped the ABA would back in 1906.

Science had proved itself a productive force to be reckoned with during World War II, and capital was determined to turn it to the purposes of private accumulation. National Academy of Sciences president D. W. Bronk's 1953 address to the second annual meeting of the Agricultural Research Institute nicely illustrates the prevailing mood:

> I cannot understand how people can ask the question, "Is there any future for private undertakings in the field of science?", when science is, above all things, a human undertaking which requires freedom and uncontrolled direction of curiosity, thought, and conclusion. If all science falls under the beneficent control of government, then I do not see any hope for the preservation of freedom in any significant form of national activity. [Bronk 1953]

Plant breeding was to be no exception. Private corn breeders argued that public funds should not be used to pursue activities that attract private investment, that public duplication of private effort was wasteful, and that a reorientation of public effort would free resources for training and basic research. The proper role of the state, and of course the public research agencies are a part of the state apparatus, was to facilitate capital accumulation, not to constrain it.

This is the message that has been transmitted continuously to the public plant breeding community since 1945. Some breeders embraced it. Jones (1950) and Sprague (1960), for example, agreed that the proper role of the scientist was to break new ground and that most inbred testing was applied and hardly met that criterion. Others, especially Hayes at Minnesota and Neal at Wisconsin, argued that the public breeder had an obligation to the farmer as well as to "Science" writ large. Extension agronomist C. R. Porter (1961:213) asked: "Should the applied research program of an Agricultural Experiment Station be eliminated or curtailed to solve a marketing problem within the seed industry?"

Though it is difficult to specify the precise mechanisms whereby it was accomplished, this is exactly what occurred. Records of the North Central Corn Improvement Conference show a sharp drop in the number of hybrids released by experiment stations after 1941. After 1949, varietal releases were no longer even recorded by the North Central Corn Improvement Conference, though they did continue to be produced in decreasing numbers. Illinois and Iowa moved out of inbred line development altogether by the early 1950s, and the USDA followed in 1958 (Hallauer 1984). At the ASTA's 1956 Industry-Research Conference, Henry Wallace expressed satisfaction that Iowa and Illinois had realized they could not "get into the commercial game" and criticized those experiment stations that were obstructing private expansion by their recalcitrance in continuing to release hybrids (Wallace 1956:112).[17] It is certainly no accident that the first states to cease hybrid release were those that had the largest seed-corn markets, the most powerful private companies (e.g., Pioneer, Funk, DeKalb), and the most influential capitalists (e.g., Henry Wallace, Eugene Funk). Although the process by which public agencies were induced to abandon the release of finished hybrids was uneven, in effect it was substantially completed by 1970. Currently, only a very few states still produce hybrids, and these for small markets in ecological niches that private enterprise has not found it profitable to penetrate.

The progressive deemphasis of hybrid line development in public corn programs had a marked effect on the structure of the seed-corn industry. Farmer-producers of hybrid seed were faced with a declining inventory of open-pedigreed public varieties with which to compete with an expanding range of proprietary lines. Individual farmers did not have the resources to initiate their own research programs, and the period between 1950 and 1970 saw the virtual disappearance of the farmer as an autonomous producer in the seed-corn business (Porter 1961; Hayes 1963; Neal 1983). Nevertheless, it should be emphasized that the farmer still is at the center of the production process. Hybrid corn companies often contract production to individual farmers. This simply underlines the fact that the dominance of the industry by capital has less to do with economies of scale or production considerations than with access to research. The market for hybrid seed is now dominated by a handful of firms large enough to afford substantial research programs. In 1980, eight companies enjoyed 72 percent of the seed-corn market,[18] with the remainder of sales spread among some two hundred smaller firms producing for specialized geographic areas.

In withdrawing from the development of commercial hybrids, public corn breeding simultaneously subordinated itself to private enterprise. By disengaging from its direct link to the commodity-form, public breeding not only ceased to discipline the market but also surrendered its autonomy. Ultimately, research has value only insofar as its fruits can be applied to production in some fashion. With seed companies alone producing commercial hybrids, private enterprise is interposed between public research and the consumer of seed. The products of public research can enter production, and thus have value, only if seed companies choose to use them. Public breeders are therefore structurally bound to set their research agendas in accordance with the goals of private enterprise. If they do not produce utility for seed companies, their work will never be used. *To control the shape the commodity-form assumes in the market is effectively to control all upstream research.* This is a point of great importance for understanding how the division of labor between public and private breeders has developed.

An excellent example of this dynamic is the development of breeding methods in corn after 1950. It is one of the great ironies of plant breeding that methods of population improvement that were neglected in the 1920s and 1930s in favor of inbreeding and hybridization have, since 1950, formed the foundation of corn breeding. In the late 1940s, corn yields appeared to have plateaued (see Figure 4.1), and it seemed that either a genetic ceiling had been reached or that breeding methods were at fault (Steele 1978:32; Hallauer and Miranda 1981:8; Sprague 1983:62). The latter was the case. In their haste to find commercially acceptable hybrids, breeders had tended to rely upon a narrow sample of elite inbreds isolated from a few superior open-pollinated varieties.[19] New and more productive inbreds were needed, but they would be obtained most efficiently if the open-pollinated source populations from which they would be derived could be improved.

Since 1950, a wide variety of highly sophisticated methods for the cyclical upgrading of open-pollinated populations has been developed (e.g., recurrent selection, reciprocal recurrent selection, full-sib selection) (Hallauer and Miranda 1981:15-18). It became clear that open-pollinated varieties could be rapidly improved. Once again, the possibility was raised of developing stable, superior commercial corn varieties whose seed could be saved and replanted by farmers. But though some suggested that "synthetics" or "composites" should yield virtually as well as hybrids and be cheaper and more reliable in areas of variable environmental conditions (Genter 1967; Jugenheimer 1976:89), trying another route besides the hybrid road was never seriously considered. Population improvement has been subordinated to the dominant hybrid paradigm. But it is a measure of the power of population improvement that the productivity of inbreds has been increased so much since 1965 as to make possible the replacement of the double-cross hybrid with single crosses. Corn breeding has come full circle to the realization of Shull's original scheme. Resumption of the dramatic increases in corn yield after 1960 (see Figure 4.1) has been due not to hybridity or heterosis but to the improvement of inbred lines via recurrent selection.

Indeed, population improvement was so effective that with the advent of the single cross, some suggested that inbred line development could well be added to industry's "responsibilities." Public breeders could concern themselves principally with perfecting breeding techniques, evaluating exotic germplasm, and upgrading populations (Huey 1962:11). This view gained some support, and North Central Corn Improvement Conference reports record a decrease in inbred releases by experiment stations during this period. However, it is extremely difficult to evaluate germplasm for commercial promise without testing it in combination with inbreds currently in use. Companies found increased testing quite expensive, so inbred release is still a public function, though several companies specializing in inbred line development have become important in this phase of the breeding process. But even these depend substantially upon public material. It is a measure of the quality of experiment station and land-grant university work, and testimony to the extent of private sector dependence on publicly subsidized research, that 72 percent of commercial hybrid corn lines had one or more public inbred parents in 1979 (Zuber and Darrah 1980:241).

Public research has done more than provide the germplasm for the development of proprietary hybrids. It has also contributed to increasing the efficiency of the hybridization process itself. Production of hybrid seed, as we have seen (Figure 5.1), involves detasseling the female parent lines. It is crucial that detasseling be effectively accomplished, because any release of pollen by female parents will result in a certain percentage of self-fertilization. The seed crop of a field that has been inadequately detasseled is composed of a combination of hybrids and inbreds. Inbreds are weak, and the resulting seed mixture, if planted, will be poor in performance. Because detasseling is done manually, the hybrid seed-corn industry was fundamentally dependent on the mobilization of a labor force during the period each year when corn is in flower and releasing pollen.

Through the 1930s, labor presented no problem; wages were low, and there was no shortage of people needing work. But as hybrid usage accelerated, mobilizing a labor force for the growing number of acres used for seed production became increasingly problematic. During World War II, German and Italian prisoners of war supplemented the women and high school students who had replaced more highly paid male laborers (ASTA 1944:122). After the war, detasseling costs jumped sharply, and the difficulty of mobilizing and organizing the work force emerged as a principal concern of seed firms, especially since this was the period when yields appeared to be plateauing. Each summer the companies had to organize, train, and supervise some 125,000 laborers for a fortnight or two of work on which the success of the whole production process depended (Mangelsdorf 1974:239). The industry was vulnerable in the extreme to any kind of labor difficulty, especially because the best results were obtained when detasseling was done within 24-36 hours of pollen-shed.[20] A mechanical solution was not technically feasible. Machinery tended to bog down in wet fields, and mechanical cutters too often took leaves along with tassels, thus reducing photosynthetic activity in the damaged plant and reducing overall yields.

In what is widely admired as a most elegant piece of applied genetics, public agricultural science provided the hybrid corn industry with a genetic solution to its labor problem. The existence of genetic factors producing male sterility in corn had long been recognized, and there had been many efforts to apply this to seed production even in the 1930s and 1940s. However, it was again D. F. Jones who succeeded in translating theoretical possibility into practical reality (Jones and Everett 1949), proposing the incorporation of cytoplasmic male sterility (CMS) into female parent lines and incorporation of "restorer" genes into male parents. All female plants in the seed producer's crossing plot would be sterile, eliminating the need for manual detasseling. Restorer genes in the male lines would ensure that the fertility required to produce grain in the farmer's field would be restored in the final cross (Figure 5.2).

The seed industry quickly adopted the process and over the next decade incorporated CMS and restorer materials into its commercial lines. By 1965, almost all hybrid seed-corn production utilized the technique, and Pioneer's Don Duvick estimated that in that year over 3 billion plants with the CMS character were grown (U.S. District Court 1970a:30). That is, 3 billion plants no longer had to be detasseled manually, and those 125,000 laborers were out of jobs. The loss sustained by these workers was not compensated for by lower seed prices for farmers. In a letter to Jones, an executive of DeKalb AgResearch had warned that "If everyone stops detasseling, and passes all the benefit on to the consumer by lower prices, then the farmer is the only one who gains" (U.S. District Court 1970b:19). From 1958, when companies first began using the process commercially, to 1965, when the process was ubiquitous, the price of a bushel of hybrid seed-corn in constant 1967 dollars *increased* from $11.95 to $12.70,[21] despite the fact that using CMS lines reduced detasseling costs by as much as a factor of *twenty-five* (Becker 1976b:22). It would appear that gains from the cost reductions permitted by publicly funded research were appropriated largely by seed companies for the enlargement of pre-tax profit margins which have in recent years been greater than 20 percent for industry leaders (Harvard Business School 1978:29).[22]

The qualifier "largely" is an important piece of the previous sentence. In fact, seed companies were not the only beneficiaries of the application of the CMS/restorer system to hybrid seed-corn production. A second set of beneficiaries included not the farmer but Donald Jones, the Connecticut Agricultural Experiment Station, Harvard University, and the Research Corporation. From 1969 to 1973 these parties shared a total of $2,859,124 in royalties accruing to United States patent 2,753,663, "Production of Hybrid

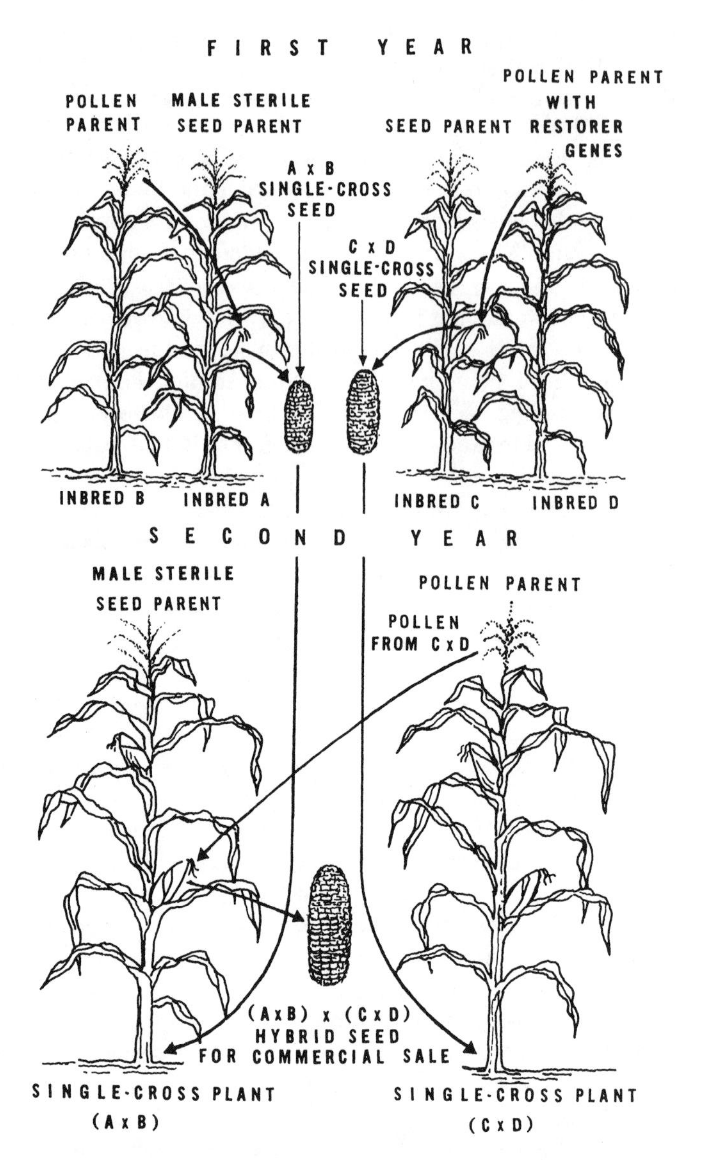

Figure 5.2. Production of double-cross hybrid seed corn using CMS. The need for manual detasseling is obviated by the CMS factor, carried by the seed parent, which prevents the shedding of fertile pollen. However, fertility-restoration genes carried in one of the pollen parent lines ensures that the double-cross seed ultimately produced for commercial sale is fertile.

Seed Corn ... by the utilization of genetic factors capable of restoring pollen fertility to the progeny of cytoplasmic pollen sterile strains."

Donald Jones patented the CMS/restorer system he developed. He assigned the patent to Research Corporation, a non-profit foundation that administers the commercialization of patents for institutions and individual inventors. In his agreement with Research Corporation, Jones specified that a portion of the royalties from the patent would go to support research at the two institutions with which he had been most closely associated: the Connecticut SAES and Harvard University. Though Jones' patent was issued in 1956, the flow of royalties came in a five-year period beginning in 1969, because only in 1969 was the seed industry forced, through litigation, to honor the patent.

That Donald Jones should apply for a patent on his breeding technique is not necessarily surprising. Recall the clear awareness of and interest in the proprietary implications of the hybrid method in his early manuscripts. He had considered patenting the double-cross method of hybridization that he had developed, but was advised against doing so by the USDA. The attempts of others to patent that process, though unsuccessful, deeply disturbed Jones (Everett 1984). His decision to claim property rights in the CMS/restorer technique was probably determined less by motives of financial gain that by a desire to secure scientific credit for his work (S. Becker 1984).[23]

Whatever his motives, Jones' action threatened the seed industry with the prospect of paying for scientific work, which previously had been freely available as a subsidized public good. Additionally, his patent claim violated the Mertonian tenets of communalism and disinterestedness that dominated the ideologies of public plant scientists. Jones simultaneously challenged the prevailing conditions of capital accumulation and the prevailing norms of scientific practice. His departure from the appointed place of the public scientist in the social division of labor and from the collegial responsibilities of the "republic of science" meant that he would confront the opposition of both capital and the scientific community to his patent.

Although seed companies began using the CMS/restorer process very quickly, none licensed the technology from Research Corporation, in spite of that organization's efforts. In addition to a refusal to commercially recognize the patent, the seed industry, through the ASTA, encouraged efforts to discredit Jones among public plant scientists. At its 1956 annual meeting, the American Society of Agronomy took the unprecedented step of censuring a member. In a special resolution, the ASA called the patent a "severe blow" to scientific cooperation. It pointed out the many individual contributions to knowledge that had enabled Jones to make his discovery, and it concluded:

> The capitalization on such information by any one individual thus becomes a breach of faith in this principle of free exchange of information and material, and seriously jeopardizes future continuation of such cooperative endeavor. [ASA 1956:603]

Faced with the intransigence of the seed industry and the moral isolation of Jones from his colleagues, Research Corporation felt it had no recourse but to defend the patent in court.

On April 10, 1963, the Research Corporation filed suit in federal court against several of the more prominent companies using the CMS/restorer technique. The next seven years were consumed in complex legal maneuvering as additional firms were brought in as defendants and as both seed companies and many public plant breeders sought to show that the patent should not be held valid. A settlement was reached in January of 1970, after an impartial adviser appointed by the court returned a report favorable to Jones, and after the presiding judge agreed to permit the opening of discovery on allegations that the seed company defendants had "violated the Antitrust Laws by reason of engaging in an unlawful combination, conspiracy and concert of action to infringe the patent, to refuse to deal with Research [Corporation] respecting the patent in suit, and to induce others to do the same" (U.S. District Court 1970a:3).[24]

The episode of the Jones patent may seem somewhat anachronistic today. But it clearly illustrates the manner in which the ideological presuppositions of public scientists reflected their position in the division of labor with private enterprise. The commodity-form must remain the domain of capital, not of the state. As far as the seed industry was concerned, the sanctity of private property rights was contingent upon who claimed them and how much it cost to violate them.

Previewing the Green Revolution

Commentators have occasionally made the point that hybrid corn constituted an American Green Revolution (e.g., Staub and Blase 1971; Becker 1976a). However, the analogy generally is not pursued any further than the broad statement that hybrid corn, like the "miracle" wheats and rices of the 1970s, greatly increased production. The last section examined two particular social impacts resulting from the development of hybrid corn: changes in the division of labor between public and private research, and the labor-displacing impact of the use of CMS in hybrid seed production. Taking a cue from the many studies treating the social impacts of the introduction of Green Revolution crop varieties into Latin America and Asia, this section explores additional socioeconomic impacts connected with the development of hybrid corn.

The development of hybridization, coupled with the emasculation of public research programs, created an important new space for the creation of a new branch of the seed industry and the accumulation of capital in agriculture. Hybrid corn was also instrumental in facilitating the expansion of other branches of the agro-inputs sector. The noted corn breeder G. W. Sprague has observed that "the objective in plant breeding is to develop, identify and propagate new genotypes which will produce economic yield increases under some *specified management system*" (Sprague 1971:96, emphasis added). From the 1940s, the specified management system for which hybrid corn was being bred presupposed mechanization and the application of agrichemicals.

In 1938, only 15 percent of American corn acreage was harvested by machine (USDA 1940:14). The genetic variability of open-pollinated corn varieties posed a serious problem for the agricultural engineer. Plants bore different numbers of ears at different places on the stalk. They ripened at different rates, and most varieties were susceptible to lodging (falling over). Mechanical pickers missed many lodged plants, had difficulty stripping variably situated ears, and tended to shatter overripe cobs. Genetic variability is the enemy of mechanization. But the principal phenotypic characteristic of hybrid corn is its uniformity. As the progeny of two homozygous inbreds, a field of single-cross hybrids is essentially a field of genetically, and thus phenotypically, identical plants. The double-cross, having four ancestral lines, is somewhat more variable than a single-cross, but far more uniform than open-pollinated varieties.

As early as the 1920s, Henry Wallace had contacted a major manufacturer of harvesting equipment to offer to develop a "stiff-stalked, strong-rooted hybrid" (Wallace 1956:111). Hybrid varieties resistant to lodging that ripened uniformly and carried their ears at a specified level greatly facilitated the adoption of mechanical pickers. The breeders shaped the plant to the machine (Mangelsdorf 1951:43).[25] In the ten years after 1935, the percentage of corn harvested by machine jumped from 15 to 70 percent in Iowa (May 1949:514). Between 1930 and 1950, the number of mechanical corn pickers and combines with corn heads increased ninefold (Cochrane 1979:198). But changes in plant architecture did not just facilitate mechanized harvest. New hybrids shaped to the machine had multiple ears and stiffer shanks connecting the ear to the stalk. These features actually *increased* the difficulty and expense of hand harvest, thereby encouraging mechanization. Insofar as adoption of mechanized harvest was linked to hybrids, the benefits of the new seeds tended to flow to those farmers with the wherewithal to purchase machinery. Despite its divisibility, genetic technology is not necessarily scale-neutral. Mechanization of the corn harvest displaced an undetermined but apparently large number of laborers who had been accustomed to finding work during the picking season (Macy et al. 1938; USDA 1940; Crabb 1947).

The volume of agrichemicals applied to corn (or any other crop, for that matter) before 1945 was negligible. But military needs during the war created an enormous production capacity, especially for nitrogen. The 1942 annual meeting of the American Society of Agronomy was held in conjunction with a conference addressing the anticipated problem of surplus fertilizer production. Increasing farmers' use of commercial plant nutrients appeared to be a profitable solution. ASA president Richard Bradfield told the assembled plant scientists that

> There seems little question but that after the war there will be available for use as fertilizer at least twice as much nitrogen as we have ever used at a price much less than we have ever paid. [Bradfield 1942:1070]

But the hybrids in use in 1944 were not suited to the higher nutrient levels made possible by the availability of cheap fertilizer. The plants responded to fertilizer application by developing weak stalks, and lodging again became a problem (Steele 1978:32). Recall also that this was the period when yields leveled off because of the overreliance on a narrow range of inbreds. Moreover, at the same time, the market for hybrid seed-corn was approaching saturation as the proportion of corn acres planted to hybrids topped 90 percent (Table 5.1, last column). Just as breeders were recognizing the need to reorient their breeding strategies, they were faced with pressure to utilize excess fertilizer capacity and to address the problem of a maturing seed market.

The result was that the new population improvement and inbred extraction programs of the late 1950s and 1960s incorporated higher fertility levels and plant populations as parameters that framed public breeding objectives. The approach is summarized by what Funk Seed's research director called the " 'high profit trio idea' – use special hybrids, plant them thicker, and fertilize heavier" (Steele 1978:33). Between 1950 and 1980, per acre seeding rates for corn nearly doubled, with the result that the volume of hybrid seed-corn sales increased by 60 percent even though the number of acres in corn rose by only 2 percent in that period. In the same time span, the tonnage of nitrogen fertilizer applied to corn jumped by a factor of *seventeen*. Whereas there were but 7 firms producing ammonia (the basis of much nitrogen fertilizer) in 1940, there were 65 firms by 1966 (Cochrane 1979:229). Higher plant populations and more luxuriant growth provided ideal conditions for insect, disease, and weed buildup, and this in turn encouraged the use of insecticides, fungicides, and herbicides. Corn now accounts for a third of

United States herbicide sales and a quarter of the market for insecticides (*Farm Chemicals* 1981b:58). It has been a major contributor to the historical increase in the intensity of chemical use in American agriculture (Pimentel et al. 1973), and thus also to the growth of the agrichemical industry.

Table 5.1 illustrates additional facets of the impact of hybrid corn on industrial structure and organization. Much of the enlarged production resulting from the introduction of hybrid seed and associated cultural practices was absorbed by the rapidly growing livestock feed and fattening industry. Continuously increasing output kept corn prices low, facilitating the development of large feedlot operations (Simpson and Farris 1982) and ultimately expanding markets for pork, beef, and poultry meat.[26] Also, corn exports increased some twenty times between 1950 and 1980, providing the government with a "food weapon," as former Secretary of Agriculture Earl Butz put it, and creating massive accumulation opportunities for merchant capital involved in the international grain trade (Morgan 1979).

If the development of hybrid corn has been important in undergirding the rise of agribusiness, it has proved to be powerfully destabilizing for many farmers. Hybrid corn was introduced just as the Roosevelt administration was trying to raise farm prices by reducing production. Intensification of land use, much of which can be attributed to the adoption of hybrids, effectively nullified the effect of Agricultural Adjustment Administration production controls on corn acreage (USDA 1940; Paarlberg 1964). In their classic study of hybrid corn adoption, Ryan and Gross (1950:675) note that the AAA was a major factor in the rapid acceptance of the new seed as farmers sought to intensify production on their reduced acreages. Indeed, the link was so clear that Henry Wallace, then Roosevelt's Secretary of Agriculture, was forced to weather charges of conflict of interest stemming from his continued association with Pioneer Hi-Bred. Burgeoning yields and production merely exacerbated endemic overproduction, and the arrival of World War II only temporarily relieved the problem. The price of a bushel of corn in constant dollars has actually *fallen* steadily since 1940 (Table 5.1). And what gains the farmer enjoyed from enlarging production have been largely eaten up by additional expenses for the inputs demanded by hybrid corn.

The introduction of hybrid corn set the technological treadmill turning at an unprecedented pace. Mechanization and chemical technology associated with the new corn varieties further accelerated the vicious cycle of innovation, increased production, depressed prices, further innovation. While farmers on the treadmill's leading edge survived and even prospered, attrition rates were high. Between 1935 and 1960, the number of farms in the North Central region (which encompasses the Corn Belt) declined by 35 percent. Tenants, in particular, were hard hit, and over that period the

Table 5.1. *Development of grain-corn acreage, yield, production, prices, and selected uses, United States, 1930–1980*

Year	Acreage harvested (million acres)	Yield per harvested acre (bushels)	Production (million bushels)	Season average price per bushel ($)	Season average price per bushel (constant $, 1967)	Used for feed (million bushels)	Exported (million bushels)	Approximate corn acres planted to hybrids (%)
1930	85.5	20.5	1,757	0.60	1.00	NA	NA	00.1
1935	82.6	24.2	2,091	0.66	NA	NA	NA	00.5
1940	76.4	28.9	2,206	0.62	1.55	NA	NA	15.0
1945	77.9	33.1	2,577	1.23	1.48	NA	NA	53.0
1950	72.4	38.2	2,764	1.52	1.48	2,482	117	78.0
1955	68.5	42.0	2,873	1.35	1.45	2,366	120	87.0
1960	71.4	54.7	3,907	1.00	1.01	3,092	292	94.0
1965	55.4	74.1	4,103	1.16	1.13	3,362	687	95 +
1970	57.4	72.4	4,152	1.33	1.33	3,570	517	95 +
1975	67.5	86.3	5,829	2.55	1.27	3,570	1,711	95 +
1977	70.0	90.8	6,357	2.02	1.05	3,744	1,948	95 +
1980	NA	NA	NA	NA	NA	4,518	2,355	95 +

Sources: Harvard Business School (1978), Leath et al. (1982), Sprague (1967).

number of tenant operations in the North Central region was reduced by 62 percent (Kloppenburg and Geisler 1985).

Just as hybrid corn had differential impacts on farmers, it affected regions differently as well. It is no accident that extensive commercial development of hybrids occurred first in the Corn Belt, where profit potential was highest. Hybrid seed entered "good" areas before "poor" ones (Griliches 1960:325). While Iowa had 90 percent of its corn land planted to hybrids by 1936, Alabama waited until 1948 before a hybrid variety adapted to its climate was available (Staub and Blase 1971:120). Although corn is harvested for grain in 41 states, the Corn Belt now accounts for over 80 percent of U.S. corn production, with half of that coming from Illinois and Iowa (Leath et al. 1982:1).

Increasing regional specialization in corn production was reflected at the farm level as well. Those farmers who survived each cycle of the technological treadmill absorbed their failed neighbors and found that the growing scale and technical complexity of their operations compelled them to specialize. Many Corn Belt farmers eliminated their livestock operations and switched completely to cash-grain production. No longer requiring roughage for livestock, such farmers replaced hay with soybeans in their crop rotation with corn. Farmers also found that with heavy fertilization, continuous corn production was possible. Fully 21 percent of corn planted now follows a previous corn crop (Ruttan and Sundquist 1982:83). The increased incidence of row crops has greatly exacerbated soil erosion (Batie and Healy 1983).

Specialization has brought corn monocultures to vast acreages. While the practice of monoculture contributes to efficiency in farming operations, it greatly increases susceptibility to pests and disease, especially at high plant populations. The threat of epidemic is further enhanced when the individual plants in the population are genetically uniform (Yarwood 1970; Sprague 1971), and this is precisely the case with regard to hybrid corn. Inbreeding is itself a process of genetic homogenization. Moreover, the pressures of competition frequently compel seed companies to utilize the same elite inbreds as parent material for their proprietary hybrids. In 1969, only six hybrids accounted for 71 percent of all acreage in the United States planted to corn (NRC 1972a:287).

The germplasm on which elite inbreds are based may itself be very narrow. For example, after 1950, the requirements of the mechanical harvester led to the elimination of practically all germplasm that did not produce a single well-developed ear (Wallace and Brown 1956:121). Before 1970, there was little attention paid to preserving germplasm that was not commercially useful. Thus, hybrid corn did not simply replace the genetically diverse open-pollinated varieties that had been developed by farmers, it actually eliminated them in a process of genetic erosion, a phenomenon directly parallel to that occurring in developing nations today.

The genetic vulnerability accompanying dependence on a narrow base of germplasm was dramatically highlighted in 1970. Some 15 percent of the corn crop in that year was lost to an epidemic of southern corn leaf blight (NRC 1972; Horsfall 1975). Corn prices rose 20 percent, and losses to consumers and farmers totaled some $2 billion (Myers 1983:17). The process of hybridization was directly implicated in this episode. The type of CMS (type T) incorporated into corn lines to eliminate detasseling was the genetic component susceptible to a new race of corn blight (*Helminthosporium maydis*). Because nearly every corn plant in the country carried this cytoplasm, the epidemic swept cornfields from Miami to Minnesota and would have been worse but for a change in the weather unfavorable to the disease organism (Harlan 1975b). In the next few years, seed companies converted their lines back to the normal cytoplasm that is resistant to the disease. As a result, they have been forced to resume manual detasseling.

The rapid response to the 1970 epidemic has been used to illustrate the quality of public agricultural research. In 1972, University of Florida administrator E. T. York (1978:270) told a Senate committee that

> Primarily because of the Land Grant's system of "genetic gadgetry" it was possible to overcome this corn blight situation to the point today that it no longer poses the serious threat so evident two years ago.

In fact, it was the land-grant system of genetic gadgetry and its subordination to private enterprise that placed the nation's agriculture in such a vulnerable situation in the first place. Moreover, there are suggestions that public and private breeders recognized the dangers to race T of the southern corn blight in 1969, yet took no action to prevent an epidemic.[27] Some breeders now look back on the corn blight and see it as the result of stupidity. At the time, a number of farmers felt something more was involved and that culpability was a more appropriate term. At least three class-action suits against hybrid seedcorn firms were initiated by farmers who had been severely damaged, even bankrupted, by the epidemic. In the face of sustained legal obstructionism on the part of the seed companies,[28] none of the suits was brought to trial. And despite the lesson of the blight, the genetic base of hybrid corn remains dangerously narrow (Zuber and Darrah 1980; Myers 1983).

The American hybrid bias in corn breeding has been transferred to other nations. The Food and Agriculture Organization of the United Nations (FAO) conducted a hybrid corn school in Italy in 1947 (Coffey 1949:128), and 50 percent of all corn grown in that nation is still a combination of two lines developed by public agricultural research agencies in the United States (Johnson 1983:157). Corn breeding programs utilizing hybridization were begun in Mexico, Guatemala, El Salvador, Venezuela, Brazil, Uruguay, Argentina, Costa Rica, Cuba, Colombia, Peru, and Chile by 1951 (Mangelsdorf 1951:46). In Kenya, in 1956, the U.S. Agency for International Development and the Rockefeller Foundation funded a hybrid corn breeding program under which purchasers of seed were obliged to buy fertilizer and agree to follow certain cultural practices (Sprague 1967:777). Where commercial agriculture provides a market for hybrid corn, the large companies have established a presence. Pioneer, for example, markets its hybrid corn in over ninety foreign countries and has research centers in nine, including Egypt, India, the Philippines, and Thailand (Gregg 1982).

In areas where extensive factor markets were yet to be created, the influence of hybrid corn was more subtle. But in shaping American breeding practices, it also helped shape the nature of the Green Revolution. The Green Revolution was implemented largely by American scientists working in the Rockefeller and Ford Foundation-funded international agricultural research centers (IARCs). These centers are reminiscent of public agricultural research agencies in the United States not only in their institutional character but also in their mission orientation and ideological commitment to client service. It is therefore not surprising that the varieties developed by the IARCs closely followed the pattern established by the development of hybrid corn in the United States: high-response varieties with stiff stalks bred for the best available lands, assuming the use of fertilizer and other agrichemicals (Jennings 1974; Plucknett and Smith 1982). Nor should it surprise us to find that the extensive social impacts of Green Revolution genetic technology are closely paralleled by distinctly similar and equally broad social impacts generated by the development and deployment of hybrid corn in the United States.

Heterosis in other crops

Long before hybrid corn had become a reality, Edward East anticipated the use of heterosis in other crops as well (East and Hayes 1912). With the commercial success of hybrid corn, public and private plant scientists outside the corn breeding fraternity looked to hybridization as the route to the botanical promised land (Duvick 1959:167; Sneep et al. 1979:204). In the 1947 *Yearbook of Agriculture*, Sears (1947:245) observed that "Hybrid corn is truly epochal in and of itself, but a greater good is the example, the impetus, and the key that it has given to all scientific breeding."

But in no other crop is the yellow brick road of hybridization as easily traveled as it is in corn. Corn is unique in that its male (tassel) and female (silk) flower parts are widely separated and are of such a size as to make hand emasculation a simple process. Moreover, because it is naturally an outbreeding species, no special arrangements for the distribution of pollen are necessary. However, nearly all other crops of economic importance (e.g., wheat, soybeans, rice, sorghum, cotton) are inbreeders that normally self-fertilize. Being homozygous and stable, such varieties are already in a sense inbreds, but the technical problem becomes that of achieving cross-pollination while eliminating self-pollination. This is a difficult proposition indeed, because the flower structures of autogamous (inbreeding) species are formed to do just the opposite. The labor-intensive techniques of hand emasculation and hand pollination are possible in a few high-value seed crops such as tomato, but are prohibitively expensive for such species as wheat or sorghum.

The most important solution has been the system that was to displace detasseling labor in corn: CMS. The first successful use of this trait in hybrid production came in 1943 with an onion cultivar (Jones and Clarke 1943). This work touched off what Thomas Whitaker (1979:360), in a historical review for the American Society for Horticultural Science's 75th annual meeting, described as "a frantic search for cytoplasmic male sterility." A CMS system was successfully applied to sugar beets two years later, and, in a 1952 address, Cornell breeder Henry Munger (1952:47) told the American Seed Trade Association that "The word 'hybrid' has magic in it at the present time." Over the last three decades, various ingenious and often very complex genetic, chemical, and mechanical strategies have been employed to regulate sexual expression and crossing so as to permit the production of commercial hybrid seed in over a score of species (Table 5.2) (Frankel 1983).

But it would seem from Table 5.2 that, apart from a handful of crops, hybridization has made less headway than might have been expected, given the widespread interest in it and the substantial resources devoted to its development. Certainly there have been important technical difficulties with hybridization, but a related problem is the failure of many inbreeding species to exhibit levels of heterosis for yield markedly superior to those for open-pollinated or pure-line varieties (Genter 1967; Simmonds 1979:160; Frankel 1983). Indeed, in recent years, barley hybrids have been replaced by better-performing non-hybrid cultivars (Ramage 1983:72). And in reviewing yield data for onions, Dowker and Gordon (1983:227) conclude that "There should be serious questioning of the faith that many authors seem to have in the advantages of F_1 onion hybrids." Proponents of hybrid wheat have resorted to reporting results of yield tests using hand-crossed seed on specially tended plots to justify continued research expenditures by demonstrating the existence of useful levels of heterosis (Edwards 1983:141). Nevertheless, "The major trend in horticultural breeding is the development of F_1 hybrids in almost all sexually produced crops" (Craig 1968:246), and

Table 5.2. *Selected crops in which hybrid seed is commercially available*

Crop	Date hybrid seed available	Hybridization system	Acreage planted to hybrids (1980) (%)
Corn	1926	CMS/hand emasculation	99
Sugar beet	1945	CMS	95
Sorghum	1956	CMS	95
Spinach	1956	Dioecy	80
Sunflower	?	CMS	80
Broccoli	?	Self-incompatibility	62
Onion	1944	CMS	60
Summer squash	?	Chemical sterilant	58
Cucumber	1961	Gynoecy	41
Cabbage	?	Self-incompatibility	27
Carrot	1969	CMS	5
Cauliflower	?	Self-incompatibility	4
Pepper	?	Hand emasculation	?
Tomato	1950	Hand emasculation	?
Barley	1970	Genic male sterility	Negligible
Wheat	1974	CMS/chemical sterilant	Negligible

Source: Author's compilation from numerous sources.

Wilson and Driscoll (1983:94) blithely assert that "As wheat is the world's second largest food crop, every effort should, and will be made to succeed with its hybridization."

Why do we keep rolling down the yellow brick road of hybridization in autogamous crops, even in the absence of the dramatic yield increases achieved with corn? Having examined the case of hybrid corn, we may anticipate the answer given by Simmonds (1979:159-60): "(1) the economic interests of breeders . . . and (2) field uniformity." These two sources of interest in hybridization are well recognized by plant breeders, and general opinion seems to be that, as far as self-fertilized crops are concerned, the more important one is the proprietary character of a hybrid (Munger 1952; Craig 1968; Whitaker 1979; Pearson 1983). The prospect of bringing growers back into the market every year for purchase of proprietary seed is a powerful incentive.

Issues surrounding the development of proprietary varieties are addressed in some detail in the next chapter. But the question of field uniformity deserves treatment here. As we saw in the case of hybrid corn, genetic (and hence phenotypic) variability is the enemy of mechanization. Mechanical harvesting equipment does not have a wide range of flexibility or selectivity when it comes to dealing with variation in a field. On the other hand, biological organisms are exceedingly plastic, and as breeding techniques became increasingly sophisticated, it became apparent that plants were amenable to being tailored to the requirements of the machine. Webb and Bruce (1968:104) assert that for successful mechanization of harvest, a research project

> must go back to the plant, and indeed, even back to the seed from which the plant comes... Machines are not made to harvest crops; in reality, crops must be designed to be harvested by machine.[29]

The degree to which the *plant* can be transformed has often been a principal factor determining whether or not capital replaces labor in the agricultural production process.

The very visible case of the tomato is a frequently cited example of the integration of plant variety and machine (Webb and Bruce 1968; Schmitz and Seckler 1970; Hightower 1973; Friedland and Barton 1976; Schrag 1978). But, historically speaking, the tomato is but the tip of the iceberg. Plant breeders and agricultural engineers have been working together systematically for more than four decades, and the union of plant and machine is so well established that Wittwer (1973:69) coined the term "phytoengineering" to refer to it. According to Warren (1969:237), the introduction of dwarf pea varieties and subsequent mechanization "took place so long ago that most of us cannot remember seeing tall-vined peas grown in wide rows to be hand harvested." As early as 1920, breeders had selected a storm-resistant cotton that retained its bolls even under the impact of High Plains winds and so was responsible for the early success of the mechanical stripper in Texas and Arizona (Oheim 1954:14). Sorghum and rice breeders had developed dwarfed varieties to fit combine harvesters by 1935 (Hambidge and Bressman 1936). The plant uniformity achieved through hybridization greatly accelerated the mechanization of sweet corn, field corn, onion, and sugar beet harvesting. Hybrids were particularly attractive to the vegetable industry, and spinach, carrots, cucumbers, and the brassicas (cabbage, cauliflower, etc.) have all been hybridized and redesigned to permit nonselective, once-over mechanical crop harvesting (Warren 1969; Litzow and Ozbun 1979). And, of course, non-hybrid crops such as beans, tomato, lettuce, and many fruits have also been bred to facilitate a wide variety of mechanical harvest techniques. Some of the more important characters that breeders seek to incorporate into machine-harvestable varieties are listed in Table 5.3.

There are no estimates of the number of workers displaced by phytoen-

Table 5.3. *Plant characteristics useful for mechanization of harvest*

Uniform maturity	Dwarfing determinacy
Early maturity	Fruit abcission
Concentrated fruit set	Disease resistance
Uniform shape of fruit	Blossom abcission
Fruit resilience	Slow seed development
Uniform plant size	High plant population

gineering over the years. Obviously this number is very large and many times the number of those put out of work by the tomato harvester. This is not an aspect of their work that breeders have questioned, however. They have simply assumed that mechanization is a necessity, as the developers of the nation's first machine-harvestable fresh *market* tomato put it, in "an age of urbanization where labor simultaneously becomes both more expensive and less available for crop production" (Crill et al. 1971:3).

The emphasis on the characters listed in Table 5.3 constrains the realization of other objectives. Tomato breeder M. A. Stevens (1974:87) admits that

> Traditionally, plant breeders have been concerned with those characteristics that relate to yield – particularly disease resistance. Nutritional quality has not been a principal objective... quality is an adjunct, and often an afterthought.

In 1971, a breeder for the Joseph Harris Seed Company promised that

> As we solve the *more pressing needs*, such as giving our growers varieties which will be healthy, mature evenly, machine pick, and merchandise properly, *we are going to go back* to refine these varieties and incorporate in them the color, tenderness, flavor and quality factors to which the consuming public is entitled." [Scott 1971:469, emphasis added]

Mandated levels of nutritional value in food crops have been strenuously resisted by both public and private breeders. In 1970 the Food and Drug Administration (FDA) attempted to bring newly developed plant varieties within the GRAS (generally recognized as safe) classification in order to prohibit any decline in nutritional value (i.e., that the quality of cultivars must not be worsened in breeding new varieties). The hue and cry from the plant science community forced the FDA to rescind the regulations (Hanson 1974; Gabelman 1975).

Conclusion: The road not taken

It should now be clear that the development of hybridization has had a great variety of social impacts. Of particular importance are the changes it galvanized in the social division of labor on the farm and in the sphere of agricultural research. At one level, self-provisioning of seed corn by farmers was increasingly displaced by seed production and marketing by capitalist firms. The commodification of seed corn was completed. At another level, these firms were able to move public breeding programs into activities complementary to, rather than competitive with, private enterprise. By pushing public science into basic research, capital was able to assume control over the shape of the commodity-form (seed corn) and thus to control upstream state-subsidized research. The emergence of a new technical form clearly stimulated significant restructuring of social relations.

This should not, however, be taken to imply a simple unidirectional causal relation between the forces and relations of production. The development of hybrid corn cannot be understood as the natural outcome of an immanently scientific technical reason. Rather, the very production of scientific knowledge that culminated in hybridization was itself shaped and directed by social relations. Henry A. Wallace played a prominent role in the selection of the hybrid road as the principal avenue of corn improvement. But he was the proximate expression of broad interests, not a necessary and sufficient stimulus. Hybrid corn would have been developed without Wallace, though certainly somewhat later. But he was in the right place at the right time, a personification of liberal business interests that had initiated the historical trend to commodification and the rationalization of agriculture and had supported the Country Life Movement.

But what of the road not taken? Was the wet well of population improvement unnecessarily abandoned? Might open-pollinated corn varieties have been developed that would have given as great yield improvements as hybrids? Might this still be done? These questions are not entirely counterfactual, for population improvement has been pursued as an important component of an overall process of breeding directed to the production of hybrids. There are a few plant breeders who still insist publicly that "maximum potential yield cannot be obtained in an F_1 hybrid, but inbred populations which yield more than the F_1 can be developed" (Genter 1982:69). And there are more who, in private, will admit that inbred lines exist *now* that are only marginally less productive than their F_1 progeny.

Representatives of the seed industry deny that open-pollinated synthetics could ever be as good as hybrids (Sneep et al. 1979:194). But Simmonds (1983a:12) comments:

> No one has ever had the time and money to push big populations thus for decades. Hybrid maize is successful but it took decades of work on a huge scale to succeed. What would happen if we put a similar effort into population improvement?

Elsewhere, he has answered his own question:

> In practice we may never get a good answer because the economic stimulus to adopt [hybrid] breeding is great; if hybrid seed can be made cheaply enough, [hybrid] varieties will be bred, *whatever the best overall strategy might be.* [Simmonds 1979:162, emphasis added]

That is, the subordination of population improvement to the service of hybrid production is not a matter of scientific discretion, but of political economy. Hybrids are amenable to commodification; open-pollinated varieties resist it.

So it is entirely possible that the road not taken would have been as productive as the hybrid route. If this is indeed the case, hybrid seed-corn sales represent a tax on the farm population. That we are not likely, as Simmonds has pointed out, to have an answer to this question is testimony to the extent to which public agricultural science has been subordinated to the service of capital. It may indeed be the case that social relations are fettering the development of the forces of production.

6

Plant breeders' rights and the social division of labor: historical perspective

> A man can patent a mousetrap or copyright a nasty song, but if he gives to the world a new fruit that will add millions to the value of the earth's harvest, he will be fortunate if he is rewarded by so much as having his name associated with the result.
>
> Luther Burbank (in U.S. House of Representatives 1930)

> The United States has just changed its plant variety protection law amidst rather bitter controversy and, by executive decision, has also accepted the International Convention for the Protection of New Varieties of Plants (UPOV). This convention reflects a global, but not uncontested, trend towards plant patents.
>
> John Barton (1982)

Hybridization furnished capital with an eminently effective technical means of circumventing the natural constraints on the commodification of the seed. But not all crops submitted to hybridization. There is, however, a second route to the commodification of the seed: the extension of property rights to plant germplasm. Plant breeders' rights (PBR) have now been an issue in the plant science community for over a century.

In 1970 the United States followed the lead of 17 Western European nations by passing the Plant Variety Protection Act (PVPA), which gave patent-like protection to developers of novel, sexually reproduced (i.e., by seed) plants. At the time, this legislation attracted little attention outside the agricultural community. However, a 1980 extension of the act that covered six previously excluded species engendered widespread and often heated debate as to the advisability of granting proprietary rights in so fundamental a resource as plant germplasm. Recent attempts to introduce PBR legislation in Australia, Canada, and Ireland have been stalled by opposition from diverse farm, labor, church, and environmental groups. As the advanced industrial nations press for the globalization of PBR, controversy has also erupted over whether or not Third World nations would benefit from the adoption of such legislation. In view of this continuing debate, there is good reason to examine closely the American experience with PVPA.

PVPA: the issues

Critics of the PVPA have called attention to a wave of acquisitions that has swept many prominent American seed companies into the corporate folds of large multinationals over the last decade. They contend that the PVPA enhances economic concentration in the seed industry, facilitates noncompetitive pricing, constrains the free exchange of germplasm, contributes to genetic erosion and uniformity, and encourages the deemphasis of public breeding (P. R. Mooney 1979, 1983; Fowler 1980). Corporate proponents of the act argue that the PVPA stimulates private investment in plant breeding, thereby providing a greater number of superior and more genetically diverse varieties for farmers and freeing public institutions to concentrate on basic research (Studebaker 1982).

Various analysts – principally economists – have attempted to assess these conflicting claims (Claffey 1981; Barton 1982; Godden 1982; Ruttan 1982a; Perrin et al. 1983; Lesser and Masson 1983). For the most part these studies have suggested that the PVPA is relatively benign. A congressionally mandated evaluation concluded that

> Increases in prices, market concentration and advertising and declines in information exchange and public plant breeding – the feared costs of PVPA – have either been nil or modest in nature. Thus, at this point in time, the evidence presented in this report indicates the Act has resulted in modest private and public benefits at modest private and public costs ... If a reasonable balance is maintained between the public and private sectors in the breeding of most crops, the present balance of benefits and costs should continue. [Butler and Marion 1985:79]

This conclusion is perceptive yet seriously flawed. A pivotal role in shaping the character and structure of the seed market is correctly ascribed to public research agencies – i.e., the Agricultural Research Service (ARS) and the state agricultural experiment station/land-grant university (SAES/LGU) complex. The implicit point is made that potentially negative impacts of the PVPA have been limited by the continued vitality of public breeding programs. Yet, there is no reason to assume that this reasonable balance of public and private effort will be maintained in the long term.

A major problem with economic analyses of the PVPA has been the fundamentally ahistorical approach they have taken. The act has been observed through a narrow window in time as an isolated, free-standing event uncoupled from historical processes. But changes at the margin that seem insignificant or inconclusive may take on new meaning when placed in broader historical and social context. In fact, there exists a clear historical trajectory toward commodification of the seed. The PVPA is but the most recent of a variety of juridical strategies taken by private enterprise to extend the reach of the commodity-form to encompass plant germplasm. The PVPA can also be understood as a mechanism for shifting the public-private division of labor in directions favorable to capital.

Setting a precedent: the Plant Patent Act of 1930

The Morrill Act of 1862 was intended, in the words of the legislation, "to assure agriculture a position in research equal to that of industry." Seedsmen were painfully aware that this was not the case. Private cereal and fruit breeders began calling for establishment of a plant patent system as early as 1885 (U.S. House of Representatives 1906:7; Harlan and Martini 1936:325). A proposal that a committee of experts should be empowered to recommend new varieties of appropriate quality for patent registration was rejected in 1901 by the American Pomological Society as "socialistic" (U.S. House of Representatives 1906:7). An enduring and ironic theme of efforts to introduce PBR legislation in the United States has been proponents' insistent assertions that enlarged private investment will result in superior varieties. At the same time, they have just as adamantly rejected the imposition of any regulatory framework intended to ensure that promised quality is in fact realized.

In 1905 the executive secretary of the newly established American Breeders Association expressed the hope that "laws or business practice can be devised which will give private individuals, animal breeders, seed firms and nursery firms practically a patent right or a royalty on new blood lines" (ABA 1905:62). The following year such legislation was introduced in Congress, but despite testimony from supporters that "every seed is a mechanism as surely as is a trolley car" (U.S. House of Representatives 1906:6), the bill was not reported out of committee. Legislators were not ready to countenance proprietary rights to genetic information.

It was another 24 years before similar legislation was reintroduced. Even so, the Plant Patent Act of 1930 covered only asexually propagated species.[1] The ASTA had lobbied to have sexually reproducing species included in the act. But while legislators were sympathetic to the elimination of what they regarded as the "existing discrimination between plant developers and industrial inventors" (U.S. House of Representatives 1930:2), they were reluctant to provide monopoly control of any variety of staple food crop. For this reason, and because of farmer opposition, they also specifically excluded tuber-propagated plants from coverage so that potato varieties could not be patented. The USDA also opposed the inclusion of sexually reproducing species on the grounds that they were not sufficiently stable genetically and that genetic drift between generations would present insurmountable difficulties in enforcement of the act.

Paul C. Stark, a prominent nurseryman who had drafted the bill, advised the ASTA's Plant Patent Committee not to press its case:

> It seemed to be the wise thing to get established the principle that Congress recognized the rights of the plant breeder and originator. Then, in the light of experience, effort could be made to get protection also for seed propagated plants which would be much easier after this fundamental principle was established. [ASTA 1930:66]

With passage of the Plant Patent Act a second precedent was established. Unlike the standard utility patent statute, the Plant Patent Act did not require that the invention be useful, only that it be new and distinct. Whether a novel plant variety was inferior or superior to existing varieties was immaterial to its patentability. Considerations of quality or utility were to have no place in the decision to grant or deny a plant patent.

Private enterprise militant

With the possibility of legally institutionalizing proprietary rights to sexually reproduced plant varieties at least temporarily foreclosed, the American seed industry appeared locked into its position as the "weak sister of agribusiness" (White 1969:66). And in fact the decade of the 1930s was one of stagnation for many seed firms. But after 1940, a series of factors combined to make private investment in research a strategically appealing proposition even in the absence of breeders' rights.

The rapid growth of seed certification programs after World War II exerted a steady limiting pressure on price levels throughout the seed market. As margins were cut almost to cost of production by the leveling effect of certification, a number of companies, in desperation, initiated marketing efforts based on uncertified seed marked with a brand name. This product differentiation paid handsomely; the key to profitability was a proprietary product and compelling advertising (White 1959:22). If firms were to avoid contravention of the Federal Seed Act of 1939 which – theoretically – prohibited use of synonyms for a single variety, the development of proprietary varieties meant research. By 1950, the prospect of research was less daunting to private companies than it had been in the past. The SAES/LGU complex had developed systematic and proved breeding techniques allowing predictable manipulation of plants. Moreover, there was a steady stream of quality germplasm flowing from public breeders that, with minor alterations, made for highly marketable "proprietary" varieties.

That the absolute sizes of potential markets were growing rapidly was another factor. Between 1939 and 1958, land planted to alfalfa rose from 13 to 29 million acres, and that for soybeans climbed by 132 percent to 25 million acres over the same period. Finally, seedsmen had the concrete example of hybrid corn to encourage them. Hybridization had provided a solution to the biological barrier to capital penetration posed by the seed. Just as important, the companies engaged in hybrid seed-corn production had been able to supplant public agencies as the principal developers of commercial varieties. The experience of hybrid corn showed seedsmen that both the biological obstacle posed by the seed and the institutional obstacle posed by the state could be overcome.

It was this latter point that was of crucial significance to seed companies involved in marketing the vegetable, forage, and field crops not amenable to hybridization.[2] In moving systematically into research and the development of private plant varieties during the 1950s, commercial seed enterprises sought to assume functions that had historically been discharged by public agricultural science. State and capital were thus brought unambiguously into direct competition, because both were directly giving shape to the commodity form – that is, the finished crop variety. As it had in corn, private industry sought to eliminate this contradiction by fostering a shift in the social division of labor characteristic of plant breeding research.

In 1954 the National Council of Commercial Plant Breeders (NCCPB) was established with the objective of promoting the interests of private breeders. In a 1956 address to the Agricultural Research Institute, a representative of the NCCPB outlined his organization's view of an appropriate allocation of responsibilities between publicly and privately supported breeding programs:

> There is considerable crowding in many plant breeding fields from government plant breeders. Their concentration upon the development of new varieties means an element of governmental competition in which scientific productivity is not accelerated... It follows that horizontal research, aimed specifically at development of commercial varieties, should largely be the responsibility of private firms. [Quisenberry et al. 1956:79-80]

This distinction, usually couched in terms of basic versus applied research, has since the 1950s been the battle cry of those companies wanting to expand their research programs. At issue is really the question of the release of finished varieties by public agencies. If this practice could be eliminated, private firms with research capabilities not only would dominate their weaker competitors who depend upon publicly produced varieties but also, by virtue of their structural position, would be able to determine public research agendas, because basic research has no value unless it can be used in applied work.

A second line of attack involved efforts to eliminate or at least weaken the regulatory programs that disciplined the market and provided some assurance of quality in commercially available plant varieties. By 1950 many state experiment stations were publishing lists of recommended varieties, and new cultivars were eligible for certification only if they were markedly superior to existing ones. During the late 1950s the seed industry argued that neither recommendation nor performance should be used to determine eligibility for certification (Loden 1963; Beard 1966). Seedsmen asserted that certification should be based on varietal purity only and that any determination of quality should be left to the consumer. Such an arrangement would uncouple certification from its established association with quality and remove the leveling effect exerted by certified seed. This would open up a fertile field for marketing based on product differentiation, because the varietal name would be the only criterion a purchaser would have for distinguishing among different varieties of seed of a particular species. It would also greatly facilitate the marketing of privately developed cultivars.

To pursue its objective of opening space for its own research and marketing efforts, private industry undertook a loosely organized but systematic lobbying effort to move public researchers and programs in desired directions. In 1956 the ASTA initiated annual Farm Seed Industry-Research Conferences designed, as a seed executive stated at the first meeting, to achieve "complete understanding, confidence and cooperation between science and industry" (Apfelbaum 1956:58). Members of the seed trade also became regular participants in the annual meetings of such groups as the Agricultural Research Institute, the American Society of Agronomy, the International Crop Improvement Association, and the American Society for Horticultural Science. In the proceedings and publications of these organizations one can clearly see the division of labor between public and private breeders being gradually and progressively negotiated and renegotiated (e.g., Christensen 1957; White 1959; Porter 1961; Kennedy 1963; Loden 1963; Beard 1966).

By the late 1950s, certified seed were rapidly losing ground to brand name products (Porter 1961). Directors of seed certification programs were reevaluating their programs, and seed company executives could express satisfaction that, "from a predominantly farmer-grower service, certification has turned its attention towards methods and procedures that better serve the seed industry" (Beard 1966:47). Substantial research investments had been made, principally by large companies and firms enjoying the high profit margins associated with hybrid corn and sorghum production (Kalton 1963:48). Smaller firms and individual growers found it increasingly difficult to compete, and as early as 1959 the industry was clearly becoming more concentrated. The seed department manager of the Tennessee Farmers Cooperative complained that the "research" that gave larger companies their advantage was "nothing more than 'Borrowing' . . . what has been developed by USDA and Experiment Station plant breeders, adding a little private stock in some instances, slapping a fancy label on it, mapping out a Madison Avenue advertising program for it, and putting it on the market" (Little 1958:131). He summed up his view of the industry by observing, "It's either grow or go."

Conditions in the seed sector reflected what was occurring in agriculture as a whole. The great structural changes that were transforming American farming from a way of life into a business had been accelerating for a decade. Secretary of Agriculture Earl Butz had, in 1955, already issued his own version of the "grow or go" ultimatum:

> Adapt or die; resist and perish . . . Agriculture is now big business. Too many people are trying to stay in agriculture that would do better some place else. [quoted in Young and Newton 1980:134]

Agribusiness was ascendant and public breeders did not need to be weathermen to know which way the wind was blowing. As a prominent plant scientist noted,

> Our objective as minions of the state is better varieties for the farmer. If these come from private sources, we are not opposed; in fact, we may have some sort of obligation to help private breeders do a good job. [quoted in White 1959:27]

Historical circumstances were ripe for the reemergence of the question of PBR for sexually reproduced species.

The struggle for a law

The European seed industry has historically been no less interested in the commodification of plant germplasm than has the American seed trade. And in fact it pursued the social solution to the commodification of the seed as vigorously and with earlier success than its American counterpart (Berlan and Lewontin 1986a). In 1961, the Union for the Protection of New Varieties of Plants (UPOV) was created by six European nations to provide an international legal framework for PBR legislation.[3] This event proved to be the catalyst that revived the issue of breeders' rights in the United States. The ASTA immediately initiated a study group to examine the European system and consider its usefulness in an American context.

But if American seed companies' European counterparts enjoyed patent-like protection, they had also been subjected to regulations specifying that new cultivars had to be demonstrably superior to be permitted to be offered for sale. Though varietal protection and "seed lists" are nominally distinct, they are functionally related in an important way: There is little point in protecting germplasm that cannot enter the market. And, in 1962, anticipating pressure to institute some form of PBR legislation, the USDA proposed amendments to the Federal Seed Act that would have required compulsory review and registration of all new varieties. The challenge facing the American seed industry was to obtain protection without losing its freedom to release varieties "of obvious *or dubious* merit" (Caren 1964:35, emphasis added). If PBR was to involve any sort of quality control, the cure might well be worse than the disease. As the president of Northrup King put it: "Compulsory registration . . . these are fighting words to most of the seed industry" (Christensen 1962:96).

The decade of the 1960s was marked by a process of negotiation between public and private breeders as to the shape that PBR legislation might most appropriately take – if, in fact, it was necessary at all. Symposia held in conjunction with the annual meetings of the American Society of Agronomy (1964) and the Crop Science Society of America (1969) considered the matter in detail. Private interests insisted that prospective legislation include no requirements for performance testing and opposed anything that would tend to restrict marketing of new varieties. The research director of a major seed company explained:

> Mandatory registration, likewise, has little to offer in a constructive way for the seedsman. It would place the government in the position of being the judge on novelty (and merit?) of any new variety or hybrid instead of the originator or customer . . . A mandatory system apparently designed to curtail expansion of varietal numbers in each crop species . . . has little appeal. [Kalton 1963:56-7]

While the seed company executives' principal justification of the need for PBR was the anticipated flow of superior plant varieties that would result from increased private investment in breeding (e.g., White 1969:63; U.S. Senate 1970:54), the seed industry steadfastly opposed the creation of any institutional mechanism for ensuring that new varieties were in fact improvements. For the seed industry, PBR was less research than marketing legislation.

A second prime motivation for the seed industry was the opportunity to use PBR to lever public agencies away from release of finished varieties, thereby also facilitating the marketing of proprietary products. One seed company executive likened breeding to a "dynamic assembly line" and asserted that

> The stage at which the private sector assumes responsibility in this assembly line operation should change as competing private firms show that they are able to assume added responsibility... Plant patents, or any other effective scheme of breeders' rights will hasten the shift of responsibilities. [Buker 1969:19-20]

To the extent that PBR encouraged private investment in varietal development, public activities in such applied work could be regarded as "duplicative" and "redundant" - an "unnecessary use of sorely pressed tax resources" (Kalton 1963:49). PBR could provide an argument for the emasculation of public breeding and its relegation to "basic" research complementary to rather than competitive with private enterprise.

For their part, public breeders found themselves in an ambivalent – not to say contradictory – position. Public breeders are, after all, fish in a capitalist sea and were and are committed to the general ideological precepts of that mode of production. When confronted with the shibboleth of "private property" and the right of industry to "fair profit," they were placed immediately on the defensive. It was continued public investment, not the expansion of property rights, that required justification. According to a patent attorney speaking before the American Society of Agronomy, the question posed by the possible enactment of PBR legislation was "whether we want to provide a motive – ultimately a profit motive – to private enterprise or whether we want to leave future development in the hands of governmental or quasi-government agencies, where profit is a subordinate consideration at best" (Dorsey 1964:28). He concluded that "the question does not survive its statement."

Potential opposition to PBR legislation on the part of public breeders was further tempered by the continued ascendancy of agribusiness in general. By 1965 industry was spending as much on agricultural research as was the public sector, though most private investment was concentrated in the physical sciences and engineering (Ruttan 1982a:23). In the 1960 *Yearbook of Agriculture*, USDA Secretary Earl Butz noted that "American agriculture is an expanding industry in every important respect except one – the number of people required to run our farms" (Butz 1960:381). The implications of this shift were not lost on public researchers. In a 1965 address to the ASTA, University of Nebraska plant breeder D. G. Hanway (1965:117) observed:

> Industry actually has replaced the farmer as the dominant part of agriculture. It must accept its responsibility for giving guidance to [public] agencies and for securing public understanding and tax support for them.

Public agricultural science was becoming increasingly dependent on agribusiness for the political muscle needed to obtain appropriations in Congress and state legislatures.

If many public breeders were sympathetic to *some* form of PBR, they could see negative consequences associated with a strong law. They naturally feared the possibility of their own marginalization and insisted that any variety protection system should be open to publicly as well as privately developed cultivars. Other concerns focused on possible constraints on willingness to exchange germplasm, the use of protected varieties for research purposes, the need for a farmer exemption clause, and the interests of seed growers and small companies without breeding programs (Myers 1964; Fortmann 1969; U.S. Senate 1970).

These issues were brought to a head in 1967, when the ASTA took advantage of patent-law revision then under way in Congress under the auspices of the President's Commission on the Patent System and introduced a bill of elegant simplicity and potentially enormous consequence. The bill would have amended the 1930 Plant Patent Act by the simple addition of the phrase "or sexually" in appropriate places. This would have brought *all* crops directly under that statute. The far-reaching implications of this proposed addition evoked substantial opposition from the USDA and from public breeders in the experiment stations and LGUs (U.S. Senate 1968). This opposition successfully killed the attempt to extend the Plant Patent Act to cover sexually reproducing crops, but also persuaded public agencies that some sort of protection system was inevitable (Weiss 1969:84).

The result of this realization was an intensive series of meetings involving representatives of the USDA, the state institutions, the ASTA, the NCCPB, and the Association of Official Seed Certifying Agencies. The ASTA in 1969 drafted a bill entitled the Plant Variety Protection Act (PVPA). This document became the basis of negotiation and the vehicle by which PBR was ultimately institutionalized in the United States. The seed industry succeeded in its principal objective of obtaining proprietary rights to new varieties unhampered by any considerations as to quality. Novelty, uniformity, and stability (consistent phenotypic reproducibility) were to be the sole criteria for protection. If these characteristics could be demonstrated, then a certificate of protection would be issued for the new variety. This gave the variety's originator the right to exclude others from using it for a period of seventeen years. The public agencies introduced language ensuring that products of their breeding plots were eligible for protection, that farmers could save and replant protected seed (and even sell to neighbors) without infringement, and that protected varieties could be used for research purposes. It was explicitly recognized that a system of variety protection would regrettably but inevitably reduce freedom of germplasm exchange (Fortmann 1969; Weiss 1969).

With a compromise agreement thus hammered out, the bill was sent to Congress, where hearings were marked by their brevity. In the Senate, testimony took less than an hour as subcommittee chairman Senator B. Everett Jordan observed:

> I see no reason why anybody would be against [PVPA] legislation... There is not much reason for a man or a company or whatever it might be to work hard for years – and it takes years, sometimes, to produce a new strain of anything – and not be able to get some benefit from it. [U.S. Senate 1970:51]

In fact, vegetable canning and freezing interests objected to the legislation, fearing that monopoly control of commercial varieties would lead to substantial rises in the price of seed. These concerns were taken into account through the exclusion of six vegetable species from coverage under the act. In contrast, suggestions from wheat growers that provisions be made for ensuring the maintenance of quality in newly released varieties had no apparent effect on the shape of legislation (U.S. Senate 1970:87-9). On December 24, 1970, the PVPA became law.

Assessing the PVPA

It should now be clear that passage of the PVPA was not an isolated event, but the outcome of a historical process involving the progressively more complete penetration of plant breeding by private industry. I have emphasized the manner in which this penetration has been shaped by two intimately related processes:

1. efforts by private enterprise to enhance the marketability of proprietary plant varieties, and
2. the continual struggle over the "proper" role of the public agricultural research complex that increasing privatization of the seed industry has necessitated.

I have suggested that the PVPA should be understood less as a "research" act than as a "marketing" act and that it could be a powerful mechanism for levering additional shifts in the public-private social division of labor characteristic of plant breeding. In this section the impacts of the PVPA are assessed in light of these arguments.[4] For this purpose the ASTA's own 1970 congressional testimony as to the PVPA's anticipated benefits provide a useful heuristic framework.

"[*The PVPA*] *will greatly stimulate private plant breeding*" (ASTA 1970:54). There can be no doubt that since 1970 there have been very substantial increases both in the number of firms engaged in plant breeding and in the absolute level of money expended for research. All sources agree on these points. Butler and Marion (1985:27), for example, found that in a sample of 51 seed companies, 30 began their research programs in or after 1970. Similarly, in their sample of 59 seed firms, Perrin et al. (1983:25) recorded a doubling of total constant dollar research expenditures between 1970 and 1979. Concomitant increases in facilities and research personnel have also been noted (Jennings 1978; Kalton and Richardson 1983). Such data are frequently cited uncritically by seed industry spokesmen as evidence that the PVPA has indeed stimulated private breeding efforts.

What is seldom noted is that these data reflect historical trends whose initial points of origin are not 1970. These trajectories of expansion can be traced back at least to 1960; Figure 6.1 shows that the trend lines on research expenditures for vegetable and forage crops are unaffected by enactment of the PVPA in 1970. Cereals and soybeans do show significant shifts after 1970, but even here the connection with the PVPA is by no means clear. Between 1970 and 1979 the acreage planted to soybeans increased by two-thirds, and that planted to wheat jumped 47 percent. The value of the annual production of wheat doubled, and the value of the soybean crop more than quadrupled to $14.25 billion. The explosive growth in seed demand associated with these trends would have attracted private investment whether there was PBR legislation or not.

Moreover, although absolute levels of private breeding research expenditures have continued to grow since 1970, the relative *intensity* of research, expressed as a relation between R&D and sales, grew most strongly over the 1960s and has actually flattened out in the post-PVPA period. Table 6.1 illustrates this phenomenon. Of the crops potentially affected by the PVPA, only forages and grasses show consistent growth in research intensity beyond the passage of the act. In 1979 the figure for soybeans had fallen to half its 1970 value, and that for the cereals had reached its apogee in 1965.

There is little evidence to support the contention that the PVPA has powerfully stimulated additional private investment in plant breeding research. Much of the investment that has been forthcoming would probably have been made even in the absence of the act. More firms are doing more research, but the intensity of their effort has, since 1970, been more or less flat.

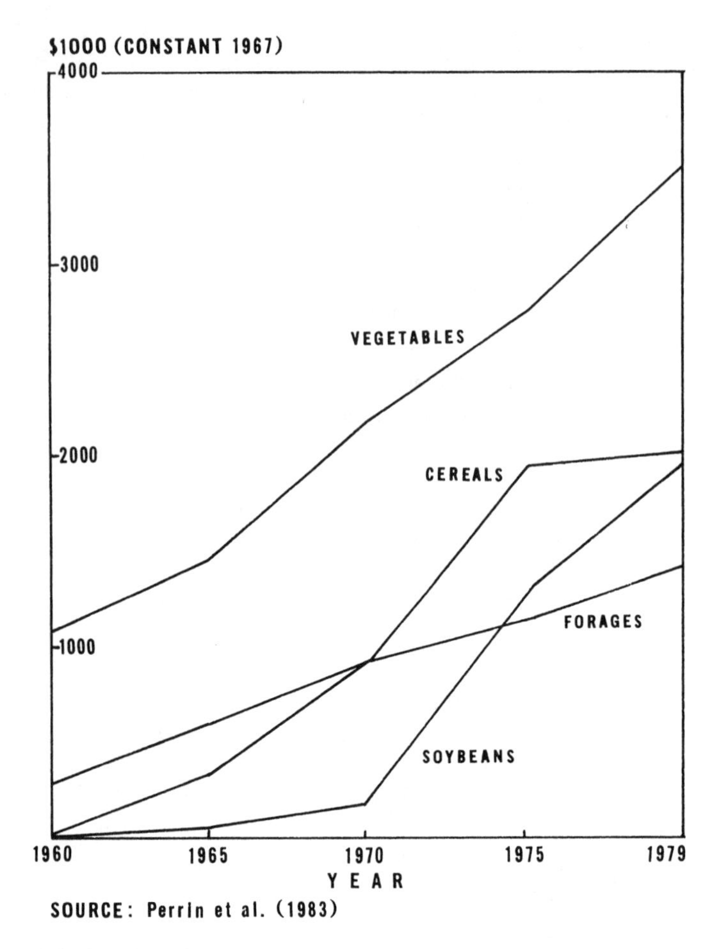

Figure 6.1. Crop breeding research expenditures by 59 seed firms for various crops, 1960–1979.

"[*The PVPA*] *will give farmers and gardeners more choice, and varieties which are better in yield or in quality*" (ASTA 1970:54). The question of yield has been specifically addressed by Perrin et al. (1983) in regard to soybeans, the species on which private effort has been most intensively concentrated. Soybean variety test results from North Carolina, Iowa, and Louisiana for the years 1960-1979 were analyzed. No statistically significant difference in rate of yield improvement was found for the post-1970 period compared with the earlier years. That is, private breeding activities, despite the growth in their magnitude since 1970, have not resulted in an increment of yield gain over the historical trend established by public researchers prior to 1970.

Table 6.1. *Research expenditures per $100 of sales by 56 seed firms, 1960–1979*[a]

Crop	1960	1965	1970	1975	1979
Soybeans	0.4	3.9	8.8	6.1	4.1
Cereals	1.2	18.7	28.0	20.7	20.7
Forage/grasses	0.7	0.9	1.4	1.9	1.7
Vegetables	3.9	4.2	5.2	3.3	4.8

[a]The figures in this table are unweighted by size of firm. Weighted data expressed as average research and development expenditures per firm are available in Butler and Marion (1985) and show a parallel trend.

Source: Perrin et al. (1983:29).

As for quality, the congressional Office of Technology Assessment (1979:72) complained that "Insect resistance has not been a significant component of commercial breeding programs." And the journal *Plant Disease* (1983:1051) reports that "The new varieties being planted have higher yield potential but also often have lower disease resistance."

As we have seen, the seed trade has historically been opposed to the imposition of any regulatory framework establishing quality as a criterion for varietal release. And ASTA lobbying has successfully eliminated varietal performance as a requirement for certification in all but a few states (Copeland 1976:314). The key word in the quotation from the ASTA testimony given earlier is not "yield" or "quality," but "choice." As one company executive put it, the

> seed industry is and always has been a merchandising industry. After all, we are only a few years away from the time that we all had the same public varieties to sell. [Kinsell 1981:64].

The PVPA was pursued by the seed industry primarily as a mechanism for permitting the *differentiation* of its products. Varietal improvement may or may not be an outcome of research, but a larger selection of "choices" for the farmer is the principal goal.

And it is undeniable that farmers do have more choices. As of December 31, 1985, a total of 1,462 certificates of protection had been granted by the Plant Variety Protection Office. However, choice is not distributed evenly across all species. Five crops account for 62 percent of new and protected varieties, and half of those are soybean or wheat cultivars (Table 6.2). Most private investment is apparently attracted to crops with high potential markets yet to be captured from the farmer and the public agencies (e.g., wheat, soybeans) or to crops in industry-dominated markets where varietal competition is more appealing than price competition (e.g., peas, beans, cotton).

Even in these crops the expanded range of "choice" may well be more apparent than real. The National Academy of Sciences has noted that most plant breeding involves a genetic "fine-tuning" of elite adapted varieties. The fact that eligibility for protection under the PVPA requires no demonstration of economic utility over existing varieties means that this fine-tuning can be used to create "pseudo-varieties." In the *Plant Variety Protection Office Journal* (1984:13), the novelty of Northrup King Co.'s soybean variety "S30-31" is described as follows:

> "S30-31" is most similar to "Pella", "Cumberland", and "Agripro 25"; however, "S30-31" has grey pubescence vs. tawny for "Pella", yellow hila vs. imperfect black for "Cumberland", and white flowers vs. purple for "Agripro 25."

It would appear that private breeding work may involve a substantial amount of unproductive effort to achieve uniqueness, and thus protectability, through transfer of non-economic traits such as flower color. A variety is changed but not improved. Seed certification officials have noted a "trend toward 'loose and vague' variety descriptions resulting from P.V.P. requirements" (Seed Certification Officials 1982:35). Seed company executive Robert Kinsell (1981:62) admits that "It almost seems that we are trying to fill the needs of the law rather than the needs of the public."

But Kinsell's ultimate (1981:65) concern is necessarily his bottom line, and he goes on to describe the marketing advantages of these sister-line "pseudo-varieties":

> As an example, my company happens to be a member of one of the group breeding efforts. We pay our royalties and produce and market one of their PVPA varieties. One of our fellow members in the next county has a sister line from the same program. I believe that performance is identical. He has a dealer within sight of my plant, and I have one between his office and the local coffee shop. I promote my variety, and he promotes his. I seriously doubt that my or any of our customers are aware that the two varieties are virtually identical, or that they care.

In fact, farmers care very much, and the fact that they do care was the source of the seed industry's historic opposition to seed certification. Indeed, the whole point of product differentiation is that consumers should *not be able* to perceive the real uniformity of products, be they soap powders or plant varieties, and so will be willing to pay higher prices.[5]

Robert Judd (1979:88-9) of the National Soybean Crop Improvement Council describes the farmer's predicament:

Table 6.2. *PVP certificates issued and selected seed market characteristics by crop*

Crop	PVP certificates issued, 1970–1983	Total seed market (1979) ($1,000)	Market supplied by seed industry (1979) (%)	Potential additional market (1979) ($1,000)	Commercial varieties of public origin (1978) (%)
Soybeans	344	816,804	60	326,434	89
Peas	162	27,012	85	4,052	15
Wheat	143	664,680	35	432,042	86
Beans	133	25,623	85	3,843	15
Cotton	128	120,450	50	60,225	31
Ryegrass	55	2,992	95	150	NA
Fescue	42	3,630	95	181	NA
Barley	32	69,876	50	34,938	95
Alfalfa	28	190,166	95	9,508	58
Bluegrass	21	900	95	45	NA
Oats	16	131,164	30	91,815	86
Tobacco	16	2,912	90	291	45
Rice	13	100,500	70	30,150	92
Vegetables	164	138,714	85	20,807	NA

Sources: Plant Variety Protection Office (1984), Butler and Marion (1985), Hanway (1978a), Leibenluft (1981).

> The number of varieties, blends and brands does present a problem for farmers today. When public varieties were about the only ones available, new varieties were not released if they didn't yield about 3 bushels more per acre or possessed a more desirable characteristic than existing varieties. At that time each farmer had a choice of perhaps 5 varieties at most. Today, it's different. Dr. Gary Pepper of Illinois listed 253 selections available to Illinois farmers for planting in 1980.

Private plant breeding research has been directed as much to the problems of marketing as to plant performance. The blizzard of pseudo-varieties to which the farmer is being subjected is not unambiguously in the farmer's interest. Indeed, it would appear to introduce substantial inefficiencies in both breeding and crop production.

It is also worth noting that it is farmers and their organizations, not competing seed companies, who have been regarded as the principal target of litigation under the PVPA (House 1981; Kinsell 1981). Of particular importance is the case of Delta and Pine Land Company v. Peoples Gin Company, settled in 1982. A United States district court in Mississippi held that an agricultural cooperative was in violation of the PVPA in acting as agent for its farmer-members in arranging sale of collectively ginned cottonseed from one farmer to another (U.S. District Court 1983). This decision effectively precludes cooperatives from facilitating seed exchanges among their membership.

In sum, the PVPA has not resulted in the development of private varieties significantly superior in yield or quality. It has been associated with a proliferation of varieties and greater choice. But that choice is more apparent than real.

"[*The PVPA*] *will allow our government agricultural research stations to increase their efforts on needed basic research . . . It would permit public expenditures for plant breeding to be deviated to important areas which industry may not pursue*" (ASTA 1970:54). The ASTA has long known the kind of division of labor it wished to establish with public researchers. The problem has been achieving it. Despite the disingenuous reference to "allowing" public research to shift its priorities, there was never any question that public breeders would have to be cajoled, pushed, and enticed away from varietal releases.

Passage of the PVPA reinforced the logic of the arguments long used by seed companies to foster the circumscription of public cultivar development. Thomas Roberts (1979:215), chairman and chief executive officer of DeKalb AgResearch, Inc., has provided a succinct and nicely paradigmatic statement of this position:

> The Plant Variety Protection Act has provided incentive for the improvement of self-pollinated species by private plant breeders. This law is effective because it encourages the private sector to invest research

> funds on crops they could not otherwise afford to breed. Further encouragement can be provided by minimizing the use of public funds for the improvement of these crops. With the need for agricultural research so great, public institutions should avoid duplicating research efforts being carried on in the private sector; they should limit their applied research to those crops where experience has demonstrated private effort to be inadequate. It is wasteful and counter-productive for public research funds to be used to compete with private research.

This message has over the last decade been broadcast in hundreds of professional meetings, congressional hearings, and other public forums. Statistics concerning the growth in private research spending and the number of proprietary varieties released since 1970 are put forth as evidence of the capabilities of private enterprise and as justification for the elimination of public "competition" (White 1976; Kalton and Richardson 1983).

The force of these arguments has been given additional weight by an astounding wave of acquisitions and mergers that has swept the American seed industry since 1970 (Table 6.3). Many of the companies acquiring seed firms are large transnational corporations with established agricultural (often agrichemical) interests. The principal factors contributing to this consolidation trend within the agricultural inputs sector include the rising commodity prices and export markets of the 1970s, the opportunity to rationalize and coordinate the marketing of agricultural inputs, and, of course, the passage of the PVPA. The seed industry is no longer the "weak sister" of agribusiness. Indeed, it is now part and parcel of the most powerful elements of agribusiness. When Thomas Roberts of DeKalb AgResearch now calls at the ASTA Congressional Breakfast for "the elimination of redundant public research" (Roberts 1979:46), his words are backed by the financial and political muscle of Pfizer, a Fortune 500 company.

Industry has increasingly used financial carrots as well as political sticks to move public breeders in desired directions. Between 1966 and 1979, private contribution to state agricultural research grew 63 percent in constant dollars, a rate of increase substantially greater than that for any other funding source (Office of Technology Assessment 1981b:58). In absolute terms, private funding represents a small proportion of the total budget available to the experiment stations and LGUs, but it has a high leverage value. Most of the monies appropriated by Congress and the state legislatures are tied to fixed items such as salaries and infrastructure. Relatively small amounts of carefully directed private support can influence the use of substantial magnitudes of public resources by providing the incremental cash needed to get a desired research project under way (Day 1974; McCalla 1978). It is not now unusual to find articles in professional plant science journals describing techniques for establishing funding relationships with private industry.[6]

Table 6.3. *Selected American seed companies by parent firm*

Parent firm	Seed companies
ARCO	Dessert Seed Co.; Castle Seed Co.
Diamond Shamrock	Golden Acres Hybrid Seed
Cargill	ACCO; Dorman; PAG; Paymaster Farms; Tomco Genetic Giant
Celanese	Celpril, Inc.; Moran Seeds; Jos. Harris Seed Co.; Niagara Farm Seeds
Ciba-Geigy	Columbiana Farm Seeds; Funk Seeds International; Germain's; Hoffman; Louisiana Seed Co.; Peterson-Biddick; Shissler; Swanson Farms
Lubrizol	Colorado Seed; Agricultural Laboratories; Arkansas Valley Seed; Jacques Seeds; Keystone Seed Co.; R.C. Young; Gro-Agri; McCurdy Seed; Seed Research Associates; Sun Seeds; Taylor-Evans Seed Co.; V.R. Seed
Monsanto	Hybritech Seed International; Jacob Hartz Seed Co.; DeKalb Hybrid Wheat
Occidental Petroleum	Excel Seeds; East Texas Seed Co.; West Texas Seed Co.; Missouri Seeds; Moss Seed Co.; Payne Bros. Seed Co.; Ring Around Products; Stull Seeds
Pfizer	Warwick Seeds; Clemens Seed Farms; DeKalb AgResearch (joint venture); Jordan Wholesale Co.; Ramsey Seed; Trojan Seed Co.
Sandoz	Woodside Seed Growers; Gallatin Valley Seed Co.; Ladner Beta; McNair Seeds; Northrup N-K; Pride Seeds; Rogers Bros. Seed Co.
Shell Oil Co.	Rudy Patrick; Tekseed Hybrids; Agripro Inc.; H.P. Hybrids; Nickerson American; North American Plant Breeders; Sokota Hybrid Producers Assn.; Ferry-Morse (Farm Seed Div.)
Stauffer	Prairie Valley Seed Co.; Blaney Farms; Stauffer Seeds
Upjohn	O's Gold; Asgrow Seed Co.; Associated Seeds; Farmers Hybrid Seed Co.
W.R. Grace	Pfister Hybrids

There is no question that private industry's efforts to shift the boundary of the division of labor in plant breeding have met with a great deal of success over the past decade. In a survey of field and forage crop breeding in plant science departments in the state research complex, Hanway (1978a) found a general shift toward the more basic activities of population improvement and germplasm enhancement. Two experiment stations reported that, as a matter of policy, they would no longer release finished varieties. Pressures on soybean and wheat breeders have been particularly intense (Leffel 1981; Johnson 1984). With regard to horticultural species, USDA breeder Clinton Peterson (1984:14) observes that "states are abandoning conventional vegetable breeding almost as rapidly as retirement or other personnel changes will permit." Under the Reagan administration, the federal Agricultural Research Service has completely acquiesced to corporate demands and is now phasing out *all* federal varietal release. ARS Administrator Terry Kinney has stated that his agency will "develop its programs on the basis of true complementarity with industry" (quoted in Leffel 1981:47) and is using federal influence to encourage the states to follow suit.

Seed industry executives such as Agrigenetics' (Lubrizol) Robert Lawrence (Qualset et al. 1983:472) welcome what they frankly see as the "changing balance of power" between private enterprise and public agencies. And well they should. Public varietal release has historically functioned to discipline the seed market in important ways. Public varieties have consistently set a standard of quality that private breeders were forced to meet. The success of public breeding and, from industry's point of view, the source of the need for its emasculation are clearly seen in the last column of Table 6.2. The percentage of public varieties in use in such crops as soybeans, wheat, oats, barley, and rice is eloquent testimony to the effectiveness of public research in producing new varieties of use to the farmer. Liberal varietal release policies, and especially the close relationships of the SAES/LGUs to the crop improvement associations, had also long served to maintain a relatively competitive market structure.

In 1980, congressional hearings convened to examine the extension of the PVPA to include coverage of the six vegetable species exempted in 1970; critics of the act expressed a number of concerns. Prominent among these were the issues of growing economic concentration and the possibilities of noncompetitive pricing that were being opened. As of December 1985, seventeen corporations, all transnationals, held 40.9 percent of all Plant Variety Protection Certificates issued.[7] Should finished varietal development be left entirely to private industry, not only would public influence over quality be forfeited, but small seed companies and individual growers without breeding programs who are now dependent upon public agencies for their products would have nowhere to turn for new varieties but to the transnationals.

In uncoupling the direct link between the shape of the commodity and public agencies, society also would have its capacity to generate options foreclosed in a significant way. If finished varieties are the exclusive province of private industry, then it follows that upstream public research must serve the goals of those who determine the final shape of the products or be vulnerable to charges of irrelevancy. For example, public breeders have for some time been interested in the development of "mixed lines" or "multi-lines" that show a wider range of genetic variability, wider adaptation, and more stable performance over a period of years than standard genetically homogeneous varieties (Poehlman 1968:671; Jain 1982). Such lines are a potential solution to the problem of genetic vulnerability, but their development is unlikely to be vigorously pursued by private interests because of the difficulty of achieving the distinctness and stability required for varietal protection under PVPA. What is profitable is not always coterminous with what is socially optimal. Preservation of an autonomous public capacity to develop new technology is a legitimate and vitally necessary objective of social policy.

Public breeders are not yet completely emasculated, however. Indeed, in a few states, experiment stations are resisting pressures from industry and the USDA to move systematically away from varietal development. The PVPA is a two-edged sword that can cut in more than one direction. A handful of experiment stations are using the act to strengthen their own breeding programs through the protection of their varieties, and some have even begun to collect royalties for the use of their lines. Overall, only 12.5 percent of Plant Variety Protection Certificates have been issued to public agencies. But in certain crops the public position is substantial: 23.8 percent of certificates issued in wheat, 28.6 percent of certificates issued in alfalfa. The major seed companies deplore the businesslike orientation developed in some experiment stations, and the ASTA, as it did with Donald Jones, has taken exception to the commodification of science by any entity other than industry (Strosneider 1984). It is well to remember that capital has carved out space for accumulation only through struggle and that this struggle continues today.

Conclusion: PVPA and the lessons of history

In one of the quotations that opened this chapter, John Barton noted the existence of a global trend to plant patenting. Besides extending coverage to six additional species, the 1981 revision of the PVPA also brought the act into accord with the Paris convention and enabled the United States to become a member of UPOV. As private companies located principally in

North America and Europe have reached out for global markets, they have also sought global extension of the legal framework that gives them proprietary rights to the new seed varieties they develop. What can the historical experience of the PVPA in the United States tell us about the likely impact of such extensions?

The passage of the PVPA may have resulted in more private research expenditures, but these have been unevenly distributed by crop. The most potentially profitable species have received added attention, while less lucrative crops have gotten no more effort than might have been expected by extrapolating existing trends. And, if there has been an increase in absolute breeding expenditures, this has been associated with increasing sales. Research intensity has in fact gone flat.

There has clearly been a tremendous increase in the number of varieties available, but much of this proliferation appears to be the result of minor changes whose purpose is less varietal improvement than product differentiation. Efficiency of breeding is reduced through diversion of effort to manipulation of non-economic traits for essentially cosmetic purposes. Despite increased levels of private research spending, the rate of yield increase has not been changed significantly.

It bears repeating that the PVPA is less a research act than a marketing act. If there is inefficient redundancy of research effort in American plant breeding, it would seem to be in the private, not the public, sector. The PVPA has also facilitated the elaboration of a social division of labor in which public research has been progressively subordinated to private interests. The evident demise of public varietal release removes the disciplinary effect that public breeders had exerted on the seed market and eliminates constraints on existing trends to concentration, rising prices, and genetic uniformity.

But the most important lesson is that the PVPA is a product of historical processes of struggle dating back to the nineteenth century. It took private industry nearly 100 years to enact PBR legislation in the United States. Rejection of PBR by Third World nations now does not eliminate the forces that have given rise to plant patenting elsewhere in the world. Indeed, historical trends to commodification of life and privatization of public functions are gathering momentum. And, as we shall see in the following chapters, the struggle has really just begun.

7

Seeds of struggle: plant genetic resources in the world system

Like every other science, the modern science of heredity is international, not only in its theoretical findings but in its practical applications in agriculture. Some of the most valuable of our present-day varieties of plants in the United States, for example, trace their parentage back to far and obscure places. Scientists search the earth for breeding material that will be useful in improving the products grown in their own country. They exchange this material, and the results of their own work, freely between one country and another. What is the net effect of all this? A great improvement, of course, in productive efficiency in our own country – but equally, a great improvement in other countries. From its rivals a nation may get the wheat germ plasm or the cotton germ plasm that enables it to supply its own needs or overwhelm those rivals in international trade ... Will nations have the wisdom to deal with this situation, or will it lead to more bitter rivalries and more deadly conflicts, as the beneficent science of chemistry has enormously increased the deadliness of war?

G. Hambidge and E. N. Bressman, *Yearbook of Agriculture* (1936)

You have heard of "Star Wars." Now there are seed wars.

Bill Paul, *Wall Street Journal* (1984)

Plant genetic resources enjoy a unique distinction: They are considered the "common heritage of mankind" (FAO 1983a:6; Myers 1983:24; Wilkes 1983:156), humanity's collective "genetic estate" (Frankel 1974). As such, PGRs have been available as a free good, the only cost associated with their acquisition being the expenses of collection. Few other resources share this honor. Certainly coal, oil, and mineral resources are not regarded as common property. Even water may become a commodity. And as the wrangling over the "Law of the Sea" treaty demonstrates, it is only with the greatest difficulty that the advanced industrial nations of the capitalist West have been persuaded to confer "common-heritage" status on resources entirely outside national boundaries. Yet there has long been universal consensus that "The major food plants of the world are not owned by any one people and are quite literally a part of our human heritage from the past" (Wilkes 1983:156).

This consensus has recently begun to dissolve. As the quotation from the

Wall Street Journal implies, access to and control over plant genetic resources has now emerged as a field of international concern and conflict. The capacity to utilize plant germplasm is now being recognized, as Hambidge and Bressman feared half a century ago, as an important dimension of national competitiveness in the world economy.

Many analyses of the international divisions in the world economy have noted the asymmetric distribution of benefits characteristic of trade between the core of advanced capitalist societies and the periphery of those that are less developed. The situation with regard to the transfer of plant genetic resources represents an "unequal exchange" (broadly construed) of a unique and extreme form. It is highly ironic that the Third World resource that the developed nations have, arguably, extracted for the longest time, derived the greatest benefits from, and still depend upon the most is one for which no compensation is paid. Indeed, it is not merely ironic, it is contradictory. And as a result of capital's own efforts at expansion, this contradiction is becoming increasingly apparent to nations of the Third World. It is my purpose in this chapter to relate historical patterns of germplasm flow in the world system to the development of capitalism and to illuminate the historical, structural, and institutional dynamics of the contemporary struggle over control of plant genetic resources.

From Columbus to Mendel: imperialism, primitive accumulation, and plant genetic resources

The spread of cultivated plants to new regions has been a constant feature of human history. Such movement was long a slow extension at the margins of adaptation or, less often, small-scale transplantation of a crop into a distant but particularly well-suited area. Such processes could be very effective; by 1300, Europe had added barley, wheat, alfalfa, and a variety of vegetables to the complement of crops with which it had been originally been endowed. But for the most part, the food complexes associated with the great centers of crop origin and plant genetic diversity (Figure 2.2, Table 2.5) remained reasonably distinct (Grigg 1974; Braudel 1979). This pattern changed dramatically with the establishment of contact between the Old and New Worlds. The last 400 years have seen global and unprecedentedly rapid movement of plant germplasm, a process that has been shaped in important ways by an ascendant capitalism committed to the creation of new social forms of agricultural production worldwide.

Marx graphically described the global character of the elemental "primitive accumulation" that undergirded the genesis of capitalism:

> The discovery of gold and silver in America, the extirpation, enslavement and entombment in mines of the indigenous population of that continent, the beginnings of the conquest and plunder of India, and the conversion of Africa into a preserve for the commercial hunting of blackskins, are all things which characterize the dawn of the era of capitalist production. These idyllic proceedings are the chief moments of primitive accumulation...The Treasures captured outside Europe by undisguised looting, enslavement and murder flowed back to the mother-country and were turned into capital there. [Marx 1977:915, 918]

Certainly some of the gold and silver thus acquired from the New World was turned into capital, but much passed through Europe and continued east to pay for plant products such as spices, tea, sugar, and drugs from the Orient (Braudel 1966; Brockway 1979). Though much attention has been given to primitive accumulation of mineral (and human) resources, little note has been taken of the appropriation of plant genetic resources. What Braudel (1966:464) calls the "hemorrhage of precious metals" from Europe was stanched only by the establishment of plantation economies in the new European possessions that could replace imports from the Orient. And while "the wealth obtained by plunder of hoards amassed over years can only be taken once" (Magdoff 1982:14), plant germplasm is a resource that reproduces itself, and a single "taking" of germplasm could provide the material base upon which whole new sectors of production could be elaborated.

The New World supplied new plants of enormous culinary, medicinal, and industrial significance: cocoa, quinine, tobacco, sisal, rubber. More than this, the Americas also provided a new arena for the production of the Old World's plant commodities (e.g., spices, bananas, tea, coffee, sugar, indigo). In what P. R. Mooney (1983:85) has called an imperial "botanical chess game," plant germplasm was appropriated and shifted across the continents and archipelagos of what is now the Third World as the European powers sought commercial hegemony. Table 7.1 illustrates the geographic extent of the game. Because most of these plantation crops were of tropical or subtropical origin, the movement of germplasm tended to be lateral, among colonial possessions, rather than between the colonies and the metropolitan center.

As the focus of the extraction of surplus-value in the colonies shifted from precious metals to agricultural products, germplasm was recognized as a crucial resource. Plant and seed transfers took on tremendous political and economic import. Elaborate measures were taken by the Dutch, English, and French to keep useful materials out of competitors' hands. The Dutch, for example, destroyed all nutmeg and clove trees in the Moluccas except those on three islands where they located their plantations. The French made export of indigo seeds from Antigua a capital offense.

Table 7.1. *Principal plantation crops, areas of origin, and areas in which plantations were established by 1900*

Crop	Origin	Plantations established
Banana	Southeast Asia	Africa, Caribbean, Central America, South America
Cocoa	Brazil, Mexico	West Africa, Southeast Asia, Caribbean
Coffee	Ethiopia	East Africa, Caribbean, South America, Central America, East Asia, Southeast Asia
Cotton	Mexico,[a] Peru[b]	East Africa, North Africa, East Asia, South America, North America, Caribbean
Oil palm	West Africa	Southeast Asia
Pineapple	Brazil	West Africa, Southeast Asia
Rubber	Brazil	West Africa, Southeast Asia
Sisal	Mexico	East Africa, East Asia, South America
Sugar cane	Southeast Asia	East Africa, North Africa, Southern Africa, Caribbean, Central America, South America
Tea	China	East Asia, Southeast Asia, East Africa

[a]Upland cotton.
[b]Sea island cotton.
Sources: Compiled principally from Grigg (1974) and Brockway (1979).

What A. W. Crosby (1972) has called the "Columbian exchange" was not limited to plantation crops. Returning in 1493 from his first voyage of exploration and conquest, Columbus brought with him seeds of the maize plant. The next year he was back in the New World bringing wheat, olives, chickpeas, onions, radishes, sugar cane, and citrus fruits to support a colony. As more voyages of exploration were undertaken and as colonization proceeded, germplasm transfers of staple food crops were made as a matter of course, principally by sailors and settlers interested in subsistence production. Maize, the common bean, potatoes, squash, sweet potatoes, cassava, and peanuts went east. Wheat, rye, oats, and Old World vegetables went west.

Maize and potatoes had a profound impact on European diets. These crops produce more calories per unit of land than any other staple but cassava (another New World crop that spread quickly through tropical Africa). As such, they were accepted, though often reluctantly, by peasantries increasingly pressed by enclosures and landlords, and by a growing urban proletariat. Braudel (1979:166) writes of maize:

> In the valley of the Garonne, in Venetia, and in general wherever it was grown, it was inevitably the poor, whether in town or country who had to take without enthusiasm to eating cornmeal cakes instead of bread ... The peasant ate maize and sold his wheat.

And at the time he wrote *Capital*, Marx (1977:867) found that the Irish factory worker depended upon "Indian [maize] meal" and "a few potatoes" for subsistence.[1] McNeill (1974) may be guilty of exaggeration when he says that Germany's industrialization would have been impossible without the potato, but new crops from the Americas certainly played an important role in feeding a European population that nearly doubled between 1750 and 1850 as the Industrial Revolution swept people off the land and into Marx's "dark, satanic mills" (Langer 1975; O'Brien 1982). Primitive accumulation of plant germplasm thus served capital in two important ways: directly, by providing the genetic foundation for the production of plantation crops, and indirectly by the introduction of crops that greatly lowered the costs of reproducing the burgeoning proletariat.

A nascent botanical science was called early into the service of capital. The creation by European powers of worldwide networks of botanical gardens in the eighteenth century was directly related to economic needs associated with agricultural development of colonial possessions.[2] Such institutions systematically collected the world's plant materials, with the object of ascertaining their commercial utility and the areas in which they might be grown.[3] In a study of the role of Britain's botanical complex, Brockway (1979:6-7) comments:

> As important as the physical removal of the plants was their improvement and development by a corps of scientists serving the Royal Botanic Gardens, a network of government botanical stations radiating out of Kew Gardens and stretching from Jamaica to Singapore to Fiji. This new technical knowledge, of improved species and improved methods of cultivation and harvesting, was then transmitted to the colonial planters and was a crucial factor in the success of the new plantation crops and plant-based served as a control center which regulated the flow of botanical information from the metropolis to the colonial satellites, and disseminated information emanating from them.

In the pursuit of this information, botanists and naturalists in the employ of the European powers did not disdain to engage in irregular and even illegal activity. In order to protect its infant industry, the government of Brazil banned the export of rubber germplasm. And Peru and Bolivia made trade in quinine, which is extracted from the bark of the cinchona tree native to those countries, a government monopoly. But in the middle of the nineteenth century, Kew Garden botanists nevertheless undertook the removal of rubber and cinchona plants from South America in operations of Bondian intrigue.

Transfer of rubber provides a graphic illustration not only of the benefits that may accrue to the appropriator of germplasm but also of the losses that may be borne by the area from which the material is extracted. At the turn of the century, Brazil dominated world rubber commerce with 95 percent of the market. Yet, as the progeny of the few hundred seedlings that survived the 1877 journey from their seedbed at Kew Garden to Ceylon and Singapore began to mature, British Southeast Asia became an increasingly important producer. Today's multi-billion-dollar rubber industry is dominated by British and American corporations like Dunlop and Firestone whose sources of supply for raw latex are in places such as Malaysia and Liberia. Brazil now has about a 5 percent share of the world rubber market (Brockway 1979:42).

As we have seen, the appropriation of plant genetic resources from other lands has been even more important for the United States than it has been for Europe. Official recognition of the crucial importance of foreign germplasm led to the formal institutionalization of germplasm collection programs in 1898. In that year the USDA established a Section of Seed and Plant Introduction to coordinate these activities. Developments in the plant sciences also contributed to a heightened awareness of the value of plant genetic material. The rediscovery in 1900 of Mendel's work on heredity opened new horizons in plant breeding. Although simple selection of best-adapted introductions was increasingly replaced by techniques that permitted creative recombination of genotypes, dependence on exotic germplasm in no way declined. W. M. Hays, Assistant Secretary of Agriculture and secretary of the newly created American Breeders Association, commented in 1905:

> Never before was there apparent greater reason for pushing the work of plant introduction . . . Those in charge of this introduction are working in closest cooperation with the breeders of plants . . . This work must continue that we may have all the needed wild forms and all forms heretofore or henceforth improved in foreign lands. [American Breeders Association 1906:160]

Taking Hays at his word, the Plant Introduction Office inaugurated what has been termed the "Golden Age of Plant Hunters" (Lemmon 1968). Between 1900 and 1930, over fifty separate USDA-sponsored expeditions spread over the globe in search of useful germplasm.[4]

The Green Revolution and plant genetic resources

The upheavals of World War II marked a hiatus in germplasm collection activities, but they also created conditions in which plant breeding was to become an explicitly political tool of American foreign and economic policy and in which the flow of plant genetic materials from the Third World to the developed nations would be accelerated and further institutionalized. The possibilities of what came to be known as the Green Revolution were first explored in a meeting between U.S. vice president Henry A. Wallace and Rockefeller Foundation president Raymond Fosdick in 1941 (Stakman et al. 1967; Cleaver 1975; Oasa and Jennings 1982). It was thought that a program of agricultural development aimed at Latin America in general and Mexico in particular would have both political and economic benefits.

Later that year the Rockefeller Foundation sent three prominent plant scientists – E.C. Stakman, Richard Bradfield, and Paul C. Mangelsdorf – on a survey of Mexico. Bradfield's views are well summarized in the presidential address he delivered to the American Society of Agronomy in 1942:

> I am convinced that the post war services of American agronomists will not be confined within the United States . . . When the war is over, there will be millions to feed, large communities of people to be resettled, and farms to be supplied with seed, fertilizer, machinery, and livestock. A roster of qualified personnel for assisting with such work is already being prepared . . . the leaders of some of our large philanthropic foundations have become convinced that the best way to improve the health and well-being of people is first to improve their agriculture. [Bradfield 1942:1068, 1071]

It was this volatile mix of business, philanthropy, science, and politics that marked the Green Revolution. In 1943 the Rockefeller Foundation initiated its Mexican Agricultural Program, concentrating principally on the improvement of wheat and corn. Over the next eight years, similar projects emphasizing hybrid corn breeding were begun in Guatemala, El Salvador, Venezuela, Brazil, Uruguay, Argentina, Cost Rica, Cuba, Colombia, Peru, and Chile under the auspices of the USDA or American land-grant universities (Mangelsdorf 1951).

The emphasis on corn is not surprising. As detailed earlier (see Chapter 5), hybridization opened a significant new space for capital accumulation in plant breeding and seed sales. Of course, before assuming the vice-presidency, Henry Wallace had been Secretary of Agriculture. And he had come to that post as the best-known champion of hybrid corn and founder of the seed-corn firm Pioneer Hi-Bred. Wallace well understood the articulation of agricultural science and business. By 1946 Rockefeller interests had conducted a survey of the market potential for hybrid maize seed in Brazil, and later that year their International Basic Economy Corporation invested heavily in the only hybrid-seed-producing firm in that country (Hoffman 1971:188). The giant grain merchant, Cargill, followed suit by initiating hybrid seed-corn production in Argentina in 1947.[5]

Creation of a class of farmers in the image of the Corn Belt and concomitant commercial penetration of new markets were not the only objectives of these programs, however. From the very first, the collection of indigenous germplasm was an important component of the Rockefeller Foundation's Mexican Agricultural Program and of the other Latin American initiatives. Indeed, agricultural development in the host country might even be a secondary consideration for American researchers. Edward May (1949:515), president of an Iowa seed-corn company cooperating with Iowa State University, explained the university's decision to establish a corn research project in Guatemala in the late 1940s:

> We know how to build resistance into the corn plant. Now we must develop techniques for finding and evaluating this germplasm. Past experience with other crops has taught us not to confine our search exclusively to our own corns. Thus it is that the Tropical Research Center has been located in Guatemala to search for genes or characters that will improve our corns and thereby contribute to greater freedom from hunger and improve the welfare and security of all nations.

In close cooperation with the Rockefeller Foundation, the National Academy of Sciences supervised a coordinated effort to collect, classify, and preserve the maize varieties of the Western Hemisphere (Chang 1979:94). By 1951 the United States had amassed a large collection of corn germplasm as a by-product of its development efforts, and the USDA had set up a system of Plant Introduction Stations in the United States to evaluate and preserve exotic plant materials collected abroad. As illustrated in Figure 7.1, these accessions came in at a rapidly increasing rate in the immediate post-war period.

Figure 7.1 also shows that the rate at which accessions are received steadily increased. The need for effective storage facilities for acquired plant genetic materials became acute. Improved understanding of seed physiology and advances in seed preservation technology made long-term storage feasible. In 1956, Congress appropriated funds for the construction of a National Seed Storage Laboratory (NSSL) at Fort Collins, Colorado. The NSSL was completed in 1958 and is the flagship of the network of gene banks that now serves as the repository for the fruits of global germplasm collection.

During the 1950s the early initiatives sponsored by the Rockefeller Foundation and the U.S. government spawned a whole series of secondary agricultural programs that encompassed an increasingly broad number of crops, countries, and funding agencies.[6] These programs spread to other continents during the 1960s (Cleaver 1975; Oasa and Jennings 1982). A series of international agricultural research centers (IARCs) was established in the

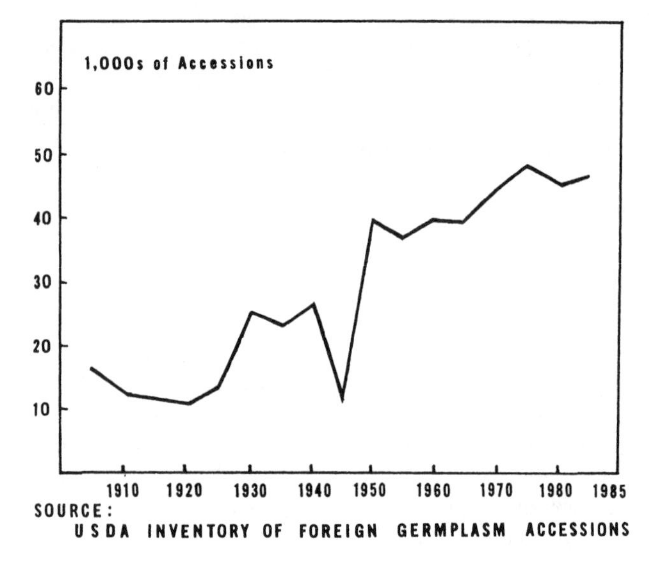

Figure 7.1. Number of germplasm accessions recorded by the USDA, by five-year period, 1898–1985.

Third World, with funding coming from an international consortium of donors from the advanced capitalist nations. Each IARC was charged with the improvement of a particular set of crops in a particular region (Table 7.2). In 1971, the Consultative Group on International Agricultural Research (CGIAR) was created by the Rockefeller and Ford foundations and other sponsoring agencies to coordinate and extend this network of institutions that has spearheaded and sustained what has come to be known as the Green Revolution.

The nature of the Green Revolution as a moment in the self-expansion of capital is well recognized. The contradictions inherent in the Green Revolution development model and the negative, in addition to the positive, consequences of the deployment of the "miracle" high-yielding varieties (HYVs) developed by the IARCs have been much debated (Cleaver 1972, 1975; Griffin 1974; Perelman 1977; Pearse 1980). Such commentary has focused on social and economic impacts in the Third World. Less well recognized has been the reciprocal impact that the Green Revolution has had on the advanced capitalist nations.

No less than the early Latin American programs that were their progenitors and on which they were modeled, the IARCs perform a dual role in the processing of plant germplasm. They necessarily collect and

Table 7.2. *Crop research centers in the CGIAR system*

Center	Acronym	Location	Founded
International Maize and Wheat Improvement Center	CIMMYT	Mexico	1959
International Rice Research Institute	IRRI	Philippines	1960
International Center of Tropical Agriculture	CIAT	Colombia	1967
International Institute of Tropical Agriculture	IITA	Nigeria	1968
International Potato Center	CIP	Peru	1971
International Crops Research Institute for the Semi-Arid Tropics	ICRISAT	India	1972
International Center for Agricultural Research in the Dry Areas	ICARDA	Syria	1976
West Africa Rice Development Association	WARDA	Liberia	1971

evaluate indigenous land races and primitive cultivars that are the raw material from which HYVs are bred. And because their "imported" agricultures are based on the very species that the IARCs are mandated to improve (i.e., corn, wheat, potato), such collection and evaluation are of direct value to the developed nations. The IARCs are not only a mechanism for encouraging capitalist development in the Third World countryside, they are also vehicles for the efficient extraction of plant genetic resources from the Third World and their transfer to the gene banks of Europe, North America, and Japan. It is not happenstance that the CGIAR institutions are located in the Vavilov centers of genetic diversity (Figure 2.2, Table 2.5, Table 7.2). The CGIAR system is, in one sense, the modern successor to the eighteenth- and nineteenth-century botanical gardens that served as conduits for the transmission of plant genetic information from the colonies to the imperial powers.

Coordinating germplasm flows: International Board for Plant Genetic Resources

The role of the CGIAR institutions, as channels of flow for genetic information from the gene-rich periphery to the gene-hungry center, was rendered increasingly important by the very success of the IARC breeders in producing

HYVs. As early as 1936, Harlan and Martini (1936:317) noted the replacement of traditional cultivars by improved varieties in the United States. They also foresaw the projection of such genetic erosion onto a global scale and realized just how valuable the germplasm stored in gene banks would ultimately become:

> In the hinterlands of Asia there were probably barley fields when man was young. The progenies of these fields with all their surviving variations constitute the world's priceless reservoir of germ plasm. It has waited through long centuries. Unfortunately, from the breeder's standpoint, it is now being imperiled... When new barleys replace those grown by the farmers of Ethiopia or Tibet, the world will have lost something irreplaceable. When that day comes our collections, constituting as they do but a small fraction of the world's barley, will assume an importance now hard to visualize.

And, advising the Rockefeller Foundation on its Mexican Agricultural Program in 1941, Dr. Carl Sauer warned that

> A good aggressive bunch of American agronomists and plant breeders could ruin the native resources for good and all by pushing their American stocks. And Mexican agriculture cannot be pointed toward standardization on a few commercial types without upsetting native culture and economy hopelessly. The example of Iowa is about the most dangerous of all for Mexico. Unless the Americans understand that, they'd better keep out of this country entirely. [quoted in Oasa and Jennings 1982:34][7]

By 1970 it was apparent that such predictions were correct and that a corollary to the adoption of the new Green Revolution cultivars was the displacement and disappearance of the land races that provided breeders with the genetic variability on which their advances were founded (Frankel 1970; Harlan 1975b). As Wilkes (1983:134) observes: "The technological bind of improved varieties is that they eliminate the resource upon which they are based."

The process of genetic erosion in the Third World is linked in an important way to the problem of genetic vulnerability in the advanced capitalist nations. The elite commercial varieties on which modern industrial agriculture is based show a high degree of genetic uniformity because they have undergone rigorous selection in breeding. Their narrow genetic makeup renders them systematically vulnerable to diseases and pest infestations in a way that heterogeneous land races are not. As the gene pool for a species is drained by genetic erosion, it becomes more difficult to find characteristics to combat the appearance of disease or pest epidemics that challenge the genetically vulnerable commercial cultivars.[8]

The material consequences of genetic vulnerability were brought dramatically home to American agriculture with the corn blight of 1970. Fifteen percent of that year's corn harvest was lost to a disease organism that attacked a cytoplasmic character carried by over 90 percent of American corn varieties. A subsequent National Academy of Sciences study found American crops to be "impressively uniform genetically and impressively vulnerable" (NRC 1972a:l). This judgment was based on the discovery that in most crops grown in the United States, a small number of varieties account for a large proportion of the acreage planted (Table 7.3). For example, in 1969, 96 percent of the acres under peas were planted to one or the other of only two cultivars. While there is disagreement over the extent to which genetic vulnerability is accurately reflected in such figures,[9] there emerged in the American agricultural community a general perception that genetic uniformity is indeed a significant problem and that broadening the crop genetic base is a worthy objective (NRC 1972a; Harlan 1980; U.S. General Accounting Office 1981; Brown 1985). This in turn has generated an awareness of the need to address the global erosion of genetic diversity, because that which is being lost is the raw material out of which responses to future pest and pathogen challenges must be fashioned and with which the broadening of the crop genetic base can be accomplished (Pioneer Hi-Bred 1983; Yeatmann et al. 1984).

Increasing attention to the issues of plant genetic resource conservation in the United States reinforced the development of a "genetic resources movement" already under way internationally (Frankel 1970; Harlan 1975b; Wilkes 1977). The late 1960s and 1970s were marked by growing concerns with human impacts on the environment. It was at two path-breaking international conferences organized under the auspices of the Food and Agriculture Organization of the United Nations (FAO) that the problem of genetic erosion was first systematically addressed. At these meetings, in 1961 and 1967, there developed a consensus that a coordinated global program of collection and conservation was necessary to ensure that the essential raw materials of plant improvement would not be lost to humanity (Frankel 1985, 1986a, 1986b).

The locus for such an international program might logically have been the FAO, which had since the early 1960s been the most active institutional proponent of genetic conservation and had in 1968 created a Crop Ecology Unit for that purpose. But the CGIAR, which had established its centers – as opposed to the FAO – as the active *research* arm of world agricultural development, argued that its network was a more appropriate medium for such efforts. A compromise was reached in 1974 that created the International Board for Plant Genetic Resources

Table 7.3. *Extent to which small numbers of varieties dominate crop acreage (1969 figures)*

Crop	Major varieties	Acreage planted to major varieties (%)
Bean, dry	2	60
Bean, snap	3	76
Cotton	3	53
Corn	6	71
Millet	3	100
Peanut	9	95
Peas	2	96
Potato	4	72
Rice	4	65
Soybean	6	56
Sugar beet	2	42
Sweet potato	1	69
Wheat	9	50

Source: National Research Council (1972a:137).

(IBPGR). In what the CGIAR itself has described as a "historic anomaly" (CGIAR 1980), the IBPGR was placed physically in FAO but was constituted as a CGIAR institution.

The IBPGR is housed in the FAO's Rome headquarters and superficially appears to be an integral part of the United Nations system. However, the board's budget is provided not by the FAO but by a group of twenty-two national governments and other organizations that are members of the CGIAR. With the exception of India, China, and the United Nations Environment Programme, all of the donors represent the advanced capitalist nations.[10] Sixty-nine percent of the IBPGR's 1984 budget was underwritten by just six of these donors: Canada, Japan, Netherlands, United Kingdom, United States Agency for International Development, and the World Bank (IBPGR 1985:103). The board's policies are set not by debate among member nations of the FAO but through decision-making processes internal to the CGIAR. The IBPGR may cloak itself in the "internationalist" legitimacy provided by its association with the FAO, but the board is not subject to the control of the United Nations. The financial heart and political soul of the IBPGR lie elsewhere.

According to IBPGR Executive Secretary J. Trevor Williams, the CGIAR has chosen to define the board's role as essentially "catalytic." The IBPGR is mandated

Table 7.4. *Numbers of IBPGR-designated global base collections by location*

Location	Number
Advanced capitalist nations	50
Advanced, centrally planned nations	4
Non-aligned/least developed nations[a]	22
CGIAR centers	14

[a]Includes China.
Source: IBPGR (1984).

> to promote and coordinate an international network of genetic resources centres to further the collection, conservation, documentation, evaluation and use of plant germplasm... [But] while the Board is to recommend overall policies and develop long-range programmes, and to estimate the annual financial requirements of those programmes, it is not basically an agency to provide finance itself for those programmes. [IBPGR 1985:iii, vii]

That is, the IBPGR has been instructed to forgo the FAO's original intention of establishing its own regional gene banks in favor of designating existing facilities as cooperating "base collections" where collected plant genetic materials are deposited for storage. Second, IBPGR has tended to coordinate and fund collection activities by third parties rather than emphasizing its own expeditions. The consequences that have followed from these policy decisions are illustrated in Tables 7.4, 7.5, and 7.6.

One consequence is that the IBPGR has relied upon existing gene banks for storage of the germplasm collected under its sponsorship. Existing gene banks are found principally in the industrialized North. And Table 7.4 shows that fifty of IBPGR's ninety designated global base collections are located in the advanced capitalist nations. It would appear that politics may be as important as technical capacity as a locational criterion for an IBPGR base collection. The advanced, centrally planned nations are highly under-represented. The CGIAR centers also account for a substantial portion of base collections located in the Third World.

A similar pattern for IBPGR grant allocations in support of collection work emerges from Table 7.5. Some 58 percent of funds disbursed by IBPGR through March 1983 have gone to advanced capitalist nations, even though these countries already have substantial national germplasm programs of their own. P. R. Mooney (1983:79) notes that, in most years, the United States has actually been a net beneficiary of foreign aid from IBPGR. Grants

Table 7.5. *IBPGR grant allocations by recipient, 1974–1983*

Recipients	$US	%
Advanced capitalist nations	5,015,612	58.1
Advanced, centrally planned nations	19,471	.2
Non-aligned/least developed nations	2,634,119	30.5
CGIAR centers	966,455	11.2
Total		100.0

Source: P. R. Mooney (1983:79).

to the advanced, centrally planned nations, on the other hand, have been virtually nonexistent.

Because most global base collections are in the advanced capitalist nations, and because most of IBPGR's collection funds have been allocated to the advanced capitalist nations, it is hardly surprising to find that the advanced capitalist nations, though poor in naturally occurring plant genetic diversity, are as rich in "banked" germplasm as the developing nations of the Third World (Table 7.6). Indeed, in a number of crops (wheat, barley, food legumes, potato) the advanced capitalist nations possess more stored germplasm accessions than do those nations that are the regions of natural diversity for the crop. The IBPGR has further institutionalized the historically asymmetric flow of genetic resources between the Third World and the capitalist societies of the Northern Hemisphere. Coupled with the continuing failure to stem the process of genetic erosion, this asymmetry has potentially ominous implications. As the well-known economic botanist Garrison Wilkes (1983:173) points out, "The centers of diversity are moving from natural systems and primitive agriculture to gene banks and breeders' working collections with the liabilities that a concentration of resource (power) implies."

Of course, concomitant with the principle of "common heritage" - which justifies the free collection of plant genetic resources – is the principle of "free availability," which mandates unrestricted exchange of banked germplasm among plant breeders and other scientists. Although the IBPGR's network of designated global base collections has no formal legal status, the norm of free exchange has been sufficient to maintain the relatively free international flow of plant genetic material stored in the gene banks of the world.

Table 7.6. *Percentages of germplasm accessions in world gene banks by crop and location, 1983*

Crop	Advanced capitalist nations (%)	Advanced, centrally planned nations (%)	Non-aligned/ LDCs[a] (%)	CGIAR centers (%)
Wheat	38	27	20	15
Rice	23	2	41	35
Maize	33	19	36	13
Barley	50	23	16	11
Sorghums/millets	24	28	29	20
Food legumes	28	22	20	30
Potato	41	33	23	4

[a]Includes China.
Source: FAO (1983b).

Reaping the benefits of free exchange

The ideology of common heritage and the norm of free exchange of plant germplasm have greatly benefited the advanced capitalist nations, which not only have the greatest need for and capacity to collect exotic plant materials but also have a superior scientific capacity to use them. The utility of plant genetic resources for the maintenance and improvement of the elite commercial cultivars of the industrial North is not mere theoretical proposition, it is historical fact. Table 7.7 illustrates some of the contributions made by exotic germplasm to crop improvement in the United States since 1900. I have given only one example for each of a wide variety of crops to make the point that *every* species of economic importance has benefited from introgression of foreign genes, and to illustrate the diverse sources from which germplasm has been drawn. I give multiple examples for wheat, however, to make the additional point that many contributions have been made to the improvement of each species.

No systematic effort has been made to estimate the monetary value of these infusions of genetic material. In a few instances some rough valuations have been reported. Several examples can be drawn from Table 7.7. A Turkish land race of wheat supplied American varieties with genes for resistance to stripe rust, a contribution estimated to have been worth $50 million per year (Myers 1979:68). The Indian selection that provided

Table 7.7. *Exotic germplasm and crop improvement in the United States*

Crop	Character	Germplasm source
Alfalfa	Stem nematode resistance	Turkey
Barley	Yellow dwarf virus resistance	Ethiopia
Bean	Fusarium root rot resistance	Mexico
Cabbage	Black rot resistance	Japan
Cauliflower	Mosaic virus resistance	Iran
Cucumber	Bacterial wilt resistance	Burma
Lettuce	Lettuce mosaic resistance	Egypt
Muskmelon	Powdery mildew resistance	India
Oat	Crown rust resistance	Uruguay
Onion	Thrips resistance	Iran
Pea	Mosaic virus resistance	Iran
Potato	Late blight resistance	Panama
Sorghum	Greenbug resistance	India
Soybean	Cyst nematode resistance	China
Spinach	Downy mildew resistance	Iran
Tomato	Increase of soluble solids	Peru
Watermelon	Wilt resistance	Africa
Wheat	Semi-dwarfing	Japan
	Leaf rust resistance	Brazil, China, Russia
	Stripe rust resistance	Turkey
	Bunt resistance	Russia, Turkey, Australia
	Septoria resistance	Brazil, Bulgaria
	Stem rust resistance	Russia, Kenya, Egypt, Ethiopia, Palestine
	Hessian fly resistance	Turkey, Greece, Uruguay
	Cereal leaf beetle resistance	Russia, China, Ethiopia
	Aluminum toxicity resistance	Brazil

Sources: Dolan and Sherring (1982), Meyers (1983), Peterson (1975), Pioneer Hi-Bred (1983), Reitz and Craddock (1969).

sorghum with resistance to greenbug has resulted in $12 million in yearly benefits to American agriculture. An Ethiopian gene protects the American barley crop from yellow dwarf disease to the amount of $150 million per annum (*New Scientist* 1983:218). Iltis (1981-2:185) reports that the value to the American tomato industry of genes from Peru that permitted an increase in the soluble solid content of the fruit is $5 million per annum. And new soybean varieties developed by University of Illinois plant breeders using germplasm from Korea may save American agriculture an estimated $100-

500 million in yearly processing costs (*Diversity* 1986b:40). Social rates of return to plant breeding expenditures are recognized by economists to be unusually high (Griliches 1958; Ruttan and Sundquist 1982). It is no exaggeration to say that the plant genetic resources received as free goods from the Third World have been worth untold *billions* of dollars to the advanced capitalist nations.

The seed industry and global reach

Thus far I have emphasized the quantitative asymmetries in the historical flows of plant germplasm in the world system over the last century. The pattern of plant genetic transfer between North and South has been largely unidirectional: from the Third World to the developed nations. Since the mid-1950s, however, there has been a reciprocal flow that began as a trickle but is assuming an increasing importance. The initiation of commercial seed exports from the industrial nations to the Third World introduced a crucial qualitative dimension to the established asymmetry of germplasm flow. Plant genetic resources leave the periphery as the common – and costless – heritage of mankind, and return as a commodity – private property with exchange-value.

Again, an example may be taken from Table 7.7. The watermelon originated in Africa, which has also been the source of important disease resistance in American varieties. In a monograph celebrating its 100th anniversary in 1956, the Ferry-Morse Seed Company (1956:32) commented:

> For the watermelon, America owes a real debt of gratitude to Africa. Ferry-Morse is helping in part to repay that debt by supplying North African and Eastern Mediterranean countries with thousands of pounds of watermelon seeds each year.

The watermelon germplasm supplied by Ferry-Morse was emphatically not, however, a free good.

Over the last two decades the seed industry has become increasingly global in scope. This process is well advanced in the industrialized nations, but has progressed more slowly in the Third World. The problem faced by companies looking to the developing nations for an increase in revenues has been not so much market penetration as market creation. Though they represent a vast potential market, most farmers in the Third World have been too poor to afford commercial seed even where they are available. However, the Green Revolution has helped to galvanize the emergence of a growing class of well-capitalized and technologically sophisticated producers who are receptive to commercial seed and able to pay for them. The maize production and export sector in Thailand, for example, has become a major new market for seed-corn (USAID 1985). The development of agro-industrial concerns has also extended the market. Though commercially supplied seed in the Third World now account for only 12 percent of global seed sales, both the volume and value of that market are expected to grow (Kent 1986:25). One need only browse through several current issues of seed industry trade journals and examine the advertisements to see graphic examples of corporate interest in the Third World market.[11]

In their attempts to achieve a global reach, the seed companies of the advanced capitalist nations must again confront the fundamental obstacle posed by the biological characteristics of the seed. Not surprisingly, the leading edge of seed market development in the Third World has been hybrid corn. As early as 1964, Pioneer Hi-Bred initiated overseas operations and now has fifteen foreign research facilities and does business in over ninety countries. But the reproducibility of non-hybrid crops has presented a substantial barrier to development and capture of Third World markets. As was the case in the United States and Europe, Plant Breeders' Rights (PBR) are viewed as a solution to this problem. UPOV, national seed trade associations, the International Federation of Seedsmen, and the International Organization of Private Plant Breeders have actively been encouraging the adoption of PBR legislation by Third World countries as well as by advanced industrial nations that currently lack a legal framework for the patenting of plants. A model PBR law has been prepared for developing countries by UPOV (P. R. Mooney 1983:141), and ASTA trade delegations are expounding upon the benefits of legal rights to plant varieties. Familiar claims are made regarding the beneficial impacts of PBR. A seed industry consultant opines that "Many private companies would work their hearts out in developing countries, if they thought there was a possibility of generating reasonable future sales and returning a reasonable margin of profit" (Underwood 1984:39). And an official of the IBPGR observes that "We need Plant Breeders Rights to support public research in the Third World" (quoted in *icda Seedling* 1984:2).

Seed wars at the FAO: North vs. South, common heritage vs. the commodity

But capital's efforts to provide the global conditions for its own expansion have had unanticipated results. The internationalization of the commercial seed industry has brought plant germplasm as a commodity and plant germplasm as a public good into unambiguous and contradictory juxtaposition in the Third World. On the one hand, governments and companies of the advanced capitalist nations have encouraged the developing nations to adopt PBR legislation – that is, to recognize private property rights in one form of plant germplasm. At the same time, they have argued forcefully for the need to collect and preserve other forms of germplasm such as primitive cultivars and land races. Although the rationale for efforts at plant genetic conservation in the Third World has emphasized the ultimate economic utility of the genetic material located there, these resources have been held to be the common heritage of humanity, a public good to be freely appropriated.

The last twenty years have seen the development of a geopolitical climate one of the central features of which has been an increasing Third World sensitivity to structural inequities in the global economy. In the political milieu characterized by demands for a "New International Economic Order," the niceties of the distinction between "elite" commercial germplasm as private property and "primitive" germplasm as common heritage seemed less persuasive. Indeed, the distinction came to appear to many Third World observers as so much ideological sleight of hand designed to maintain the subordinate position of the South in the global economy. Third World nations found their own genetic resources, albeit transformed by plant breeders, confronting them as commodities. This pattern has been seen as doubly inequitable because the commercial varieties purveyed by the seed trade have been developed out of germplasm initially obtained free from the Third World.

Third World sensibilities regarding the established patterns of global plant germplasm exchange were further offended by revelations concerning restrictions placed on the availability of germplasm stored in the CGIAR's system of global base collections. There have always been exceptions to the principle of free exchange. The national programs of a number of Third World nations have from time to time apparently restricted either the collection or exchange of germplasm of certain crops. Such restrictions have been relatively rare and have been applied to industrial crops of special economic importance to the economies of the countries imposing the limits.[12] But with regard to the materials held in IBPGR's designated base collections, the principles of common heritage and free exchange were thought to be inviolate.

But such an assumption is not necessarily tenable. Publication in 1980 of a 1977 letter from the USDA/ARS Administrator to the chairman of IBPGR made it clear to the world that the United States was willing to violate the norm of free exchange in pursuit of political objectives. The IBPGR chairman had written to the ARS to request that the National Seed Storage Laboratory participate in the IBPGR's network of global base collections. The ARS Administrator responded as follows:

> We are willing to accept selected collections for long-term maintenance at Fort Collins. *They would become the property of the U.S. Government*, would be incorporated with our regular collections, and made available upon request on the same basis as the rest of the collection... As you know it has been our policy for many years to freely exchange germplasm with most countries of the world. *Political considerations have at times dictated exclusion of a few countries*. [Agricultural Research Service 1977, emphasis added]

Through deposition in the NSSL, germplasm collected under the auspices of the IBPGR as common heritage was transformed into national property of the United States. Further, free access to these materials was not guaranteed, even for those from whom the plant genetic resources had been collected. Subsequent investigation revealed that the United States had refused germplasm to Afghanistan, Albania, Cuba, Iran, Libya, Nicaragua, and the Soviet Union (P. R. Mooney 1983:29). It became apparent that the IBPGR base collection network's lack of concrete legal status had an important implication: There were no means of enforcing the free exchange of the global common heritage other than moral suasion.

Growing unease with the global germplasm system among Third World politicians, diplomats, and scientists was reinforced through the activities of environmental, consumer, and other activist groups opposed to PBR legislation and to growing concentration in the seed industry. Pat Roy Mooney's 1979 book *Seeds of the Earth: A Private or Public Resource?* was widely distributed and was instrumental in focusing worldwide attention on the questions surrounding control of plant genetic resources.[13] Mounting Third World dissatisfaction found expression in political action in the United Nations system. At the FAO's 21st biennial conference in 1981, a resolution was passed instructing FAO's director general to prepare a draft of an international agreement that would provide a legal framework for controlling the flow of genetic resources (FAO 1983a).

The introduction of Resolution 8/83, International Undertaking on Plant Genetic Resources, occasioned sharp debate among the delegates to the 22nd biennial conference of the FAO in November of 1983. Pre-conference diplomatic maneuvering had made a voluntary "Undertaking" of what was originally to be a legally binding "Convention." That bit of compromise was to be all the agreement achieved between the advanced capitalist nations and the remainder of the FAO membership on the issue of plant genetic resources. Delegates from Third World and industrialized socialist countries called for the application of the principles of common heritage and free exchange to *all* categories of germplasm. While recognizing the scientific contributions of the IBPGR, they questioned its lack of a juridical personality and suggested that its activities would be strengthened if they were formally carried out under the legal auspices of the FAO. In response, representatives of the developed capitalist nations reiterated their commitment to the principle that plant genetic resources are the common heritage of mankind. However, they steadfastly maintained that inclusion of elite or commercial varieties in any agreement was flatly unacceptable (FAO 1983c). Further, they defended the IBPGR as an effective, decentralized, purely scientific entity and refused to countenance its "politicization" by incorporation into FAO.[14] Three days of heated and often acrimonious debate failed to produce any significant narrowing of differences. Finally, in a rare departure from the consensus decision-making preferred at FAO, the Third World forced and won a vote carrying the Undertaking.[15]

The Undertaking is premised on the familiar and universally accepted principle that "plant genetic resources are a heritage of mankind and consequently should be available without restriction" (FAO 1983d:5). But Article 2 of the Undertaking makes a crucial addition to the range of materials that have conventionally been included under the rubric of "plant genetic resources." To the primitive cultivars, land races, and wild and weedy relatives of crop plants that have long been objects of collection and have been appropriated free of charge for preservation in gene banks and for use in plant breeding programs, the FAO Undertaking explicitly appends "special genetic stocks (including elite and current breeders' lines and mutants)" (FAO 1983d:5). That is, commercial cultivars and breeding lines are claimed as no less the "common heritage of mankind" than peasant-developed land races. Article 7 of the Undertaking is also anathema to the advanced capitalist nations. It mandates the development of

> an internationally coordinated network of national. regional and international centres, including an international network of base collections in gene banks, *under the auspices or jurisdiction of FAO*, that have assumed the responsibility to hold, for the benefit of the international community, and on the principle of unrestricted exchange, base or active collections of the plant genetic resources of particular species. [FAO 1983d:7, emphasis added]

The enlarged conception of what constitutes the plant genetic "heritage of mankind" directly challenges the commodity-form. And the proposed institutional restructuring threatens the established web of control over the exchange of plant genetic resources. As such, the Undertaking is patently unacceptable to those nations with highly developed private seed industries that are engaged in breeding proprietary crop varieties for commercial sale.

Adherence to an Undertaking is voluntary, and national governments have been asked to inform FAO as to the extent they are able to comply with the measures specified. The United States, Denmark, Finland, France, Federal Republic of Germany, Netherlands, Norway, Sweden, United Kingdom, and New Zealand have all officially indicated that they are unable to support the Undertaking or are able to do so only with restrictions.[16] Conversely, virtually every non-aligned or Third World member nation of the FAO that has provided an official response has expressed "support without restriction" for the Undertaking (FAO 1985a).

There has been little movement toward accommodation since the 1983 FAO conference. The first meeting of the Commission on Plant Genetic Resources, created by the FAO to oversee implementation of the Undertaking, in March 1985 was largely taken up by the reiteration of the respective positions of the opposing camps in the debate (FAO 1985b). The developing nations insisted upon free access to proprietary lines, repeated their allegations that current patterns of germplasm transfer constitute the exploitation of a "gene-rich" South by a "gene-poor" North, and asserted that placement of the germplasm exchange system under the jurisdiction of the FAO was a prerequisite to the achievement of a more equitable plant genetic world order. The developed capitalist nations saw such arguments, in the words of the American Seed Trade Association's executive secretary, who attended the commission meetings as an observer, as an attempt to

> wrest control of the international germplasm system from IBPGR-CGIAR; use the Commission to manipulate a supposedly voluntary Undertaking on Plant Genetic Resources into a mandatory, legalized system which, through political domination in and patronage by FAO, they can control, and use the Commission as a visible forum to advance their prejudices against private enterprise and intellectual property-breeders' rights. [Schapaugh 1985]

The 23rd biennial conference of the FAO concluded in November 1985 with the Third World and the developed nations as far apart as ever on the question of implementing the Undertaking and no prospect of a rapprochement on the horizon (Sun 1986a; Witt 1986).

Although the germplasm controversy has received a substantial amount of attention in the scientific, political, and business communities, the mode of debate in FAO and other forums has been characterized more by polemic than careful analysis (e.g., FAO 1983c; P. R. Mooney 1983; ASTA 1984; U.S. Department of State 1985; Arnold et al. 1986). In the remainder of this chapter, I first supply an empirical framework designed to objectively reveal the degree to which regions of the world now depend upon one another for plant genetic resources. Second, I critically examine the principal arguments made in response to the concerns of the Third World regarding the current structure and pattern of global germplasm use, exchange, and control, and in justification of the distinction that is made between commercial cultivars as private property and other forms of plant genetic resources as common heritage.

The common bowl: plant genetic interdependence in the world economy[17]

Historical processes of appropriation and transfer of plant genetic resources have directly shaped contemporary patterns of the distribution of the crops now produced throughout the world. Inter- and intra-hemispheric transfers of germplasm have created a world in which domestic agricultures are often based on genetic materials with origins well beyond domestic borders. Any assessment of the political economy of plant genetic resources must take into account this "genetic geography."

The point of departure for the study of such geography is the work of N. I. Vavilov, who has given his name to the "Vavilov centers of genetic diversity" (Figure 2.2, Table 2.5). Though Vavilov's studies were seminal, subsequent research has shown that centers of diversity are not necessarily coterminous with the area in which a crop originated and that both crop domestication and the subsequent patterns of development of crop genetic diversity were more dispersed in time and space than Vavilov realized. The concept of a center itself has been questioned (Harlan 1971; Hawkes 1983), and the term "regions of diversity" is now generally used to account for the variability generated as crops spread from their original points of origin. Zhukovsky (1975), for example, identifies twelve "mega-gene-centers" of diversity that encompass almost the entire globe.

In Figure 7.2, all the nations of the world are divided into ten regions of genetic diversity on the basis of current scientific understanding of the location and extent of plant genetic variability. The twenty food crops and twenty industrial crops that lead global tonnage of production are identified and listed under their respective regions of diversity in Table 7.8.[18] Melding regions of genetic diversity, political boundaries, and crops permits empirical assessment of the plant genetic contributions and debts of particular geopolitical entities.

Using statistics from FAO's *Production Yearbook, 1983* (FAO 1984), several types of measures were computed. First, for each region the proportion of production accounted for by crops for which that region is the locus of genetic diversity was calculated. For each region, the proportion of production accounted for by crops associated with each of the other regions of diversity was also calculated. Computations for food crops are based on metric tons. However, because of the skewing introduced by tremendous

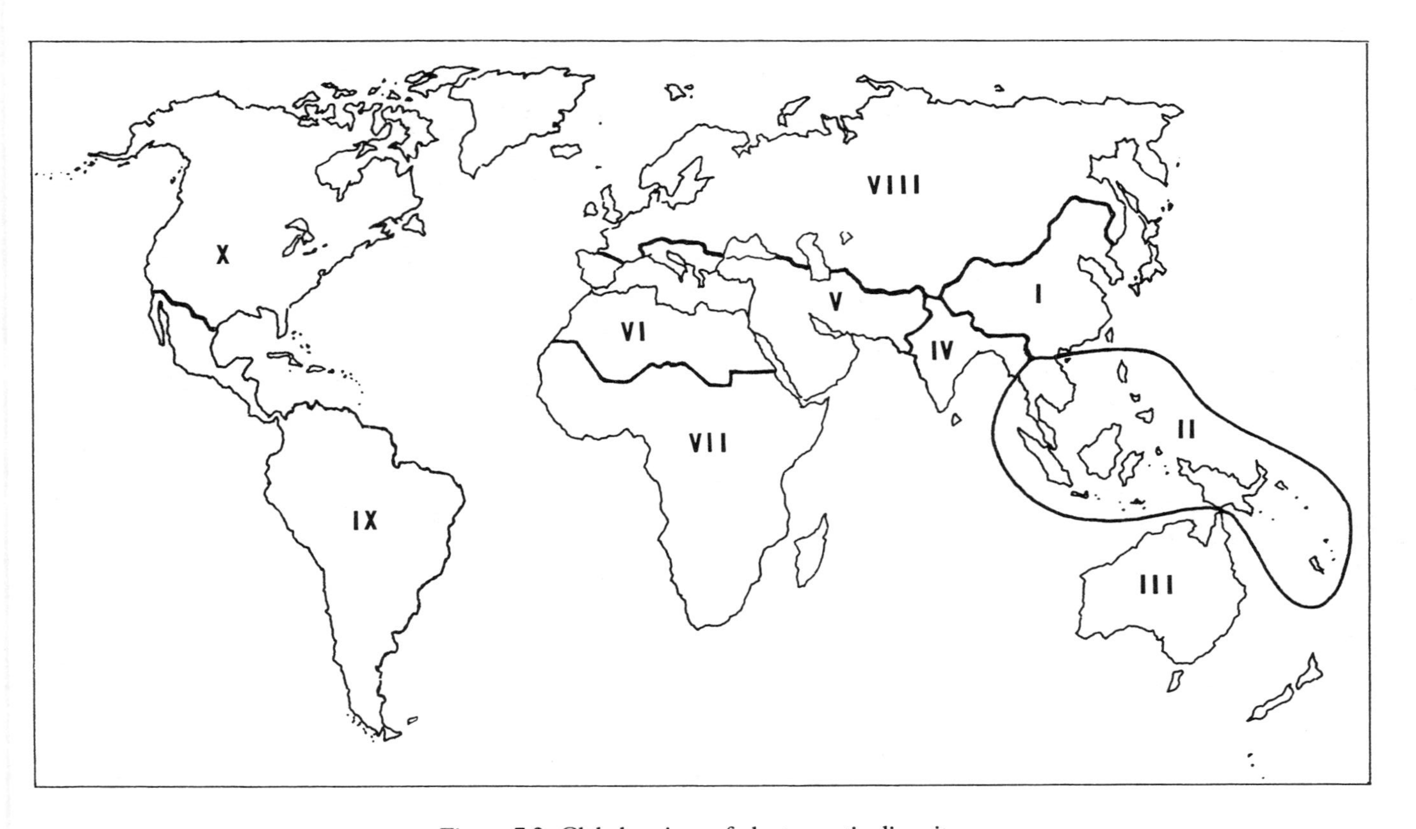

Figure 7.2. Global regions of plant genetic diversity.

Table 7.8. *Regions of genetic diversity and their associated crops*

I. *Chino-Japan*	V. *West Central Asiatic*	VIII. *Euro-Siberian*
Soybeans	Wheat	Oats
Oranges	Barley	Rye
Rice	Grapes	
Tea[a]	Apples	IX. *Latin American*
	Linseed[a]	Maize
II. *Indo-Chinese*	Sesame[a]	Potato
Banana	Flax[a]	Sweet potato
Coconut (copra)[a]		Cocoa[a]
Coconut	VI. *Mediterranean*	Cassava
Yam	Sugar beet[a]	Tomato
Rice	Cabbage	Cotton (lint)[a]
Sugar cane[a]	Rapeseed[a]	Cottonseed (oil)[a]
	Olive[a]	Seed cotton (meal)[a]
III. *Australian*		Tobacco[a]
None	VII. *African*	Rubber[a]
	Oil palm (oil)[a]	
IV. *Hindustanean*	Oil palm (kernel)[a]	X. *North American*
Jute[a]	Sorghum	Sunflower[a]
Rice	Millet	
	Coffee[a]	

[a]Industrial crops.

Source: Kloppenburg and Kleinman (1987b).

differences in weight among some industrial crops (e.g., sugar cane and cotton), industrial crop figures were calculated on the basis of hectares in production rather than tonnage. The results of these operations are reported in Tables 7.9 and 7.10.

Tables 7.9 and 7.10 give two types of information. Read horizontally, the figures show the percentages of production within a given area derived from crops whose regions of diversity are listed in the column headings. For example, 40.3 percent and 2.8 percent of food crop tonnage produced in North America come from crops whose regions of diversity are Latin America and Euro-Siberia, respectively. Reading vertically, these figures can be interpreted as indicators of the importance of crops associated with a given region of diversity to the agricultures of the areas listed in the row headings.

In a sense, reading Tables 7.9 and 7.10 horizontally provides a measure of the "genetic debt" of a given area's agriculture to the various regions of

Table 7.9. *Percentages of regional food crop production accounted for by crops associated with different regions of diversity*[a]

Regions of production	Regions of diversity										Sum (%)[b]	Total dependence
	Chino-Japanese	Indo-Chinese	Australian	Hindustanean	West Central Asiatic	Mediterranean	African	Euro-Siberian	Latin American	North American		
Chino-Japanese	37.2	0.0	0.0	0.0	16.4	2.3	3.1	0.3	40.7	0.0	100	62.8
Indo-Chinese	0.9	66.8	0.0	0.0	0.0	0.0	0.2	0.0	31.9	0.0	100	33.2
Australian	1.7	0.9	0.0	0.5	82.1	0.3	2.9	7.0	4.6	0.0	100	100.0
Hindustanean	0.8	4.5	0.0	51.4	18.8	0.2	12.8	0.0	11.5	0.0	100	48.6
West Central Asiatic	4.9	3.2	0.0	3.0	69.2	0.7	1.2	0.8	17.0	0.0	100	30.8
Mediterranean	8.5	1.4	0.0	0.9	46.4	1.8	0.7	1.2	39.0	0.0	100	98.2
African	2.4	22.3	0.0	1.5	4.9	0.3	12.3	0.1	56.3	0.0	100	87.7
Euro-Siberian	0.4	0.1	0.0	0.1	51.7	2.6	0.4	9.2	35.5	0.0	100	90.8
Latin American	18.7	12.5	0.0	2.3	13.3	0.4	7.8	0.5	44.4	0.0	100	55.6
North American	15.8	0.4	0.0	0.4	36.1	0.5	3.6	2.8	40.3	0.0	100	100.0
World	12.9	7.5	0.0	5.7	30.0	1.4	4.0	2.9	35.6	0.0	100	

[a]Reading the table horizontally along rows, the figures can be interpreted as measures of the extent to which a given region of production depends upon each of the regions of diversity. The column labeled "total dependence" shows the percentage of production for a given region of production that is accounted for by crops associated with non-indigenous regions of diversity.

[b]Because of rounding error, the figures in each row do not always sum exactly to 100.

Source: Kloppenburg and Kleinman (1987b).

Table 7.10. *Percentages of regional industrial crop area accounted for by crops associated with different regions of diversity*[a]

Regions of production	Regions of diversity										Sum (%)[b]	Total dependence
	Chino-Japanese	Indo-Chinese	Australian	Hindustanean	West Central Asiatic	Mediterranean	African	Euro-Siberian	Latin American	North American		
Chino-Japanese	8.3	4.7	0.0	1.4	7.4	27.5	0.1	0.0	45.4	5.1	100	91.6
Indo-Chinese	5.0	43.5	0.0	7.1	2.9	0.0	22.6	0.0	18.8	0.0	100	56.4
Australian	0.0	51.2	0.0	0.0	1.8	3.3	0.0	0.0	15.4	28.3	100	100.0
Hindustanean	2.6	14.2	0.0	7.2	20.5	17.2	0.9	0.0	35.2	2.1	100	92.7
West Central Asiatic	1.5	14.7	0.0	0.0	4.5	14.2	0.1	0.0	56.6	8.4	100	95.5
Mediterranean	0.0	3.9	0.0	0.2	2.4	25.3	0.0	0.0	31.8	36.5	100	74.9
African	1.3	16.3	0.0	0.1	10.6	0.4	22.4	0.0	46.0	3.0	100	77.7
Euro-Siberian	0.4	0.0	0.0	0.1	12.8	41.3	0.0	0.0	17.5	27.9	100	100.0
Latin American	0.2	30.4	0.0	0.4	5.9	0.4	25.7	0.0	28.0	9.1	100	72.1
North American	0.0	3.7	0.0	0.0	8.3	33.1	0.0	0.0	39.6	15.3	100	84.7
World	2.1	13.7	0.0	2.0	10.8	18.2	8.3	0.0	34.4	10.5	100	

[a]Reading the table horizontally along rows, the figures can be interpreted as measures of the extent to which a given region of production depends upon each of the regions of diversity. The column labeled "total dependence" shows the percentage of production for a given region of production that is accounted for by crops associated with non-indigenous regions of diversity.

[b]Because of rounding error, the figures in each row do not always sum exactly to 100.

Source: Kloppenburg and Kleinman (1987b).

genetic diversity. Reading the tables vertically provides measures of the "genetic contribution" made by a particular region of diversity to other areas. To the extent that productivity improvement through plant breeding in any crop depends crucially on continued access to the genetic resources in that crop's region of diversity, the figures in Tables 7.9 and 7.10 also provide indices of what can be termed plant genetic "dependence" of various regions on each other. Thus, in Tables 7.9 and 7.10, a "total dependence" index is included. This is simply the total percentage of production for a region that is accounted for by crops associated with other regions of diversity.

On the other hand, the numbers along the principal diagonal in Tables 7.9 and 7.10 measure the proportion of production accounted for by crops whose region of diversity is indigenous to the given area. These figures may be viewed as an index of plant genetic self-reliance or "independence." In the analysis that follows, the terms "dependence" and "independence" are used in these senses. Although these indices are rough measures, they are useful first approximations that can illuminate the broad parameters of global genetic interdependence. They should prove valuable in bringing an empirical content to a crucial issue on which debate has been confined largely to polemic and unsubstantiated assertion.

Interdependence in food crops

Tables 7.9 and 7.10 reveal that the world is strikingly interdependent in terms of plant genetic resources. Yet within the overarching web of interdependence are important patterns of variation in regional relationships. Certain areas have been the sources of the germplasm that undergirds a substantial portion of global agricultural production. Other regions have been, and continue to be, principally recipients of this genetic largesse.

The high degree of global genetic interdependence in food crops is reflected in the figures reported in Table 7.9. Of the ten regions defined, only three (Indo-Chinese, Hindustanean, West Central Asiatic) have indices of "total dependence" below 50 percent. Even West Central Asia, the region with the lowest dependence index (30.8 percent), obtains nearly a third of its food crop production from crops whose sources of genetic diversity are in other regions.

The general global rule is not crop genetic self-sufficiency, but substantial, and even extreme, dependence on "imported" genetic materials. The Mediterranean, Euro-Siberian, Australian, and North American regions all have indices of dependence over 90 percent. Indeed, Australian and North American genetic dependence in food crops is virtually absolute. Ironically, the agricultures of what are regarded as two of the principal breadbaskets of the world are almost completely based on plant genetic materials derived from other regions.

That none of the world's twenty most important food crops is indigenous to North America or Australia is reflected in the last row of Table 7.9. This row reports the percentages of world food crop production accounted for by species from each region of diversity. Zeros are recorded for North America and Australia. The Mediterranean (1.4 percent), Euro-Siberian (2.9 percent), and African (4 percent) regions have individually made only marginal contributions to the genetic base of global food production. Plant genetic materials from the Hindustanean (5.7 percent), Indo-Chinese (7.5 percent), and Chino-Japanese (12.9 percent) regions account for a somewhat larger component of the world's larder.

But it is clearly the West Central Asiatic and Latin American regions whose germplasm resources have historically made the largest genetic contribution to feeding the world. Crops originating in these regions together account for 65.6 percent of global food crop production. Latin America is the region of diversity for maize, potato, cassava, and the sweet potato, and West Central Asia is the region of diversity for wheat and barley. These two regions have given us six of the world's seven leading food crops; hence their stature in the global plant genetic system.

The data in Table 7.9 provide a means of empirically assessing one of the principal issues in the current controversy over plant germplasm. The six regions that contain nearly all of the world's less developed nations (Chino-Japanese, Indo-Chinese, Hindustanean, West Central Asiatic, African, Latin American) together have contributed the plant genetic material that has provided the base for fully 95.7 percent of the global food crop production. By contrast, those regions with dependency indices greater than 90 percent (North American, Australian, Mediterranean, Euro-Siberian) contain all of the world's advanced industrial nations (with the exception of Japan), yet have contributed species accounting for only 4.3 percent of world food crop production. Thus, there is empirical justification for the characterization of the North as a rich but "gene-poor" recipient of genetic largesse from the poor but "gene-rich" South.

However striking this relation may be, it should not be permitted to mask the great complexity in the patterns of interdependence between individual regions. Though the North may be more or less uniformly "gene-poor" when it comes to food crops, the South is by no means uniformly "gene-rich." The figures in Table 7.9 permit exploration of the nature of specific inter-regional relationships.

The regions containing the advanced industrial nations have been characterized as genetically "dependent." This is nowhere more evident than in the relation in which they stand to Latin America and West Central Asia.

Fully 76.4 percent of North American, 87.2 percent of Euro-Siberian, and 85.4 percent of Mediterranean food crop production comes from crops for which Latin America and West Central Asia are regions of diversity. Such numbers reflect the importance of wheat, maize, potatoes, and barley in the agricultural economies of the advanced industrial nations. The dominant role played by wheat in the Australian region is evident in the 82.1 percent of that region's food production accounted for by West Central Asiatic crops.

But genetic "dependence" is not exclusively a characteristic of the northern regions. The African region has a dependency index of 87.7 percent. Africa depends more on Latin American crops (maize, cassava, sweet potatoes) than it does on indigenous plant genetic materials. Indeed, there are few regions – North or South – that have not drawn upon Latin America or West Central Asia for a significant proportion of the food crops they grow.

Though the Latin American and West Central Asiatic regions have clearly been preeminent as far as overall genetic contributions are concerned, other regions have made important contributions as well. The regions encompassing most of the rest of the developing world (Chino-Japanese, Indo-Chinese, Hindustanean, African) together are the regions of diversity for crops accounting for 30.1 percent of global food production. North America looks to Chino-Japanese crops (especially soybeans) for 15.8 percent of its food crop production. Even Africa makes a significant contribution to regions as diverse as Hindustan (12.8 percent), Latin America (7.8 percent), and North America (3.6 percent) with its millets and sorghums.

There is no such thing as plant genetic independence for either the regions of the North or the South. Even the most genetically self-sufficient of regions, the West Central Asiatic (69.2 percent of production from indigenous crops) and the Indo-Chinese (66.8 percent of production from indigenous crops), rely on crops with a Latin American origin for large proportions of their food production (17 percent and 31.9 percent, respectively).

Interdependence in industrial crops

Table 7.10 reinforces the picture of a genetically interdependent world. Indeed, the degree of genetic interdependence in industrial crops is even more marked than it is in food crops. The lowest index of dependence is 56.4 percent for the Indo-Chinese region. In all of the other regions, more than 70 percent of the hectareage planted to industrial crops is planted to non-indigenous species.

The Latin American region retains its position as the prime donor of genetic material to other regions, with 34.4 percent of the globe's industrial crop area devoted to crops that originated there. On the other hand, the Australian and Euro-Siberian regions have contributed no industrial crops. Between these extremes, contributions are more evenly distributed among regions than was the case with food crops. Indo-Chinese, West Central Asiatic, Mediterranean, and African crops each account for between 8 and 19 percent of global industrial crop hectareage. Even North America, home of the sunflower, weighs in with a contribution of 10.5 percent.

One very important difference between Tables 7.9 and 7.10 lies in the nature of the North-South relationship. In food crops it was clear that a "gene-poor" North draws broadly and systematically upon the resources of a relatively "gene-rich" South. With regard to industrial crops this relation holds, but it is weaker and involves a greater degree of region and crop specificity. Table 7.10 shows that both North America and the Mediterranean have roughly a third of their industrial cropland planted to species of Latin American origin. The Australian and Euro-Siberian regions also depend significantly upon Latin America. In addition, over half (51.2 percent) of the Australian region's industrial crop production is accounted for by sugar cane, a species of Indo-Chinese origin. Apart from these relationships, the regions containing the advanced industrial nations draw little from the South. Indeed, the most salient feature of industrial crop genetic geography is not the North-South relation, but the interdependence of regions within each hemisphere. In contrast to food crops, industrial crops have tended to be transferred laterally rather than vertically across the face of the globe.

Over a third (36.5 percent) of Mediterranean industrial crop hectares are planted to North America's sunflower. The Mediterranean's sugar beet, olive, and rapeseed germplasm has traveled the opposite direction and together accounts for 33.1 percent of North American industrial crop hectareage. In turn, the Euro-Siberian region looks to the Mediterranean and North America for the planting of a full 69.2 percent of its industrial crop area.

A similar pattern of intra-hemispheric interdependence prevails among the regions of the South. The consistently high dependence ratios associated with regions of diversity in the Third World are in large measure historical artifacts of the colonial era. They reflect the extent to which crops such as cocoa, coffee, cotton, oil palm, rubber, and tea were shifted across the continents and archipelagos of what is now the Third World as European powers sought commercial dominance (Braudel 1979; Brockway 1979).

Crops of Latin American origin (cocoa, cotton, rubber, tobacco) were of particular importance. The data in Table 7.10 show that the industrial agricultures of all regions – North and South – depend substantially, and often crucially, upon crops that originated in Latin America. Yet, despite its original genetic endowment, Latin America plants only 28 percent of its industrial cropland to indigenous crops. Over half of Latin America's industrial crop hectareage is accounted for by sugar cane (30.4 percent) introduced from the Indo-Chinese region and by coffee (25.7 percent), which originated in Africa. Similar interdependencies can be found throughout the South in a variety of crops and regions. If it is true that maintaining inter-hemispheric flows of plant germplasm is of central concern with regard to food crops, maintenance of intra-hemispheric flows is the central issue with regard to industrial crops.

Tables 7.9 and 7.10 clearly show that in the global agricultural economy there is really no such thing as "genetic independence." No region can afford to isolate itself – or to be isolated – from access to plant germplasm in other regions of diversity. If the controversy surrounding the FAO's International Undertaking on Plant Genetic Resources goes unresolved, there is a real danger that the sporadic instances of restriction on germplasm exchange now evident will become ubiquitous and comprehensive, to the detriment of both North and South.

Value in the seed

Ironically, in a world economic system based largely on private property, each side in the germplasm controversy wants to define the other side's possessions as "common heritage." The advanced capitalist nations wish to retain free access to the developing world's storehouse of genetic diversity, while the South would like to have the proprietary varieties of the North's seed industry declared a similarly "public" good. It is this that has *Wall Street Journal* reporters speculating about "seed wars":

> In seed wars, Third World nations are pitted against seed companies – and their supporters – in developed nations. At issue is whether the Third World nations should have to pay for new seed varieties developed by Western seed companies from seed obtained in the Third World. [Paul, 1984]

What is it about the germplasm in commercial varieties as opposed to the germplasm in land races that justifies classification of the former as a commodity and the latter as a free good?

Seed companies have been looking for a persuasive answer to this query. In a plenary session at the 1983 "Plant Breeding Research Forum" sponsored by Pioneer Hi-Bred,[19] Dr. Josef Schuler of the Swiss company Ciba-Geigy[20] laid out the problem as follows:

> Those active at the international level very frequently run into the following problem: Industrialized nations, in the East and West, are accused by some people from developing countries as well as by special

> interest groups within the industrialized nations, of taking resources from developing nations and not paying for them. It has become a political issue, and it is affecting activities of companies working transnationally. I would like some good arguments from you in reply to those accusations ... What could we say, as scientists from the industrialized world, that we do *not* and could not deprive the developing countries of their resources? [Pioneer Hi-Bred 1984:114]

Schuler's question was fielded by Dr. J. T. Williams, Executive Secretary of the IBPGR. Williams pointed out that one could argue that "It is not the original material which produces the cash returns. Hence, the principle that germplasm resources constitute an international heritage" (Pioneer Hi-Bred 1984:115). The forum's executive summary elaborated this point:

> Some insist that since germplasm is a resource belonging to the public, such improved varieties would be supplied to farmers in the source country at little or no cost. This overlooks the fact that "raw" germplasm only becomes valuable after considerable investment of time and money, both in adapting exotic germplasm for use by applied plant breeders and in incorporating the germplasm into varieties useful to farmers. [Pioneer Hi-Bred 1984:47]

Curiously, this argument relies implicitly on a labor theory of value. It is asserted that only the application of scientists' labor adds value to the natural gift of germplasm.

But in fact, the land races of the Third World are most emphatically not simple products of nature. Traditional agriculturalists have made very great advances in crop productivity. Domesticated forms of a species are frequently very different in form from their wild or weedy relations. Harlan (1975a:233) credits the American Indian with a "magnificent performance" in the improvement of maize, potato, manioc, sweet potato, peanut, and the common bean. It should be remembered, too, that until the 1930s, scientific breeding consisted primarily of selection among land race introductions. Thus, Robert Leffel, Program Leader of the USDA National Research Program for Oilseed Crops, told the 1980 Soybean Research Conference that "In our more modest moments, today's soybean breeders must admit that a more ancient society made the big accomplishment in soybean breeding and that we have merely fine-tuned the system to date" (Leffel 1981:36). Plant breeder Norman Simmonds (1979:11), in his widely used text *Principles of Crop Improvement*, observes that "Probably, the total genetic change achieved by farmers over the millennia was far greater than that achieved by the last hundred or two years of more systematic science-based effort."

Nor was this labor performed completely in the past. In their day-to-day agricultural activities, contemporary peasant farmers the world over constantly produce and reproduce the genetic diversity that is the raw material of the modern plant breeder. It is important to recognize, as Marx did,[21] that the germplasm of domesticated species is not a free gift of nature, but is the product of millions and millions of hours of human labor. Whatever the contribution of plant scientists, they are not the sole producers of utility in the seed. The unique status of commercial varieties as bearers of exchange-value cannot be justified on that basis.

And, as we saw earlier in this chapter, exotic germplasm has indeed made tremendous economic contributions to the agricultures of the developed capitalist nations. Dr. J. P. Kendrick (quoted in P. R. Mooney 1979:3) of the University of California warns that:

> If we had to rely only on the genetic resources now available in the United States for the genes and gene recombinants needed to minimize genetic vulnerability of all crops into the future, we would soon experience losses equal to or greater than those caused by southern corn leaf blight several years ago – at a rapidly accelerating rate across the entire crop spectrum.

Even the American Seed Trade Association is willing to admit that "our national interests are dependent on continued access to the world's germplasm" (ASTA 1984:3). So genetic resources are not valueless in the sense that they do not have utility. Nor are they valueless in the sense that they have no economic impact. Clearly, they have enormous *social* value. Plant genetic resources are, however, valueless in the sense that *market* values have not been attached to them.

A second argument made to justify the differential status of breeders' lines and land races is that "raw" germplasm *cannot* be priced: "Collections of so-called 'exotic' germplasm may and often do contain useful genes, but until the accession is evaluated and its traits identified, it is an unknown quantity" (Brown 1986:4). That there are difficulties associated with the determination of such a price is undeniable. The utility of plant genetic resources lies in their variability. But this very feature means that the utility sought by a plant breeder may reside in a very few accessions out of hundreds or thousands of samples. When plant genetic resources are collected, there is no way of knowing whether or not any of the genes contained in the sample will be of any use. Only after expensive and time-consuming evaluation and characterization of the materials does their use to current breeders become apparent. Because some traits may become useful only at some time in the future, it may be decades before their latent utility is revealed by changing conditions in agricultural production. Moreover, because genes from a variety of nations may be incorporated into a single cultivar, crediting the original supplier of a particular gene would require an impossibly large program of genetic monitoring. For these reasons, it is argued that "raw germplasm" simply cannot be priced. While specific genes are admittedly valuable, they are needles in a haystack of no worth (Brown 1985:47).

It is true that genetic materials present the market with some unique problems in pricing. But to permit the value of the whole to mask the value of the part is to beg the question. Useful genetic material is in fact identified and used, to the great benefit of those who have appropriated it. The inability to set a price through the "natural" operation of the market is not in itself justification for failure to assign a value to something with recognized utility. There are various non-market strategies that could be used to establish compensation schedules for appropriation and use of raw genetic materials if there was a willingness to do so. And while there are technical problems associated with monitoring the movement of genes in breeding programs, private breeders are developing tools to provide "genetic fingerprinting" for the purpose of keeping track of their own patented genes. Market failure is an excuse rather than a logical justification for current practice. It speaks of lack of will to make compensation; it is not a legitimate reason for failing to do so.

Given their ideological commitment to what an ASTA position paper calls "the sanctity of private property" (ASTA 1984:6), businessmen, politicians, and plant scientists of the advanced capitalist nations are particularly sensitive to the charge that they have "robbed" the Third World of its plant genetic patrimony. A third line of argument defends the appropriation of plant genetic resources as a public good by asserting that

> No plant exploration team removes all of a specific source of germplasm when making a collection. This would be impossible as well as impractical... The original source area is not deprived of anything. In fact, duplicate samples are usually deposited with that country's agricultural agency. [Pioneer Hi-Bred 1984:47][22]

When plant collectors sample a population, they acquire only a few pounds of seed or plant matter. The vast bulk of the material is left untouched and in place. Unlike the extraction of most natural resources, the "mining" of plant germplasm results in no significant depletion of the resource itself. And collectors do customarily deposit duplicates of the collected materials with agricultural officials of the country in which they are operating. If the donor nation is not giving up anything, if it is not losing any utility, why should it demand compensation?

That the logic behind such an argument is not immediately recognized as faulty is testimony to the unique characteristics of genetic information. Germplasm differs from resources such as coal and copper, and even from such renewable resources as timber or fish, in a very fundamental way. With most natural resources, the utility acquired through their extraction is directly proportional to the physical quantity of the resource extracted. But with germplasm the resource of interest is physical matter only insofar as it is the carrier of *genetic information*. The utility is contained not in the seed *per se* but in the DNA sequences encoded in the cells that compose the seed. Collection of a small sample of seed is sufficient to transfer the genetic utility contained in very large populations of plants. With plant germplasm, the entire utility of the whole is in the part, and this masks the magnitude of the transfer of use-value that is nevertheless occurring.

Now the international transfer of such utility – or use-value – does not deprive the donor country of the capacity to use the resource. But the appropriating nation clearly benefits because it has acquired something useful that it did not have before. And if the exchange is unrecompensed, as is the case with the collection of plant genetic resources, the donor has in fact lost something insofar as it has forgone the opportunity to demand a reciprocal flow of benefits in exchange for its largesse. That is, the donor has forgone the opportunity to charge what economists term the "pure economic rent" that accrues to monopoly control over a resource.

An example may clarify my meaning. In 1977, a wild relative of maize, thought to be extinct, was rediscovered in Mexico by botanists associated with the University of Wisconsin. Seeds of the plant, called *teosinte* (*Zea diploperennis*), were returned to the United States and distributed widely to both university and private breeders. *Teosinte* was found to possess resistance to seven major diseases of maize in addition to a variety of other useful characteristics. Moreover, while maize is an annual, *teosinte* is a perennial. And because *teosinte* is sexually compatible with maize, these characters are potentially transferable to commercial corn lines. An economist has estimated that perennial corn bred from *teosinte* could be worth as much as $6.8 billion per year (Wiley 1986:46). If such gains do indeed accrue to introgression of *teosinte* germplasm into commercial corn varieties, Mexico will have provided the United States with a resource of enormous utility and enormous economic value without obtaining anything in return. While "robbery" is not the term to appropriately describe the relation, neither is it accurate to say that "the original source area is not deprived of anything." There is a clear transfer of use-value and a subsequent asymmetry in the distribution of benefits accruing to use of the resource.[23]

Finally, the failure of the advanced capitalist nations to agree to the Undertaking is often justified with the observation that "The FAO Undertaking is inconsistent with plant breeders' rights as protected by law in the United States and other nations that grant proprietary rights" (U.S. Department of State 1985). The provisions of the FAO Undertaking that mandate the unrestricted exchange of "elite and current breeders' lines" do in fact contradict established legal practice in many of the advanced capitalist nations. Those countries that are members of the International Union for the Protection of New Varieties of Plants (UPOV) have adopted national legislation expressly designed to provide proprietary rights in plant germplasm. The United States' Plant Variety Protection Act of 1970 is an example of such plant breeders' rights legislation. Such laws would have to be rescinded or altered to allow for the operationalization of the concept of common heritage for all types of plant germplasm. But the mere existence of such laws does not in itself justify the differential treatment of peasant land races and elite commercial varieties. Law is a social creation, not an immutable reflection of the natural order.

The arguments put forward by the seed industry and the functionaries of developed nations' governments to justify distinguishing some germplasm as valueless (and therefore free) common heritage and other germplasm as a valuable commodity and private property are baseless. That such a distinction exists has nothing to do with the essential character of the germplasm itself and everything to do with social history and political economy.

Conclusion

The crop population of the United States is as immigrant in character as its human population. Much the same is true of Europe. The development of modern industrialized agriculture in the advanced capitalist economies has been predicated on the systematic and continuous appropriation of plant genetic resources from source areas of genetic diversity that lie principally in the Third World. This primitive accumulation of plant germplasm is an enduring feature of the historical relationship between the capitalist core and its global periphery.

Much attention has been given to the post-World War II agricultural development initiatives that culminated in the Green Revolution. Western science made the seed a catalyst for the transformation of pre-capitalist agrarian social formations and their integration into the web of commodity relations that characterizes the contemporary world economy. The institutional and scientific network of the Green Revolution has also served as a mechanism for the collection of plant genetic information in the Third World and its transmission to the industrial, capitalist North.

Free appropriation of these resources has been justified by assertions that they represent the common heritage of mankind. Such assertions have recently been called into question by events associated with capitalists' own efforts to extend the reach of private property and to enlarge their sphere of accumulation. Internationalization of the commercial seed industry has brought common heritage and the commodity-form into unambiguous confrontation. Third World nations are increasingly coming to recognize a simple truth that has long been well understood by the capitalist nations of the North: Plant genetic resources are a strategic resource of tremendous value. With such knowledge has come an awareness of the asymmetric character of gene flows in the world system and of the implications of political and institutional control over plant genetic resources. Through the medium of the FAO's International Undertaking on Plant Genetic Resources, Third World nations are directly challenging the patterns of genetic transfer established and maintained through four centuries of capitalist development.

These patterns may be ripe for a profound realignment. While it is clear that the status quo cannot be maintained, the shape of a resolution to the germplasm controversy is by no means self-evident. One objective of this book is to contribute to the development of a "New International Genetic Order" for plant germplasm. But in order to assess the manner in which such an outcome might be fashioned, we must understand the ways in which new forces of production – new plant biotechnologies –are altering the terrain of debate and the range of political possibilities.

8

Outdoing evolution: biotechnology, botany, and business

> I think that the world seed trade figure is probably something like 30 billion dollars annually. And this number, if you stop to think about it, simply represents a heck of a lot of DNA, the primary annual genetic input to the agricultural sector. Numero Uno. Not exactly a discretionary item. No agriculture without seeds.
>
> David Padwa,
> Agrigenetics Corporation (1982)

> In the future of ecology or bionomics, that science concerned with the determination of new varieties and species, when all the mystery of biological forces and of adaptation shall have been unraveled, it is quite probable that the exact characteristics of a new variety or species may be predicted and predetermined at the will of the operator.
>
> Hyland C. Kirk
> (in U.S. House of Representatives 1906)

> The solutions are coming very fast now. In three years, we'll be able to do anything [with gene manipulations] that our imaginations will get us to.
>
> Mary-Dell Chilton,
> Ciba-Geigy Corp. (in Rossman 1984)

It would be interesting to know the source of the Wellsian prescience apparent in the quotation from Hyland C. Kirk. Kirk's comments came in the course of congressional testimony during a hearing on proposed legislation that would eventually become the Plant Patent Act of 1930. Kirk doubtless wanted to emphasize the extent to which new plant varieties were the product of human manufacture. Rediscovery of Mendelian genetics a few years earlier must have formed the basis for his anticipation of rapid advance in plant breeding.

And according to Ciba-Geigy plant molecular biologist Mary-Dell Chilton, some eight decades later Kirk's vision of the future is close to being a reality. The recent emergence of the new biotechnologies promises qualitative improvements in the techniques of genetic manipulation used to produce new plant varieties. Conventional plant breeding has in fact been, quite literally, breeding. Recombination of genes has been achieved through the sexual mating of whole plants. Now, however, techniques such as protoplast fusion and recombinant DNA transfer allow direct access to a discrete piece of a plant's genome at the cellular and even the molecular level. It is becoming possible to change gene frequencies with a wholly unprecedented specificity, and such recombinations are no longer limited to organisms that are sexually compatible. New plant varieties are being engineered in the strongest sense of that word's connotations of precision and foresight.

Chilton's buoyant "in three years, we'll be able to do anything" is a species of overstatement common among corporate biotechnologists. But general optimism as to the potential of the new genetic technologies is widely shared in the plant science community. Iowa State University plant breeder Kenneth J. Frey made biotechnology the subject of his 1984 presidential address to the American Society of Agronomy. He told the assembled ASA members,

> Biotechnology procedures that permit the easy asexual transfer of genes among microorganisms, when and if mastered for higher plants, hold the potential for transferring desired genes across species, genera, and perhaps family barriers. Let your imagination roam – the high lysine trait from the pigweed might be used to improve the quality of maize protein. The resistance of maize to wheat stem rust might be used to make wheat resistant to this disease. The gene for tolerance to Al[uminum] toxicity in wheat might make maize tolerant to Al[uminum] ... The future for agronomy is not only bright, but it has no foreseeable bounds. [Frey 1985:188-9]

And, for Frey, a bright future for agronomists implies a bright future for the rest of us. For among the fruits of the new plant biotechnology will be "an enormity of crop production that may dwarf the accomplishments of the 'Green Revolution' " (Frey 1985:187).

So we are confronted once again with the familiar language of miracle varieties and scientific revolution. And we are now promised that greener pastures lie just over the next petri dish. But if the past is indeed systematically connected to the future, we should now be in a position to critically analyze the current phase of technological innovation in light of prior experience. The threads that we have thus far followed through the history of plant improvement – the commodification of the seed, the changing division of labor between public and private research institutions, and the appropriation of plant genetic resources – remain the principal themes around which this analysis can be structured.

Biotechnology: an overview

This book focuses on plant improvement. But what is now occurring in the seed sector is one instance of a much broader technological transformation that is galvanizing changes in the social organization of all production processes in which organic substances or life forms play a significant role. This section is intended to situate plant improvement in that larger context.

The roots of the current "biorevolution" go back some three decades. In 1953, James Watson and Francis Crick succeeded in specifying the helical structure of deoxyribonucleic acid (DNA), the genetic material found in all living organisms that guides biological development and constitutes the vehicle for the transmission of hereditary information between generations. The structure of the DNA molecule implied how it functioned, and subsequent research revealed a remarkable property. Genetic information is encoded in DNA by the sequencing of pairs of four different nucleotide bases that lie along the backbone of the helix. Combinations of these four basic "letters" form "words," or codons, which code for the production of one of twenty amino acids. Amino acids combine to form the proteins that are the building blocks out of which all organisms are constructed.[1]

Hence, while the language of biology is capable of enormously complex expression, it is chemically quite simple. It is not only simple, it is also universal. It might be likened to a sort of biological Esperanto common to all organisms: "No tower of Babel here: the same 'words' that instruct the *E. coli* to add another particular amino acid to a protein chain would order up the same amino acid in a honeybee or a man" (Sylvester and Klotz 1983:28). The four bases become a kind of universal equivalent that erodes the incommensurability of species and that provides an entirely new vision of the fundamental unity of all life. Monsanto's advertisement boasting that "all life is chemical" is merely the vulgar commercial version of what Francis Crick christened the "Central Dogma" of molecular biology: Genetic information moves from DNA to RNA to protein and thus guides all biological processes (Judson 1979:336-7).

This reductionist approach has a distinctly utilitarian face as well. If organisms are programmed with a genetic code written in common terms, then it should be possible to read that code, rewrite the program, and even shift useful bits of program between organisms. Progress in mapping genes during the 1960s was accompanied by the discovery of restriction enzymes capable of cutting and resplicing segments of DNA at specified sites. In 1973, Stanford University's Stanley Cohen and University of California-San Francisco's Herbert Boyer succeeded in splicing a DNA sequence from one organism into bacterial plasmid DNA, and then using the properties of the plasmid to insert the gene into an *Escherichia coli* bacterium, where it was successfully expressed. The significance of their achievement is succinctly described in United States patent no. 4,237,244, issued to them in 1980 on their "Process for Producing Biologically Functional Molecular Chimeras":

> The ability of genes derived from totally different biological classes to replicate and be expressed in a particular microorganism permits the attainment of interspecies genetic recombination. Thus, it becomes practical to introduce into a particular microorganism... functions which are indigenous to other classes of organisms. [U.S. Patent Office 1980:1]

Genetic engineers in commercial and university labs are currently pursuing the transfer of genetic information from both higher and lower organisms into microbes suited to the exigencies of industrial culture. Wide arrays of low-volume/high-value products, ranging from pharmaceuticals (e.g., hormones, insulins, interferons, vaccines, antibiotics) to specialty chemicals (e.g., food additives, enzymes, pheromones, amino acids, pesticides), are expected to be amenable to microbiological production. Other active properties of microorganisms are also of interest. Work is under way on designing plant symbiont microbes with enhanced nitrogen-fixation capabilities and frost-inhibition effects. Genetically engineered microorganisms are expected to be used for mineral leaching in mining operations, in facilitating oil recovery, in degrading pollutants and toxic wastes, and in transforming biomass feedstocks into substrates for the production of commodity chemicals.

Microorganisms are not the only targets of the genetic engineer; plants, animals, and humans will come under the genetic scalpel as well. The techniques of superovulation and embryo transfer are already changing the genetic composition of American dairy herds, and animal scientists look forward to designing new breeds using the expanding universe of genetic information available to them. Plant breeders expect to incorporate foreign genes into economically important species for the improvement of such characteristics as photosynthetic capacity, stress tolerance, nitrogen fixation, and herbicide resistance. The technique of plant tissue culture even offers the possibility of moving production of certain crops out of the fields and into the factory. Epistemological and ethical considerations may limit the rapidity and scope with which the new genetic technologies are applied to humans, but their impact will be profound. Gene therapy will give us a means of confronting – perhaps curing – the more than 1,000 genetic diseases to which the human species is subject. The development of monoclonal antibody technology, a sort of genetic tagging system of great sensitivity and specificity, is making the process of diagnosis more efficient and certain.

The point of this brief excursion into the promise of biotechnology is to emphasize the broad nature of its applicability and the tremendous variety of products it could generate. The industrial sectors that will feel the impact of biotechnology account for some 70 percent of the annual American gross national product (McAuliffe and McAuliffe 1981:28). It is estimated that by the year 2000, annual worldwide sales of bioengineered products will reach $40 billion (Russell 1983:14). Thus, the prospect of outdoing evolution also offers, in commercial terms, the prospect of outdoing competitors. The structure of DNA is not merely the Rosetta Stone that furnishes the key to the hieroglyphics of the genetic code, it is also the Philosopher's Stone of the new alchemists of modern industry: It has the power of turning not base metals but base life to gold.

This property was not clearly apparent until 1973, when Cohen and Boyer developed practical procedures for creating functioning rDNA molecules. From that achievement, events moved with astounding rapidity. It was a mere four years from the time a rat insulin gene was cloned to the time human insulin made by bacteria into which human genetic material had been spliced with rDNA technology entered clinical trials. A striking feature of biotechnology is this near identity of basic and applied research. A second feature of note is that the new genetic technologies were developed largely within research programs focused on biomedical research and supported by public funds, most of which were provided by the National Science Foundation and the National Institutes of Health (NIH). The "basic" character of biotechnology research and the public nature of its funding meant, in the words of University of California Vice Chancellor Roderic Park, "that it initially resided totally within the universities and institutes and when industry became aware of its importance they had only one place to go in order to get the human resources to help them out" (U.S. House of Representatives 1982a:127). Firms like Monsanto, Exxon, and Hoechst found that the scientific labor power they needed to realize the commercial promise of biotechnology was not in their corporate labs, but in the halls of institutions like Washington University, Harvard, M.I.T., and Cold Spring Harbor.

Such labor power was not easily extracted from academia. There is, after all, little incentive for prominent scientists to leave their well-funded labs and tenured positions in order to submit themselves to corporate discipline. However, with offers of equity positions and management freedom, venture capital proved more suited to the task of eroding the ties that bind scientists to the university. In 1976, University of California-San Francisco's Herbert Boyer and venture capitalist Robert Swanson founded Genentech, a research company devoted to commercializing the advances in genetic technology. Over the next seven years, over 110 such marriages of venture capital and university scientists were consummated. The resulting firms have taken such evocative names as Agrigenetics, Advanced Genetic Sciences, DNA Plant Technology Corp., Hybritech, Molecular Genetics, Calgene, and Repligen.

As excitement over biotechnology grew in industry and investment circles, these new biotechnology firms (NBFs)[2] began to go public, with stock offerings that made Wall Street history. In 1980, Genentech's public offering set a record for the fastest increase in price per share ever experienced on the New York Stock Exchange: from $35 to $89 a share in twenty minutes. A year later, Cetus Corporation set the Wall Street record for the largest amount of money raised in an initial stock offering: $115 million (OTA 1984:4). As of 1984, fully $2.5 billion had been invested in the new companies (*Business Week* 1984a:84).

Stimulated by the apparent commercial promise of biotechnology, the multinational pharmaceutical and petrochemical giants that dominate the sectors in which the new genetic technologies will be most immediately applicable have adopted a four-pronged approach to the shaping of the new field. About a fifth of the investment in the NBFs has come from established multinational corporations (MNCs) purchasing substantial equity positions, and hence a significant measure of management control. Research contracts afford the MNCs a further means of influencing the direction and character of research activity in the new start-ups. Second, the multinationals have concluded a series of unprecedentedly large research contracts with universities that are at the cutting edge of molecular biology. Third, the large corporations have committed themselves to the rapid development or improvement of in-house capabilities in biotechnology. Finally, in an effort to facilitate the transfer of knowledge from academia to industry, corporate interests have initiated a campaign to shape federal policies and legislation in regard to such areas as patent rights, product regulation, and the legal structures in which business may be pursued.

The NBFs have attempted to maintain their independence from the MNCs by diversifying research contracts, by obtaining financing in innovative fashion (e.g., research and development limited partnerships), and by establishing their own formal and informal ties to universities. With both monopoly and competitive capital vying for the services of their faculty members, universities have found themselves in a Faustian dilemma, torn between Mammon and the Ivory Tower.[3] And in a time when federal support for universities is declining, Mammon speaks of sums the magnitudes of which, according to Harvard president Derek Bok, "stir the blood of every harried administrator struggling to balance an unruly budget" (Russell 1983:17). Biotechnology offers not only a means of increasing extramural support but also a vehicle that an institution can ride to enhanced prominence and status in the academic community. University administrations have almost without exception embraced biotechnology and are actively soliciting the more intimate relations with industry that are dictated by corporate funding.

There has been established a Byzantine web of formal contractual obligations and informal connections between universities, the NBFs, and the multinational giants. Enthusiasm over biotechnology has carried over into the circles of government, where "high tech" is seen as the solution to the problem of economic stagnation. Biotechnology means profits to the corporations, enhanced status and new facilities to the universities, and a revitalized economy and a more competitive posture in international markets to the federal government. The climate of opinion among these institutions might be characterized as "laisser innover," and there is little enthusiasm for critical examinations of genetic engineering that might slow the pace of the biorevolution.

Yet the development of biotechnology has not escaped the scrutiny of those who recognize that scientific advance and technological innovation have rarely been wholly benign, but always carry, Janus-like, the two faces of benefit and liability. To their credit and their ultimate regret, it was the scientists closest to genetic engineering who first expressed concerns relating to the potentially negative consequences of the emerging technology. Speculation arising from discussion of proposed rDNA experiments at the 1973 Gordon Conference focused on the hazards associated with the possible escape and proliferation of novel forms of life. A working group of prominent scientists took the unprecedented step of calling for a moratorium on certain types of research. Such concerns culminated in the Asilomar Conference of 1975 at which the scientific community attempted to agree on the need for regulation of biotechnological research (Krimsky 1982). Only a weak consensus was possible, because many scientists had already begun to fear that they had opened the Pandora's Box of regulation by bringing the implications of rDNA into the public eye.

And Asilomar marked the initiation of sustained criticism of, and even overt opposition to, not only the conduct of rDNA research itself but also the social impacts that it had already generated or could possibly generate. The most persistent theme of criticism has been the fear of adverse ecological and epidemiological consequences that might stem from the accidental or deliberate release of self-propagating genetically engineered organisms into the biosphere (King 1978; Krimsky 1982; Perrow 1984; Rifkin 1986). Such concerns have constituted the core of most public opposition and have provided the logic for enactment of legislation regulating of rDNA research at local, state, and national levels. Some critics, prominent scientists among them, have argued that the very power of the new technology outstrips our capacity to use it in safety, that neither nature's resilience nor our own social institutions are adequate protection against the unanticipated impacts of genetic engineering. Given our recent experience with atomic energy and past flirtation with eugenics, can we wisely proceed with the development of new life forms and what Robert Sinsheimer has called the "genetic redefinition of man" (1975:151)?

Much depends on the competence of our social institutions in coping with the issues thrown up by scientific advance. While the federal government had, by 1976, established a Recombinant DNA Advisory Committee (RAC) within NIH and shortly thereafter set up guidelines for rDNA research, these guidelines have been substantially relaxed in recent years, and neither NIH nor any other department has yet been given a clear and authoritative mandate to regulate genetic engineering. The congressional Office of Technology Assessment has, as evidenced by its most recent report on biotechnology, viewed its task as one of assessing the competitive position of American business rather than considering broad issues of social impact (OTA 1984).

Indeed, the government's principal role in rDNA research has been not regulatory, but fiscal in nature. The progress in molecular biology and biochemistry that made genetic engineering possible was developed largely with public funds by researchers whose educations often had been underwritten by public monies. To the extent that private industry is able to appropriate this knowledge for commercial purposes, critics like Jonathan King argue that "The public is being forced to buy back what the public itself initially financed" (1982:40). Because tax revenues were instrumental in generating the new biotechnologies, should not the public have a major role in determining their manner of deployment?

The proliferation of contractual arrangements linking academic researchers and universities to corporations involved in the commercialization of biotechnology has raised important questions regarding the changing role of the academy in modern society. The university has never been the Ivory Tower of myth, but it has enjoyed at least relative autonomy from external commercial pressures. Now, however, even this modicum of insulation is being stripped away by what Congressman Albert Gore (D-Tennessee) has called the "selling of the tree of knowledge to Wall Street." It is not clear that the public interest is served by the splicing of commercial interests into the body politic of the republic of science. Conflicts of interest may easily emerge for university faculty who have commercial connections with or equity positions in their own or another's genetic engineering firm. Dick Russell (1983:20) provides a succinct summary of critics' principal fears: A university scientist with direct or indirect commercial interests

> simultaneously may be serving on government granting panels, testifying at Congressional hearings, publicly discussing the risks and benefits of new products. At universities, graduate students may be turned into an

> unpaid labor force to serve commercial ends. Institutions become overdependent on money from a single large corporation, and professors distracted from their proper duties. And secrecy may become almost a disease.

Insofar as the research agendas of public institutions are influenced by the leveraging of privately supplied research funds, society will be losing a portion of what positive control it does have over the types of products that are developed by the new genetic technologies.

In the United States, commercial development of products is the virtually exclusive province of private enterprise. Corporate perceptions of market potential are, however, by no means coterminous with objective social needs. Sheldon Krimsky (1983:56) asks, "Is there any justice in allowing the free market to determine whether and to what extent gene splicing improves people's living conditions by determining what products are introduced into the market place?" There may well be any number of opportunities for the production of socially valuable products offered by bioengineering that will never be pursued because they would not be profitable. On the other hand, any number of new products of biotechnology that are privately profitable may see the light of day even though they may have negative consequences for certain segments of the population or for the environment. Further, even when technologies are neutral, they may be introduced in a manner that reflects the relative power of various classes in society.

In sum, it has been clear to all observers that biotechnology is a profoundly powerful new technical form that contains the possibility, indeed the probability, of transforming society in important ways. Biotechnology has catalyzed significant changes in the institutional framework of biological research, in industrial structure, in the social relationships characteristic of the university, and in our system of property rights. Biotechnology is, therefore, *already* associated with social impacts. However, there is no consensus as to whether these impacts represent on balance a benefit or a liability for society.

I have outlined some of the principal concerns expressed by those who perceive dangers as well as promise in the biorevolution and who call for enlarged social control over the development of the new genetic technologies. These concerns have been elaborated over the last eight years in a voluminous stream of articles and books that have themselves registered an important social impact: They have facilitated the mobilization of citizens' groups willing to join words with action in opposition to what they believe are threats to the public interest. Weakening of the NIH guidelines and the more or less unhampered course taken by the commercialization of biotechnology need not blind us to the success of many popular initiatives. As Stuart Newman (1982:56) observed, genetic engineering "was the first new technology to stimulate a national discussion of risk prior to its widespread implementation." That in itself is a signal advance.

But the struggle for a real measure of social control over the implementation of genetic engineering will be protracted and difficult. This struggle will be more effective insofar as there is a goodness of fit between the analytical frameworks selected by critics and the social realities under scrutiny. There is now a need for a shift in our strategic approach to this problem. Up to now, much of the debate centering on the development of biotechnology has been highly generalized, often abstract, and heavily speculative. Necessarily so, because the technology itself is so new, and both material technology and social relations are in a state of tremendous flux. Biotechnology and its potential impacts needed to be grasped and understood as a whole, and critiques were appropriately couched in broad terms.

Now, however, biotechnology is entering a new phase. The creation of new genetic engineering firms has slowed to a trickle, and a new caution informs corporate investments in both universities and the biotechnology boutiques. The bio-hype of the late 1970s has given way in the 1980s to a more traditional concern with the bottom line. Most important, the products of gene splicing and other novel genetic technologies are now coming to market. Eli Lilly now markets human insulin produced in fermentation vats by genetically engineered bacteria. Molecular Genetics Inc. sells Genecol 99, a vaccine for calves. Over forty *in vitro* diagnostic products using monoclonal antibody technology were approved in the United States as of June 1983. A host of other products, including bio-produced human growth hormone and alpha interferon, is close to FDA approval. In a January 23, 1984, cover story entitled "Biotech Comes of Age," *Business Week* welcomed biotechnology as a "real" - i.e., product-generating and profit-making – industry.

As biotechnology matures, it is essential that critical and progressive modes of analysis keep pace with its development. As biotechnology moves into production spheres and is increasingly disseminated throughout modern industry, we need to refine the scope of our analyses to encompass the ramifications of its increasingly diverse concrete impacts. This does not mean the abandonment of the broad concerns that have already attracted attention and analysis. However, as a complementary undertaking, we need to ask how broad problems (e.g., patenting life, university/industry relations) manifest themselves in particular sectors of production.

Production may be organized in different ways in different industries and subsectors. In effectively analyzing and anticipating the social impacts of new technologies, it is important to be sensitive to this variation. Agriculture, for example, is organized in very different fashion from industrial production. It follows that the social contexts into which biotechnology will be deployed will exhibit different characteristics. Differences in social relations mean different political possibilities, different sorts of potential allies, and different strategies and points of leverage. There will also be different technical promises to be pursued, as well as dangers to be avoided.

Additionally, it is necessary that critical analysts develop proposals for *positive* programs for biotechnology, rather than concentrating exclusively on negative or regulatory approaches. The new genetic technologies *do* contain liberatory possibilities and tremendous positive potential. We may not, for some time at least, succeed in seriously circumscribing or controlling the activities of corporate capital in the development of biotechnology. The technology *will* be developed; and it should be. We are faced with a dual challenge in our struggle: We must decide what we do not want and do our best to avoid it, and we must decide what we do want and try to achieve it. Detailed analysis of biotechnological possibilities in the context of particular production sectors will be crucial to the setting of both positive and negative agendas. If we are to deal adequately with biotechnology as it matures, we must complement the efforts that have been made up to now with more specific analysis. To this end, this chapter and the next consider the case of plant improvement.

First the seed: nexus of the production process

With biotechnology we have gained manipulative access to the basic molecular building blocks of life itself. But if the rearrangements of these blocks by genetic engineers are to have any practical utility, they must be expressed at the whole-plant level and, ultimately, must somehow be integrated into agricultural production processes. In arable agriculture, it is the seed that provides the essential material link between research and the market. The seed is the endpoint of the research and development process, and it is in that form that the new plant variety becomes a commercial product. The advent of the new biotechnologies will not change this fundamental parameter.

Now, a seed is, in essence, a packet of genetic information, an envelope containing a DNA message. In that message are encoded the templates for the subsequent development of the mature plant. The content of the code crucially shapes the manner in which the growing plant responds to its environment. Insofar as biotechnology permits specific and detailed "reprogramming" of the genetic code, *the seed, as embodied information, becomes the nexus of control over the determination and shape of the entire crop production process*.

Both public and private plant biotechnologists are now focusing their research on an array of agronomic traits including plant architecture, harvestability, maturation, photoperiod, photosynthetic efficiency, stress (temperature, moisture, chemical) tolerance, nutrient utilization, nutritional quality, and disease resistance. As the new technologies enhance control over the specification of the particular form in which these features are expressed, genetic engineers will be able to determine where a plant may be grown, under what environmental conditions, the requisite inputs, the timing of cultural and labor activities, the mode of harvest, the manner of processing, and the characteristics of the plant product. To paraphrase Daniel Bell (1973:378) again, the seed is the monad that contains within it the imago of future agriculture.

The information in the seed is the gateway to control over the production process as a whole. And because the seed is also the material entity in which molecular biology is articulated to plant breeding, and in which science (research) is articulated to commerce (the commodity), control over the seed becomes a matter of considerable importance. An understanding of this fact must inform all analysis of plant biotechnology. The aphorism "First – the seed" is more resonant than ever before.

Biotechnology and plant breeding: revolution or evolution?

The initiative for the application of the emerging biotechnologies to plant improvement came not from plant breeders but from molecular biologists who had been working principally with microbial systems (Carlson et al. 1984:24). Familiar neither with organismic-level biology[4] nor with the techniques and achievements of plant breeding, these scientists held rather inflated views about what genetic engineering would accomplish for plant improvement. At the same time, the burgeoning bio-boom swept plant science into the economic maelstrom of investment and business start-ups, along with most other biologically oriented production sectors. There was stock to be sold, and the venture capitalists and scientist-entrepreneurs who founded the NBFs were not shy about trumpeting their claims of scientific revolution.

This combination of scientific naivete and commercial hyperbole generated what plant breeders regarded as some arrogant and rather extravagant claims on behalf of the new technologies (Sprague et al. 1980; Bingham 1983; Duvick 1984). Molecular and cellular techniques, some asserted, would rapidly displace the cumbersome whole-plant approach to breeding. Not only would rDNA transfer open the entire world gene pool to the plant genetic engineer, it would also make plant breeding qualitatively more efficient. Precise transfer of specific genes would replace the combination of entire genomes entailed by sexual crosses. Instead of evaluating plants in the field, scientists would screen and select among millions of individual

Table 8.1. *Techniques of genetic transformation associated with molecular, cellular, and whole-plant biology*

Molecular biology (genetic engineering)	Cell biology (*in vitro* horticulture)	Whole-plant biology (conventional breeding)
rDNA transfer	Meristem culture	Sexual cross
Transposable elements	Embryo culture	Wide cross
Microinjection	Callus culture	Chemical mutation
Monoclonal antibodies	Cell culture	X-ray mutation
Gene mapping	Pollen/anther culture	
Gene repression	Somatic embryogenesis	
Electroporation	Somaclonal variation	
	Protoplast fusion	
	Protoplast culture	
	In vitro infection	
	Leaf-disc transformation	
	Gamete transformation	
	Chloroplast engineering	

cells – each a potential plant – in a single petri dish, with vast saving of time and space. In short, the plant molecular biologist and the tissue culture specialist would displace the conventional plant breeder as the producer of new cultivars. These new varieties would be radically improved types: corn that fixes its own nitrogen, salt-tolerant cultivars that could be irrigated with sea water. Such observations as "in 5 to 10 years, Saudi Arabia may look like the wheat fields of Kansas" (Mintz 1984:49) have not been uncommon. Is plant breeding about to be revolutionized?

Plant breeders are fond of pointing out that insofar as they are in fact moving genes from organism to organism, they are already plant genetic engineers. The point is well taken, but it is useful to use the term "genetic engineering" to refer specifically to the new techniques deriving from advances in molecular biology, biochemistry, and genetics. It is also useful to distinguish these molecular-level techniques from the new cellular procedures based on the older technology of tissue culture. Simmonds (1983a:21) has appropriately termed these cellular procedures "*in vitro* horticulture." Genetic engineering and *in vitro* horticulture are often subsumed under the generic term "biotechnology" and are in turn distinguished from conventional plant breeding, which operates principally at the whole-plant level. Table 8.1 lists the principal techniques associated with each of these levels.[5]

Genetic engineering is potentially an extremely powerful technique when applied to plant improvement. According to Dr. Raymond Valentine, founder of the NBF Calgene, the plant genetic engineer's motto is "any gene from any organism into plants" (quoted in California Agricultural Lands Project 1982:13). In theory this is quite possible, but the practical difficulties are legion. Simmonds (1983b:68) has identified the following steps as necessary for the useful realization of rDNA transfer in plants:

> (a) the desired 'foreign' gene must be identified and the DNA isolated; (b) the DNA must be multiplied in some suitable (probably bacterial) system, a process called 'cloning' or, perhaps better, 'molecular cloning'; (c) the 'cloned' DNA must be transmitted to recipient crop cells by a suitable 'vector' which might be a plasmid, a virus, a liposome, a bacterial cell, or a micro-syringe; (d) the DNA must be incorporated into the recipient DNA, whether nuclear, chloroplast, or mitochondrial; (e) the altered cells must be made to regenerate whole plants which must then; (f) be shown to express the new gene and transmit it sexually and; (g) have the genetic potential to be 'worked up' by conventional plant breeding methods to be in an agriculturally useful form. A formidable programme.

A formidable program indeed, and one that encompasses all three levels detailed in Table 8.1.

The ability to circumvent natural barriers of sexual compatibility via genetic engineering will certainly be useful. There are numerous characteristics that might be transferred into crop plants, and no shortage of organisms as potential donors. Even where sexual recombination is possible, genetic engineering may prove useful. The capacity to transfer only genes of interest might eliminate the time-consuming, multi-generational process of backcrossing now used in conventional breeding to eliminate the extraneous genes that necessarily accompany the desired genes in a sexual cross. The problem is that molecular biologists now have great difficulty identifying the genes they wish to move. Desirable traits can be observed at the phenotypic (organismic) level, but locating the gene – or, more likely, the gene complex – controlling that trait is, according to Harvard's (and Biotechnica's) Lawrence Bogorad (quoted in NRC 1984:21), "one of the most difficult and challenging operations in molecular biology." Much fundamental knowledge remains to be gathered, and a great deal of genomic mapping must be done before scientists can routinely specify the DNA sequences they want to move.[6]

When such sequences are identified, the problem becomes one of transmitting them into the DNA of the recipient plant cell. That this can be done was demonstrated in 1983 by three different groups of researchers at Washington University, the Max Planck Institute, and Monsanto Company. All three teams made use of the Ti plasmid of *Agrobacterium tumefaciens* as a vector system for the transfer of bacterial DNA into plant cells (Chilton 1983; Schell and Van Montagu 1983). However, only small percentages of cells were successfully transformed. Moreover, the Ti plasmid is capable of penetrating only nuclear (as opposed to cytoplasmic) DNA and does not work at all in monocotyledons, a class of plants that includes the principal cereal species. In an effort to develop more efficient and ubiquitously applicable vector systems, scientists are exploring the use of other bacterial plasmids, viral plasmids, pollen, transposable genetic elements, and even transformation through brute force as in the case of microinjection.

Even when the DNA of a cell is successfully transformed, that cell must still be regenerated into a plant if the gene transfer is to have any practical value in crop production. Regeneration is founded on the phenomenon of totipotency, the capacity of cells in culture to recapitulate the entire organism. Here genetic engineering interfaces with *in vitro* horticulture. Plant tissue culture is itself not a novel technology. Meristem (shoot tip) culture has long been used for the propagation of certain high-value ornamental plants such as orchids, or for production of virus-free seedlings. Many plant species are capable of being cloned in this fashion. What is new is the level of cellular organization at which tissue culture is being attempted. As one moves down the hierarchy of cellular organization to simpler structures (i.e., from meristem to single cell), regeneration becomes increasingly difficult (Ammirato et al. 1984; NRC 1984).

Should a genetically engineered cell be regenerable as a whole plant, the alien gene must still be expressed. Even if the foreign genetic material is expressed, one does not necessarily have a new elite variety appropriate for commercial distribution. Most major traits are multigenic (e.g., the much discussed NIF – nitrogen-fixing – character in legumes involves 17 genes) and, when incorporated into an alien genome, will constitute a "profound modification of the biochemical architecture" (Simmonds 1983a:22) of the host. Conventional breeding procedures will be required to put the new variety into commercial shape, and at worst the deleterious effect of the foreign genes on the expression of native genes will prove to be a limiting trade-off.[7]

Bingham (1983:223) asserts that, in most cases, "It is going to take as long to breed a molecular engineering gene into a successful cultivar as it takes for a natural gene." This may well be the case even for single-gene, intra-specific transfers via genetic engineering. It has become apparent that one consequence of tissue culture is somaclonal variation – the regenerated plantlets exhibit novel genetic variability.[8] Such changes might eliminate the elite status of any line, no matter now simple and non-disruptive of biochemical architecture the introduction of foreign genetic material had been. Genetic engineering will not eliminate the need for testing and evaluation of converted lines through established breeding procedures.

What is a problem in one context may be an opportunity in another. Thus, the generation of somaclonal variants in tissue culture promises to provide useful genetic variability in a form that can be quickly incorporated into ongoing conventional breeding programs (Evans et al. 1983; Earle 1984; Ammirato et al. 1984; Carlson et al. 1984). *In vitro* horticulture is also capable of generating novel variability through the technique of protoplast fusion. Protoplasts (cells with their walls removed) of two different organisms may be fused chemically or electrically to create a new somatic hybrid (Evans 1983). This process, like genetic engineering, permits inter-specific and even inter-generic recombination, but, like conventional sexual breeding, it combines entire genomes and so has the disadvantage of lack of specificity. Moreover, few major crops have yet been successfully regenerated from protoplasts.

In vitro horticulture can also involve selection in culture. Selection agents such as herbicides, saline solutions, or pathogens can be applied to 3-5 million cells in a 6-inch petri dish rather than to 3-5 million whole plants on several acres of land (NRC 1984:33). But only traits expressed at the cellular level can be screened; no bioassay is yet available to permit *in vitro* selection for traits expressed at the whole-plant level. And, again, somaclonal variation means that tissue-cultured plants may be changed in more traits than just the one for which selection is being made. So, for all of these *in vitro* procedures, conventional breeding techniques will still be necessary for the evaluation of regenerated materials and for their incorporation into elite breeding lines or their development as commercial varieties (Evans 1983:259; Duvick 1984:14).

Plant breeding is emphatically not about to be replaced by plant molecular or plant cellular biology. The new biotechnologies promise to be extremely powerful adjuncts to conventional techniques, but will not wholly displace them. Indeed, it is not too much to say that realization of the potential of biotechnology in plant improvement will still depend on manipulations at the whole plant level. Simmonds (1983b:69) has concluded that "The relationship between plant breeding and genetic engineering will be coevolutionary, I think, each enhancing the other. Neither revolutions nor takeovers are likely." There is an emerging consensus among both plant breeders and molecular biologists that this is an accurate analysis (Bingham 1983; Duvick 1983, 1984; Evans 1983; Fraley 1983; Padwa 1983; Phillips 1983; NASULGC 1984; NRC 1984; Fraley et al. 1986).

But if plant genetic engineering and *in vitro* horticulture still need plant breeding, the converse is even more true. Though the rates of yield gain in major crops have shown no sign of leveling off, they are achieved at increasingly greater cost (Duvick 1984:1). About half of the historic increase in crop yield since 1930 can be attributed to the use of fertilizer, pesticides, and mechanical equipment (Russell 1974; Duvick 1977). There is evidence that the productivity potential of these chemical and mechanical technologies has been largely exploited (Sundquist et al. 1982:xiii-5). The burden of maintaining yield trajectories will fall more and more on the genetic component of plant improvement. According to Pioneer Hi-Bred's Director of Research, "More efficient and more powerful breeding techniques are needed. Biotechnology promises to give such assistance to plant breeding, but major improvements will not come as soon or as spectacularly as has been popularly expected" (Duvick 1984:1). Plant breeders increasingly recognize that molecular and new cellular methods will be used in plant improvement (Bingham 1983; Phillips 1983; Simmonds 1983a). It is very difficult to determine how quickly this will occur. Molecular biologists are, naturally, more sanguine than plant breeders. In general, the new plant biotechnologies are making rapid advance (Bliss 1984; Lawton and Chilton 1984; Fraley et al. 1986), and John Hesse, of Plant Resources Venture Fund, has commented that "All the right things are being done better and faster than expected" (quoted in Sirkin 1984b:24).

Whether or not a technical revolution is occurring in plant breeding is largely a rhetorical question. Two essential points emerge from the foregoing excursion into technical possibilities. First, the new biotechnologies will provide us with an increasingly sophisticated ability to specify the genetic composition of plants. Second, the new molecular and cellular techniques will complement, rather than supersede, plant breeding. The new technologies cannot do without whole-plant manipulation. At the same time, a combination of biotechnology and plant breeding will be markedly superior to conventional techniques used alone. The question is not so much whether or not one of the levels detailed in Table 8.1 will displace the others, but how all three levels will be integrated.

From competitive to monopoly capital

Whatever the ultimate technical impact of biotechnology may be, the new technology has *already* galvanized important social and institutional transformations. This is nowhere more clear than at the level of industrial structure. We saw in Chapter 6 that during the 1970s a confluence of factors – rising grain prices, declining rates of profit in the chemical industry, passage of the Plant Variety Protection Act, the opportunity to rationalize agro-input marketing – initiated a wave of seed company acquisitions by large multinational corporations, many of which had significant agrichemical interests (see Table 6.3). The emergence of biotechnology reinforced this trend toward centralization of the agricultural-inputs sector. Those agrichemical producers that purchased seed companies found themselves uniquely poised to take advantage of biotechnological advances in plant improvement. Agricultural biochemistry, which had of necessity always been closely adapted to biological processes, has found in biotechnology a common technical base with plant breeding. Research and development in the seed companies and their parent corporations has profound synergistic potential.

There is no question that corporate managers recognize the seed, and seed production and distribution facilities, as the crucial nexus for the commercialization of plant biotechnology. In the foreseeable future, commented Agrigenetics' David Padwa (1982:99),

> There's no way, aside from one or two vegetatively propagated species, of getting an improved crop plant to agriculturalists without going through the seed... The point is that the big ongoing values will belong to the party who can handle large scale production and market the improvement to farmers.

This realization has undergirded a continuing bull market for seed companies (*Business Week* 1984a:69; Kidd and Teweles 1986). It is the deep-pocket multinationals with major agricultural biotechnology interests – such as Monsanto, Upjohn, and Lubrizol – that are the most active participants. Such corporations are willing to make very substantial bids for key seed firms. For example, Central Soya was itself in the market for seed companies when Upjohn made an offer for its O's Gold subsidiary (fifth largest hybrid corn producer) that Central Soya was unable to refuse (*Wall Street Journal* 1983:21).

These purchases are being made with the expectation that seed companies are the conduit through which genetically engineered plant varieties will be made available to the consumer. It is important to understand that those seed companies that have been acquired are principally that elite 10 percent or so of the trade that maintain plant breeding research programs. Thus, in purchasing the Joseph Harris Seed Company, Celanese acquired not only an extensive production and marketing network but also the services of some thirty experienced plant breeders and technicians, as well as a large collection of proprietary germplasm. Access to adapted, elite breeding lines is an extremely important consideration for any company looking to penetrate seed markets. A major factor in Monsanto's acquisition of the Jacob Hartz Seed Company was that firm's collection of proprietary soybean germplasm. By buying the top seed companies, corporations obtain both essential raw material (germplasm) for their genetic engineers and a labor force capable of putting engineered genes into marketable form.

If the seed companies are meant to be the outlet for the finished commodity, the parent corporations are providing the biotechnology input. Every parent firm listed in Table 6.3 has a significant in-house biotechnology research effort. These efforts can be very large. Monsanto recently completed a $150 million Life Sciences Research Center to provide physical infrastructure for its biotechnology research. These facilities will house 1,200 researchers by 1987 (Keppel 1984:3G).[9] Ciba-Geigy opened a new $7 million Agricultural Biotechnology Research Center in 1984 to serve as the locus of its efforts.

Of course, not all agricultural biotechnology research in the multinationals is directed exclusively to plant improvement. Most of these companies have multiple interests and are also involved in exploring such areas as veterinary pharmaceuticals, plant and animal growth regulators, and microbial crop symbionts. But even when the subject of research is chemicals or microbes, there are important synergies with plant science. Developing a new herbicide or improving a nitrogen-fixing rhizobacterium requires an understanding of the genetics of the plant to which the product will be applied. And, as we shall see, the seed is increasingly being viewed as the ideal delivery system for chemical and biological inputs. So the plant is also the critical nexus of much research.

The potential market for the seed of biotechnologically altered plant varieties is itself a considerable stimulus to plant research in the multinationals. L. William Teweles and Co., agricultural consultants, have estimated that by the year 2000 this market will be in the neighborhood of $6.8 billion (*Agricultural Genetics Report* 1983:3). Monsanto expects to be selling some $500 million worth of seeds in the late 1990s (Sanford 1984:Gl). According to the company's president, Richard J. Mahoney, seed is one leg of "a triad of high-performing areas" to which Monsanto's emphasis is being shifted (quoted in Keppel 1984:3G). Atlantic Richfield (ARCO) has established a subsidiary, Plant Cell Research Institute (PCRI), to pursue biotechnological approaches to plant breeding. Jim Caldwell, ARCO's Vice-President of Operations, explains that "over the next 20 years, there's going to be a lot of technology brought to bear on agriculture. It's a fallout from the ability we now have to manipulate plants genetically at the molecular level . . . we think it's going to be a *business* that we'd like to be involved in" (quoted in Pramik 1982:14).

And it is a business to which the multinationals are not the only prospective aspirants. One of the most striking social impacts associated with the emergence of biotechnology has been the proliferate establishment of research companies dedicated expressly to the commercialization of the new genetic technologies. Since the formation of Genentech in 1976, over 100 NBFs have appeared in the United States (OTA 1984:93). A substantial number of these have plant improvement as a prime, or as an important, objective. Information regarding the principal plant-oriented NBFs is presented in Table 8.2. These companies constitute an important challenge to the multinationals.

A salient feature of the NBFs is that they are the progeny of the marriage of venture capital and university professors. Biotechnology was developed in the university, and it is the intellectual resources of academia upon which bio-industry is really founded. The biotechnology start-ups have had marked success in attracting the best and the brightest among the limited pool of scientists with expertise in the new technologies. By starting their own firms in partnership with venture capital, entrepreneurial academics have been able to combine commercial gain with a university-like research atmosphere. It is probably true that the quality of science in the small research firms is, on the whole, superior to that in the corporate labs of the multinationals (Howard 1982:93). The MNCs have found it useful to establish research contracts with and, in many cases, to purchase equity interests in the NBFs (Table 8.2).

Yet these connections, which demonstrate the NBFs' main strength, also reflect their critical weakness. The start-ups are, in fact, research and development (R&D) companies. They sell knowledge, not final products. Current income for most of the NBFs in Table 8.2 comes from research contracts, from providing services to the multinationals rather than manufacturing a product. The start-ups are not unmindful of David Padwa's (1982:99) injunction, quoted earlier, that "the big ongoing values will belong to the party who can handle large scale production and market the improvement to farmers." Sungene's Vice President for Finance observes, "We won't be a rich company if we concentrate on contract research" (quoted in Gebhart 1984b:9). However superior and lucrative new plant varieties created by genetic engineering or the newer techniques of *in vitro* horticulture may be, they are still some years away. The start-ups are fond of praising the patience of their venture capital investors, but this patience is not unlimited. Nor is there any assurance that even contract research money will continue to be made available, as Allied Chemical's withdrawal from a nitrogen fixation project with Calgene illustrates.

Even when a product in the form of improved germplasm is available, many of the plant-oriented NBFs will confront a second critical weakness – their lack of seed production and marketing facilities. Except for Agrigenetics and Plant Genetics, none of the firms in Table 8.2 has a seed subsidiary, and some do not even have a classical plant breeding capability.

Table 8.2. *Selected characteristics of principal plant-oriented NBFs*

Biotechnology start-ups	Date founded	Equity investments from	Research contracts with	Subject of contract research	Proprietary research
Advanced Genetic Sciences[a]	1979	Rohm & Haas (15%) Hilleshog (15%)	Rohm & Haas Hilleshog Plant Genetic Systems Du Pont	Microbial crop symbionts Lignin processing Plant improvement	Hybrid rapeseed Herbicide-resistant potatoes Microbial frost inhibitors Soybean tissue culture
Agracetus	1981	W. R. Grace (51%) Cetus (49%)			N fixation in plants Microbial crop symbionts
Agrigenetics	1980	Lubrizol (100%)	Hoffmann-LaRoche	Plant antiviral agent	Clonal propagation breeding lines Ti vector development Transposable elements in corn N fixation in plants Gene sequencing Hybridization techniques Tissue-culture corn inbreds
Biotechnica International[a]	1981		Monsanto H. J. Heinz	N fixation in plants Food processing	N fixation, tomatoes, corn, alfalfa Herbicide resistance in soybeans High protein forage Microbial symbionts

Table 8.2. (*cont.*)

Biotechnology start-ups	Date founded	Equity investments from	Research contracts with	Subject of contract research	Proprietary research
Calgene[a]	1980	Plant Resources Venture Fund FMC Corp. Continental Grain	Kemira-Oy Campbell's Soup Nestle Rhone-Poulenc DeKalb-Pfizer Cokers Seed Co. Phytogen Procter & Gamble Roussel-Uclaf Ciba-Geigy Philip Morris	Herbicide-resistant rape High-solid tomato Herbicide-resistant soybean Herbicide-resistant sunflower Herbicide-resistant corn Herbicide-resistant tobacco Herbicide-resistant cotton Specialty plant oils Stress tolerance Disease resistance Tobacco improvement	Herbicide-resistant cotton Herbicide-resistant corn Tissue-culture corn inbreds High-protein wheat
DNA Plant Technology, Inc.[a]	1981	Campbell's Soup (24%) Kopvenco (8.4%) John Brown Ltd. (6%)	Campbell's Soup Koppers General Foods Brown & Williamson Tobacco Co. Hershey Foods	High-solid tomatoes Plant disease diagnostic kit Processing technology Tobacco improvement Clonal propagation of	Embryo encapsulation Herbicide-resistant plants Clonal propagation Cytoplasmic male sterility

			Pepperidge Farm Archer-Daniels Arthur D. Little Monsanto United Fruit	cocoa Vegetable snacks Hydroponic herb production Synthetic seed system Sweeteners Clonal propagation of oil palm	Tissue culture of secondary metabolites
International Plant Research Institute	1978	Bio-Rad (70%)	Davy-McKee Sime-Darby Berhad General Foods Eli Lilly	Fermentation systems Clonal propagation of oil palm Processing technology	Salt-tolerant wheat Disease-free cassava
Molecular Genetics[a]	1979	Martin-Marietta (21%) American Cyanamid (18.5%)	American Cyanamid Rhone-Poulenc	Herbicide-resistant corn Corn diseases	High-protein corn Herbicide-resistant plants Tissue-culture corn inbreds
Native Plants, Inc.[a]	1973	Martin-Marietta (10–20%)			rDNA-altered corn rDNA-altered tomatoes Potato tissue culture Clonal propagation Tissue-culture secondary metabolites
Plant Cell Research Institute	1981	ARCO (100%)	H. J. Heinz	High-solid tomatoes	High-protein wheat Male sterile triticale Onion tissue culture Vector development Herbicide-resistant tomatoes

Table 8.2. (*cont.*)

Biotechnology start-ups	Date founded	Equity investments from	Research contracts with	Subject of contract research	Proprietary research
Plant Genetics	1981	Plant Resources Venture Fund	FMC Corp.	Seed/chemical encapsulation	Somatic embryogenesis
					Synthetic seed system
		INCO Securities	Ciba-Geigy	Seed/chemical encapsulation	Hybrid celery
		Whitehead Assoc.	Chevron	Seed/chemical encapsulation	
Phyto-Dynamics	1980		Lubrizol	Oil crop improvement	Tissue-culture corn inbreds
Phytogen	1981	J. G. Boswell (majority interest)			Cotton tissue culture
					Herbicide-resistant cotton
					Regeneration systems
Sungene	1981	Lubrizol			High-oil sunflower
		Mitsubishi			Tissue-culture corn inbreds
		Hambrecht & Quist			Gene sequencing
		Morgenthaler Assoc.			

[a]Publicly held corporations.

Sources: Author's compilation from numerous sources.

Sungene's ambitions include capturing 5-10 percent of the U.S. hybrid seed-corn market within 10 years (Gebhart 1984b:9). In order to realize this goal, it is considering purchase of a seed company. But even small seed companies can be expensive for small research firms without steady income, which is what the NBFs are. Sungene's president, Thomas Hiatt, admits that a more likely scenario may be production and marketing of Sungene varieties through Lubrizol, a company that has substantial equity in Sungene as well as ownership of a bevy of seed companies (Gebhart 1984b:7). Start-up firms Biotechnica International and Advanced Genetic Sciences also anticipate having to market their germplasm through other companies (*Genetic Engineering News* 1982:16; Harvard Business School 1982:19). Though strong on the new technologies and in the laboratory, many NBFs are only weakly articulated to plant breeding and the marketplace for seed.

An exception to this rule was Agrigenetics Corporation. Founded in 1975, Agrigenetics' corporate strategy was premised on the necessity of linking biotechnology to conventional breeding in laboratory and market. Enjoying $55 million in venture capital funding, Agrigenetics had the financial resources to purchase 12 seed companies between 1975 and 1983. Then-president David Padwa noted, "A major part of our game plan is to position ourselves correctly in the marketplace, so that we can effectively sell what we make. That's why we do no contract research" (Padwa 1982:97).[10] Annual income of $80 million from established seed operations (Agrigenetics 1984:4) provided financial space for the development of proprietary research in biotechnology. Attempting to duplicate Agrigenetics' strategy is the Plant Resources Venture Fund (PRVF), a group of venture capitalists who intend to build a competitive position through the acquisition of an integrated team of small seed companies and the provision of advanced genetic research services (Sirkin 1984b). PRVF has recently acquired three seed companies and an equity interest in the biotechnology firm Plant Genetics.

A problem with such a strategy now, however, is the reduced number of seed companies available for purchase. To be sure, there are plenty left. But of the 700 or so seed firms in the United States, less than a quarter combine research, production, and marketing, and most of these are already in one corporate stable or another (ASTA 1983) (see Table 6.3). Moreover, biotechnology firms wanting to gain access to marketing channels and classical breeding expertise via seed company acquisition must compete for the best firms with the multinationals who are not only interested in further seed company purchases but also are willing to reach deep into their pockets to position themselves in what they see as a crucial market. The price at which the agrichemical and pharmaceutical MNC Sandoz agreed to buy Northrup King, one of the premier American seed companies, was $110 million over the firm's book value (Doyle 1985:95). The MNCs continue to actively pursue the purchase of seed companies. In the last two years, Imperial Chemical Industries, ARCO, Shell, Upjohn, Rohm and Haas, and Ciba-Geigy have all made purchases. Such acquisitions are regarded as a sign of corporate strength, and Shell's seed subsidiary NAPB boasts in its *Seedsmen's Digest* advertising that "we've acquired two seed companies in the last six months alone."

Even if a NBF does succeed in establishing a viable plant breeding and seed production and marketing capability, this does not ensure that it will survive as an independent entity. Late in 1983, the MNC Bio-Rad acquired 70 percent of International Plant Research Institute (*Genetic Engineering News* 1983:14). In August of 1984, Cetus sold 51 percent of its plant biotechnology subsidiary (formerly Cetus Madison, now Agracetus) to W. R. Grace for over $60 million (*Journal of Commerce* 1984:22B). This was followed in September by J. G. Boswell's (a California agribusiness corporation) acquisition of a majority interest in Phytogen (*Biofutur* 1984:69). And on January 1, 1985, Lubrizol assumed complete ownership of Agrigenetics. Large-scale capital has demonstrated its willingness to use its financial strength to acquire NBFs as well as seed firms. Absorption into multinational conglomerates rather than independent development may well be the fate of other start-ups as they ripen, and as the new technologies mature commercially.[11]

A third group of firms, independent seed companies, also needs to be considered in addressing the impacts of biotechnology on industrial structure. After the wave of acquisitions in the last decade, the independents are predominantly small firms with localized markets that, having no research programs, depend on public breeders for development of the varieties they grow and market. Ironically, just as many biotechnology start-ups have no access to seed companies, these small seed firms have no direct access to the new genetic technologies. Even those independents with modest breeding programs of their own will find it very difficult to move into biotechnology. Support costs for scientists engaged in the application of genetic engineering to crop improvement are a multiple of those for conventional breeding (Duvick 1982a:34). As Agrigenetics' pithy Padwa observes, "The game of molecular biology is not a game for three people over a garage" (quoted in *Business Week* 1984b:69). As biotechnology begins to have a material impact on crop improvement, the independent seed companies will increasingly find themselves at a competitive disadvantage (*Business Week* 1984b:69; Duvick 1984:23). Whether or not such firms will survive in the coming age of synthetic biology will depend largely on whether or not public agencies are able to integrate biotechnology with traditional plant breeding and to maintain breeding programs that culminate in varietal release. There is reason to doubt that both conditions will be met.

Not all independents are small, however. The acquisition trend of the last decade has resulted in a disappearing middle among independent seed companies, and the apex of the industry's structure is very narrow indeed. It consists of only two firms, DeKalb AgResearch and Pioneer Hi-Bred. Significantly, both have embraced biotechnology, though in very different fashions. DeKalb has entered into a joint venture with Pfizer in order to take advantage of that company's research facilities – "the greatest new two-way cross in the seed industry" as DeKalb-Pfizer Genetics promotional literature has it (*Seed Leader* 1982:1). As such, DeKalb is no longer really an independent. In contrast, Pioneer has chosen to go it alone by building a $2.7 million Department of Biotechnology Research (Pioneer Hi-Bred, n.d.). Pioneer's Research Director explains his company's perspective: "Alert biotechnologists working in tandem with alert plant breeders will continue to find new practical aids to plant breeding. Individual uses usually will be unspectacular and small, but cumulatively they will be large and important" (Duvick 1984:17).[12]

Pioneer's decision to pursue the new technologies corroborates some points I have been trying to make throughout this chapter. Practical applications of genetic engineering and the more exotic techniques of in vitro horticulture to plant improvement will come slowly. But they will come, and they will be important. Biotechnology will move commercial plant improvement to a new plane of competition.[13] At the same time, biotechnology will not replace conventional plant breeding, but will transform it. The seed will still be the form in which the commodity enters the market. Commercial success will depend upon *integrating* plant breeding and biotechnology.

That said, we may ask who is in the best position to achieve this integration. As we have seen, the plant-oriented NBFs are weakly articulated to classical breeding programs and seed markets, though they are scientifically strong. Conversely, independent seed firms, with few exceptions, lack access to the new technologies. A number of multinationals, however, appear to be uniquely positioned to dominate plant improvement. The nature of their strength is illustrated in Table 8.3. Not only do such corporations as Monsanto and Ciba-Geigy have extensive in-house biotechnology research programs, they also own seed companies through which to bring improved germplasm to the market. Moreover, through contracts or equity interests, they have windows into the activities of the biotechnology start-ups.[14]

The MNCs listed in Table 8.3 already hold preeminent positions with regard to the American (and world) seed market. Together with Cargill and Pioneer, Upjohn, Ciba-Geigy, Sandoz, and Pfizer enjoy 68 percent of the hybrid corn market (Table 8.4). Because seed-corn accounts for about half of annual U.S. seed sales, these companies alone control over a third of the American seed trade. Actually this figure is higher, for these companies are also active in other species. The entries in Table 8.3 under "U.S. seed

Table 8.3. *Some characteristics of selected multinational corporations engaged in the seed industry and biotechnology*

Parent firm	U.S. seed subsidiaries	Biotech subsidiaries/ in-house capability	Equity in/ contract with
ARCO	Dessert Seed Co. Castle Seed Co.	Plant Cell Research Inst.	International Plant Research Institute Bioengineering Center Ingene
Ciba-Geigy	Funk Seeds Intl. 7 others	Agricultural Biotechnology Research Center ($7M)	Plant Genetics
FMC Corp.	Seed Research Assoc.	Extensive in-house	Centocor Immunorex Plant Genetics
Lubrizol	Jacques Seeds Sun Seeds 12 others	Agrigenetics	Sungene Phyto-Dynamics Genentech
Monsanto	DeKalb Hybrid Wheat 3 others	Life Sciences Res. Center ($150M)	Genex Biogen Genentech Collagen Biotechnica International
Pfizer	DeKalb AgResearch (joint venture) 5 others	Central Research Lab ($22M)	
Sandoz	Northrup King 7 others	Zoecon	Genetics Institute Ingene
Shell	North American Plant Breeders 5 others	Extensive in-house	Cetus Ingene
Upjohn	Asgrow Seed Co. O's Gold Associated Seed	Extensive in-house ($26M)	Synergen Biotechnica International
W. R. Grace	Pfister Hybrids	Agracetus	

Sources: Author's compilation from numerous sources.

Table 8.4. *Market shares of leading seed-corn companies, 1981*

Company	Market share (%)
Pioneer Hi-Bred	33
DeKalb-Pfizer Genetics	17
Funk (Ciba-Geigy)	7
Northrup King (Sandoz)	4
O's Gold (Upjohn)	4
Cargill	3
Total	68

Sources: Davenport (1981), Leibenluft (1981).

subsidiaries" are a compendium of the country's largest and best-known seedsmen. Not only are the multinationals continuing the purchase of new seed subsidiaries, they are rationalizing their holdings by buying companies and crop-specific research operations from each other. Thus, DeKalb (Pfizer) sold its wheat operation to Monsanto, Northrup-King (Sandoz) sold most of its wheat program to Rohm and Haas,[15] Paymaster (Cargill) bought Pioneer's cotton operation, Funk Seeds (Ciba-Geigy) acquired Ring-Around's (Occidental Petroleum) corn, sorghum, and cotton research programs, and NAPB (Shell) purchased Ferry-Morse's (Limagrain) farm seed operations.

L. William Teweles, an industry consultant who has brokered some 60 percent of the MNCs' seed company acquisitions, suggests that "only the strongest and most nimble independent seed companies or those that are subsidiaries of multinationals with their own plant science departments will be factors in the future" (quoted in McDonnell 1986:43). Speaking of his company's plans to sell $500 million worth of seed annually by the 1990s, James Windish, president of Monsanto's Hybritech Seed Co., says, "We doubt that there are 10 to 15 other firms with the financial muscle to carry their programs to that extent" (quoted in Sanford 1984:G6). The NBFs recognize the difficulties they face. But, as one genetic engineering firm executive observes, "If we are simply playing a dollar game – more dollars give more results, we'd all better give up now" (quoted in Harvard Business School 1982:9). There may well be an IBM or two among the flock of NBFs, but at the very least, competitive capital faces an uphill battle and substantial structural obstacles in its quest to maintain its position in the plant genetic supply sector. Given the appearance of a variety of plant-oriented NBFs, the casual observer might well predict a trend away from concentration and the emergence of an increasingly competitive structure in the seed industry. This analysis suggests that something quite different is actually occurring.

The campus and the corporation: biotechnology and changes in the technical division of labor

According to Dr. Arnold J. Levine, Chairman of Princeton's Department of Molecular Biology, "The kinds of questions now being asked in the life sciences are broadbased and lead to a merging of disciplines" (quoted in Sirkin 1984a:14). This is as true for the plant sciences as for any other area. Realization of the promise the genetic technologies hold for plant improvement must be predicated on the establishment of multi-disciplinary research teams capable of integrating genetic engineering, *in vitro* horticulture, and whole-plant breeding into a unified process. If there is now substantial agreement that this is true (Bingham 1983; Duvick 1983; Evans et al. 1983; Fraley 1983; Phillips 1983; Qualset et al. 1983; NRC 1984; Day 1986), it is nevertheless a difficult thing to achieve in either private or public labs.

In the halcyon days when molecular biologists first glimpsed the potential power of their discipline, they assumed a rather patronizing approach to the plant breeders whom they expected to replace. The optimistic claims that sold stock and garnered research grants for the genetic engineers also alienated many plant breeders (Bingham 1983:222; Duvick 1984:12; Day 1986). Healthy skepticism of the new techniques has sometimes been turned to outright hostility.[16] As it has become clearer that plant molecular biology and plant breeding need each other, ruffled feathers are slowly smoothing on both sides. But, as Bingham (1983:222) has observed, the relation of these disciplines still "has many of the characteristics of a polarity."

Integration is also rendered problematic by the very different characters of the disciplines themselves. Simmonds (1979:337) has called plant breeding "an applied science that is devoted to changing nature rather than understanding her."[17] In contrast, molecular biology has historically been a theoretical science that is devoted to understanding nature rather than changing it. Now, enhanced understanding implies enhanced capacity to engineer change. But the merger of the polar approaches of the molecular engineer and the whole-plant manipulator is a joining that is not easy. Duvick (1983:221) comments,

> Established plant breeders, not trained in molecular biology, generally do not understand molecular biology well enough to see its possible present-day uses or its potential future applications. Established molecular biologists generally do not understand plant breeding well enough to appreciate what plant breeders really need and realistically can use from molecular biology.

What is needed are hybrid scientists.[18]

The division of science into disciplines is, of course, a social artifact rather than a reflection of some distinctions intrinsic to nature. But this does not imply that disciplines necessarily cross-fertilize any more easily than species do. Molecular biology and biochemistry exhibit reasonably good combining ability, but molecular biology and plant breeding is a very wide cross indeed. Private industry may have certain advantages over the university when it comes to facilitating such hybridization. In a sort of social protoplast fusion, the rigid institutional walls surrounding the different disciplines can be dissolved in the corporate laboratory, and scientists fused into research teams by managerial fiat. The structure and composition of a number of agricultural biotechnology research units in private companies are detailed in Table 8.5.

Judging from Table 8.5, it would seem that genetic engineering and *in vitro* horticulture are reasonably well linked in corporate research operations. But articulations to plant breeding appear problematic. Ciba-Geigy and ARCO do not include classical breeding programs within their biotechnology research units, though they do have extensive capabilities in that area through their seed subsidiaries.[19] The NBF Advanced Genetic Sciences does not have that luxury, and its structure illustrates the weakness of many start-ups in whole-plant manipulation. In contrast, the NBF DNA Plant Technology has included a plant breeding section in its team. The biotechnology work of both Agrigenetics and Pioneer is actually embedded within an overarching structure of plant breeding and seed production. Even so, meshing biotechnology and conventional plant improvement is not easy. Pioneer's Director of Corn Breeding says there have been difficulties integrating the new section into ongoing breeding work. The biotechnologists think in different terms and have different expectations: "They don't know how to breed corn and we don't know what they do" (Seifert 1984).

It is too early to say how well these programs will achieve their objectives, but it is clear that industry can initiate inter-disciplinary research by brute force, as it were. For universities, disciplinary walls are not so easily dissolved. Vice presidents for research cannot simply order people about, or fire dead wood. The problem of inter-disciplinary research in biotechnology is especially acute with regard to plant improvement. Molecular biology developed principally in private universities on the East and West coasts (e.g., Harvard, Yale, M.I.T., Rockefeller University, University of California-San Francisco, Stanford), where it is unusual to find someone who knows what a plant breeder looks like, or what the acronym "LGU" stands for. Even in those land-grant universities with strong molecular biology or biochemistry programs, these departments are located in the colleges of arts and sciences, which are physically and administratively (and often intellectually) distinct from the colleges of agriculture in which plant breeders are located. Even within the agricultural college community, integrated research is the excep-

Table 8.5. *Structural organization of agricultural biotechnology research in selected companies*

Company	Research sections
ARCO: Plant Cell Research Institute, Dublin, CA	Biochemistry Molecular Biology Natural Products Chemistry Genetics and Tissue Culture
Ciba-Geigy, Agricultural Biotechnology Research Center, Raleigh, NC	Biochemistry Plant Molecular Biology Microbiology Plant Tissue Culture
Advanced Genetic Sciences, Berkeley Research Laboratory, Berkeley, CA	Biochemistry Molecular Biology Genetics Bacterial Genetics Rhizobacteria
DNA Plant Technology, Inc., Cinnaminson, NJ	Developmental Genetics Tropical Crop Genetics Plant Breeding
Agrigenetics, Advanced Research Laboratory, Madison, WI	Biochemistry Genetics Microbiology Cell Biology
Pioneer Hi-Bred, Department of Biotechnology Research, Johnston, IA	Molecular Biology Cytogenetics Tissue Culture Molecular Plant Pathology Microbiology

Sources: Agrigenetics (1981), Harvard Business School (1982), Pramik (1982), DNA Plant Technology (1983), Rossman (1984), Pioneer Hi-Bred (n.d.).

tion rather than the rule, and deans and experiment station directors have very limited capacity to set research agendas and to focus efforts on particular problems (McCalla 1978; Ruttan 1982b).

What industry can do by mandate, the university must achieve by indirect means. And the promise of biotechnology is of sufficient magnitude to motivate many universities to put something solid behind what heretofore has been largely lip-service to inter-disciplinary work. In the past few years, wide varieties of institutional frameworks designed to facilitate inter-disciplinary work related to biotechnology have been established on the nation's campuses. The LGUs have been represented in this trend as well, and agricultural and crop applications of biotechnology are important components of these programs (Table 8.6). Lacking industry's capacity to use a stick, universities use the carrot of funding to encourage the cooperative work required by the new technologies. It is too early to assess the effectiveness of these programs; you can lead a molecular biologist to a plant breeder, but you cannot make them talk.

The campus and the corporation: biotechnology and changes in the social division of labor

For both industry and the university there are at least two fundamental dimensions to the question of the interaction between the genetic/cellular engineer and the plant breeder. First, there is the technical division of labor, the problem of integrating the techniques characteristic of different disciplines into a unified process of plant improvement. But science, like any human activity, takes place in a particular social context. There exists also a social division of labor that encompasses and is yet larger than the technical division of labor.

Writing of the relationship between molecular biology and plant breeding, Bingham (1983:222) observed that there may be tensions between two disciplines when they share the same technical goals. We may extend his point to say that there may be tensions when two institutions share the same goals. This is especially true if these goals are commercial in nature. When both institutions are private firms, the tension is expressed as competition in the marketplace. The situation becomes problematic when one institution is a private firm and the other is a university. We are familiar, from our historical discussions of plant breeding, with the implications of the public-private division of labor in plant improvement. We need now to examine changes in this relationship in the context of the new technological climate engendered by the emergence of biotechnology.

One of the most interesting features associated with the new biotechnologies is the way they have blurred conventional distinctions between levels of research. As DNA Science's (E. F. Hutton) Zsolt Harsanyi (1981:118) put it,

> Much of the research in biotechnology cannot simply be categorized as being basic as opposed to applied research. As one of our scientists said recently, a study that he is working on in which he is looking at the genetic code is basic in the sense that it gets to the very core of what the genetic code is about, but if he can solve this problem it may be of

Table 8.6. *Examples of biotechnology centers at land-grant universities*

UC-Berkeley, USDA, California Agricultural Experiment Station	Plant Gene Expression Center	Goal is advancement of crop genetic engineering; $4 million funding from USDA; industrial matching grants being sought
UC-Berkeley, Stanford University	Center for Biotechnology Research	Research group of faculty from both universities; funding $2.4 from Engenics Corporation; center has 30% equity in Engenics (remaining ownership distributed among UCB/Stanford professors, General Foods, Bendix, Koppers, Mead, Noranda, and Elf Aquitaine)
Cornell University	Cornell Biotechnology Institute	A dynamic interface with industry; funding of $2.5 million per annum per company for 6 years from General Foods, Union Carbide, Kodak
	New York State Center for Advanced Technology for Biotechnology in Agriculture	Linked to Cornell Biotechnology Institute; intended to be a mechanism for transfer of new technologies to collaborating New York corporations; $20 million in state funding for construction of facilities
University of Georgia	Program in Biological Resources and Biotechnology	Coordinated approach will be a draw to biotechnology firms looking for university support in research efforts; $250,000 funding from Georgia Power Co.
University of Illinois	Genetic Engineering Center	Industrial Affiliates Program, membership fee $10–20,000 per annum
	Center of Excellence for Crop Molecular Genetics	$2 million funding from Sohio
University of Maryland, National Bureau of Standards	Center for Advanced Research in Biotechnology	Planned to serve as a resource to industry by providing sophisticated equipment and basic research projects that industry could not do on its own; $3.5 million funding
Michigan State University	Biotechnology Research Center	The center is slated to coordinate biotechnology research in 6 colleges and provide a central office to work with biotechnology companies; funding $6 million from the state and Kellogg Foundation

	Neogen Research Corporation	A private company founded by MSU (MSU Foundation has a 30% equity) to support and commercialize biotechnology research at the university; $25,000 investment units available to the public
New Mexico State University	Center for Semi-arid Plant Biotechnology	Research into genetic engineering of desert plants; $7 million in funding from state legislature
Pennsylvania State University	Cooperative Program in Recombinant DNA Technology	$15,000 per annum membership fee for companies; Amax, Gibco, Wyeth Laboratories, Gulf Oil, Schering-Plough, Westinghouse, IBM now members
	Biotechnology Institute	$8.8 million in funding, including $100,000 from Rohm and Haas.
Rutgers University	Center for Advanced Food Technology	Designed to foster cooperative research and development between academia and industrial communities; $584,200 in state funding
University of Wisconsin	Wisconsin Biotechnology Center	Promote formation of research units focused on specific areas of biotechnology; facilitate interactions between university and industrial scientists; funding $500,000 from Wisconsin Alumni Research Foundation, expecting $1.3 million from the state

Sources: Author's compilation from numerous sources.

> tremendous industrial use . . . The point is, the basic scientist on campus doing basic research may end up doing some extraordinarily important industrial research.

By industrial research, Harsanyi means research that is linked closely to the commodity-form. And here is the pivot of a contradiction: Scientists in the university, which in capitalist society is ideologically and structurally *not* constituted as a profit-making institution, are now producing knowledge that is directly constitutive of exchange-value.

As M.I.T. president Paul Gray (1981:54) has observed, "We are now in a situation in which the rules of the game have changed in a way because the value added is larger at the basic research end of the research development application than it has been traditionally." Suddenly, private industry needs to do the same work that is being done in universities. For example, there is no multinational, NBF, or university interested in applying the new techniques to plant improvement that does not have a research program on the Ti plasmid (Qualset et al. 1983:476). And every crop tissue culture lab in the country, from Agrigenetics to Pioneer to Cornell to Stanford, is working on corn regeneration from protoplasts.

As was the case with hybrid corn, two decades of publicly funded research in genetics have led to scientific advances that have enlarged the space for privately profitable research. And, as with hybrid corn, capital is confronting the competing activities of university scientists as an obstacle to accumulation. University researchers are producing knowledge that is valuable but is difficult to privately appropriate. The social division of labor between the public and private sectors must once again be redefined as the lines between basic and applied research dissolve.

In the best of circumstances, such a redefinition is not necessarily an easy thing for capital to accomplish. And with biotechnology there is an additional complicating factor. While industry might wish to shift the division of labor, it has found itself lacking the scientific labor power to realize the benefits of such a shift. It was academic scientists who led the way to the molecular level, and if business has followed them there, it still needs their services in negotiating the new terrain. Most of the expertise in the new technologies is located in university rather than corporate laboratories.

The most direct solution to such a problem is for corporations simply to purchase the scientific labor power they need through the market. But the highly specialized knowledge that the companies wish to obtain is concentrated among a relatively narrow set of university-based researchers, especially in the field of plant molecular biology. With a few prominent exceptions, professors have been reluctant to leave academia for the research programs of the corporate giants. They have proved much less resistant to the blandishments of venture capitalists who have promised ownership or equity interest, managerial control, and a university-like atmosphere for professorial cooperation in establishing NBFs.

The loss of faculty to private industry has been a problem for universities, and this has been true for the plant science departments of the land-grants as well (U.S. House of Representatives 1981, 1982b; NASULGC 1983; Lower 1984). The University of Wisconsin, Kansas State University, the University of Illinois, Michigan State University, the University of Minnesota, and Ohio State University have all had leading plant scientists leave for the private sector. The beneficiaries of this trend have principally been NBFs such as Agrigenetics, Allelix, Calgene, Crop Genetics International, Molecular Genetics, and DNA Plant Technology. Given the relative scarcity of scientists with agricultural backgrounds who have developed expertise in the new biotechnologies, such losses are painful to the universities involved.

Though this brain drain may reduce the university's capacity, it does not pose a direct threat to its institutional integrity. Of more concern than the simple loss of personnel is the manner in which the university is being transformed by new links that private enterprise has forged with researchers who remain in their university posts. Unable to buy labor power out of the university to the extent they would like, companies have bought into the university in unprecedented fashion. While industry provides 3 to 4 percent of total research funds spent in institutions of higher education, it now supports an estimated 16 to 24 percent of biotechnology research conducted in the nation's universities (Blumenthal et al. 1986, "Industrial Support"). Table 8.7 illustrates the types of relationships that have been developed in the last few years between companies and professors engaged in biotechnological approaches to crop improvement.

One type of arrangement is that in which a faculty member assumes substantial managerial or ownership positions in outside firms. The celebrated case of Dr. Raymond Valentine – University of California-Davis professor of biochemistry and founder of the NBF Calgene – woke the academic community to the conflicts of interest that could emerge from simultaneously holding positions in business and academia. In his capacity as a member of the California Agricultural Experiment Station, Valentine received a $2.3 million grant from Allied Chemical for research on nitrogen fixation. Allied also purchased a 20 percent interest in Valentine's Calgene. Questions arose as to the distinction between Valentine's research for the station and his work for Calgene. There were also allegations of unethical management of graduate student research that was relevant to research projects under-way at Calgene.[20] Faced with an ultimatum from the university administration, Valentine subsequently relinquished his position with Calgene, though he remains a consultant to the company.

There have been other instances where potential conflicts of interest be-

Table 8.7. *Examples of formal relationships between plant-oriented NBFs and university-affiliated scientists*

Company	Scientist	Position with NBF	Position with university
Calgene	Raymond Valentine	Science Advisory Council[a]	Prof. of Agronomy, Univ. of California–Davis
	George E. Bruening	Science Advisory Council	Prof. of Plant Pathology, Univ. of California–Davis
	Tsune Kosuge	Science Advisory Council	Prof. of Plant Pathology, Univ. of California–Davis
	Donald Helsinki	Science Advisory Council	Prof. of Biology, Univ. of California–San Diego
	Peter Geiduschek	Science Advisory Council	Prof. of Biochemistry, Univ. of California–San Diego
DNA Plant Technology, Inc.	Philip Ammirato	Manager, Developmental Genetics	Chair, Dept. of Biology, Barnard College
	David Evans	Vice Pres., Asst. Dir. Research	Adjunct Prof. of Biology, Rutgers University
	Norman Borlaug[b]	Scientific Advisory Board	Ex-Director CIMMYT, Prof., Texas A&M
	Melvin Calvin[c]	Scientific Advisory Board	Univ. of California–Berkeley
	Jules Janick	Scientific Advisory Board	Prof. of Horticulture, Purdue University
	Merle Jensen	Scientific Advisory Board	Prof. of Plant Science, Arizona State University
	Elton Paddock	Scientific Advisory Board	Prof. of Genetics, Ohio State University
	Oved Shifriss	Scientific Advisory Board	Prof. of Horticulture, Rutgers University
Advanced Genetic Sciences	Lawrence Bogorad	Director, Scientific Board	Prof. of Biology, Harvard University
	Howard Goodman	Scientific Board	Prof. of Genetics, Harvard University
	Milton Schroth	Scientific Board	Prof. of Plant Pathology, Univ. of California–Berkeley
Molecular Genetics	Charles Green	Research Director	Prof. of Agronomy, University of Minnesota
	Anthony Faras	Chairman of the Board	Prof. of Microbiology, University of Minnesota
Agracetus	Winston Brill	Director of Research	Adjunct Prof. of Bacteriology, University of Wisconsin
Biotechnica Intl.	Frederick Ausubel	Senior Research Consultant	Prof. of Genetics, Harvard University
Soil Technologies	Stanley Katz	Board of Directors	Prof. of Biology, Iowa State University

[a] Valentine is also a co-founder of Calgene.
[b] Recipient of Nobel Prize for peace, 1970.
[c] Recipient of Nobel Prize in chemistry, 1961.
Source: Author's compilation from numerous sources.

tween academic duty and corporate responsibility have been so clear that university administrators have been unable to shrug them off as nothing more than a new iteration of the established tradition of consulting. Walter Gilbert at Harvard and Timothy Hall at the University of Wisconsin were asked to choose between the university and the private sector; both opted to stay with the NBFs they had helped found (Biogen and Agrigenetics, respectively). Addressing the issues raised by such cases, Yale's president A. Bartlett Giamatti (1982:1279) commented in a *Science* article,

> The burden of mounting a teaching program and two separate research programs, where the results of one research program are to be widely disseminated and the results of the other may have to be kept secret in the pursuit of commercial success, is more than even the most responsible faculty member can be expected to shoulder.

Without a national policy, however, the determination of what is acceptable is left to each institution. Some faculty members are permitted to retain their university posts while discharging their duties as line managers or principals in NBFs (Table 8.7). More circumspect academics may pursue full-time outside work but retain a formal connection to the university through such arrangements as adjunct professorships.

Relatively few faculty members have the opportunity to directly participate in the management of an NBF or other private firm. A more ubiquitous form of faculty-industry connection in biotechnology is faculty service on what are termed (with some variation, see Table 8.7) scientific advisory boards (SABs). Most of the new biotechnology firms have created SABs and have stocked them with prominent scientists whose affiliation with the company raises its credibility with investors and whose expertise provides the company with important channels of access to strategic information. Though Advanced Genetic Sciences' SAB is actually expected to perform managerial tasks (Harvard Business School 1982:15), most SABs function principally as intelligence units. In the knowledge-intensive field of biotechnology, interaction with the university community is vital if a company is to maintain contact with state-of-the-art science. SABs provide a firm with a window on the latest developments. Board members' contacts are also useful in recruiting staff, identifying consultants, facilitating the procurement of research contracts, and smoothing the way to the establishment of cooperative research with university faculty.

Membership on a SAB requires a substantial commitment of time and effort. Board meetings are generally held quarterly, and, in addition, conferences with individual members are arranged as needed. These services are well recompensed. Members of DNA Plant Technology's SAB are guaranteed a minimum of $5,000 per year in consulting fees and also enjoy options to purchase the company's stock (DNA Plant Technology 1983:32). Such options are common (Kenney 1986), and to the extent that they are exercised, the academic becomes something more than a mere advisor. In some cases, the company becomes a funder of an SAB member's research, thereby further reinforcing the professor's commitment to the success of the firm.[21] The Harvard Project on University-Industry Relationships in Biotechnology[22] found that, in 1985, 21 percent of non-Fortune-500 biotechnology companies reported funding university faculty members who held significant equity in the companies (Blumenthal et al. 1986:243, "Industrial Support").

The most common expression of the intensified interest industry is now showing in university life sciences faculty is neither the professor-entrepreneur nor the SAB, but the extension and elaboration of the time-honored practices of consulting and extramural funding. The Harvard Project found that 23 percent of faculty engaged in biotechnology research were recipients of industry funding. This private support constituted 34 percent of these scientists' total research budgets. Moreover, 15 percent of the faculty receiving private funds for biotechnology research obtained at least 75 percent of their support from industry. Blumenthal et al. (1986, "University-Industry Research") comment further, "Controlling for other factors, faculty in our sample who were receiving industry support tended to publish more, patent more, earn more, serve in more administrative roles, and teach just as much as faculty without industry funds." Could it be that it is the best and the brightest who are most connected to industry?

In regard to plant biotechnology, this does in fact appear to be the case. Winston Brill, Research Director at Agracetus (W. R. Grace) and Adjunct Professor of Bacteriology at the University of Wisconsin, has testified before Congress that "the number of plant molecular biology and biochemistry experts in the U.S. is limiting. Most of the well-known professors in that area are now consulting for one or another corporation" (Brill 1982:14). Table 8.8 provides some insight into the concrete realities behind the Harvard Project's global statistics and Brill's generalization. In Table 8.8 are listed faculty members who acted as consultants for the NBF Agrigenetics in 1983. That year, Agrigenetics was funding research at eighteen universities, including twelve LGUs, with each project receiving between $500,000 and $2 million.

For plant science departments, sums of this size are orders of magnitude greater than any historical precedent. The $567,233 that two Cornell plant breeders obtained from Agrigenetics was ten times the size of any other private grant received by the experiment station in that year. The $2.3 million that Allied Chemical conferred upon the California Agricultural Experiment Station in 1981 was equal to all other grant and contract funds received by

Table 8.8. *Agrigenetics Corporation consulting scientists*[a]

Consultant	Institution
Wolfgang Dietzgen Bauer, Ph.D.	Kettering Research Laboratory
Andrew Binns, Ph.D.	University of Pennsylvania
Nicholas Brewin, Ph.D.	John Innes Institute
Adrienne Clarke, Ph.D.	University of Melbourne
Peter Dart, Ph.D.	Australian National University
Leon Dure, III, Ph.D.	University of Georgia
Elizabeth D. Earle, Ph.D.	Cornell University
Harold Evans, Ph.D.	Oregon State University
David W. Galbraith, Ph.D.	University of Nebraska
Vernon E. Gracen, Ph.D.	Cornell University
Peter M. Gresshoff, Ph.D.	Australian National University
Thomas J. Guilfoyle, Ph.D.	University of Minnesota
Richard Hallick, Ph.D.	University of Colorado
Maureen Hanson, Ph.D.	University of Virginia
Hauke Hennecke, Ph.D.	Zurich Microbiological Institute
Thomas K. Hodges, Ph.D.	Purdue University
Paul Kaesberg, Ph.D.	University of Wisconsin
Brian Larkins, Ph.D.	Purdue University
Sharon Long, Ph.D.	Stanford University
Alfred Puhler, Ph.D.	University of Bielefeld (FRG)
Ralph Quatrano, Ph.D.	Oregon State University
Barry Rolfe, Ph.D.	Australian National University
John Shine, Ph.D.	Australian National University
Jack Widholm, Ph.D.	University of Illinois

[a]"Consulting Scientists. Research Program management and evaluation is assisted by cellular and molecular biologists who consult with the Company and the Partnership on an exclusive basis in defined areas of their expertise. These consultants provide independent advice with respect to promising developments and areas for research, and assist management in the evaluation of research program results, as well as setting goals and direction for future research. These consultants meet regularly with the Company together with scientists conducting sponsored research in their respective areas of expertise. The Company believes that the periodic interaction of these scientists gives valuable evaluation and direction to the Research Programs. Under the agreements between the Partnership and the consultants, any intellectual property arising from the consultancy is the sole property of the Partnership, which provides the compensation for all consultants."

Source: Agrigenetics (1984).

the station that year (Meyerhoff 1982:50). And Sohio's grant of $2 million to the University of Illinois Department of Agronomy for the creation of a Center of Excellence for Crop Molecular Genetics and Genetic Engineering is of a similar relative scale.

This qualitative increase in the magnitude of private funding of plant science research has not come without strings attached. University of Nebraska wheat breeder Virgil Johnson (1983:146) noted that "it seems clear that money from the private sector now is dictating to a degree we have not seen before the kind of research that will be done." Half of the membership of the committee that decides how Illinois' Sohio money is to be used is drawn from company personnel (Laughnan 1984). Most research contracts involve conferral of some form of proprietary rights – patent assignment or exclusive licensing – in the results of research (Advanced Genetic Sciences 1983; Agrigenetics 1984). Corporate jargon reflects the existence of such arrangements: When speaking of the research their companies are funding at a university, corporate executives will commonly say, "we own that research."

Proprietary considerations also imply the need to limit access to information. Many consulting arrangements and research contracts also provide for the restriction of information flow among colleagues. Calgene (1986:23), for example, requires the execution of confidentiality agreements by all consultants. A memorandum from Dr. Emanuel Epstein to the University of California-Davis administration regarding the implications of such constraints on information exchange is worth quoting at some length:

> Any UCD scientist with a promising new slant for the improvement of nitrogen fixation or the enhancement of salt tolerance for crops will think twice before talking about it to anyone who is connected with either of the Davis crop genetic private enterprises [Calgene, Plant Genetics], or even with colleagues who in turn might speak to any such person. I know that this type of inhibition is already at work on this campus... In addition, graduate students of faculty members connected with these businesses are in danger of being directed in ways more tailored to the requirements of those enterprises than the students' educational and professional advancement... In effect, what this intimate connection of some of our faculty with corporate business is doing amounts to no less than an insidious but nonetheless real abridgement of the academic freedom of all members of our college and their graduate students. [quoted in Meyerhoff 1982:50]

Not only may the free flow of information be restricted, but there may be a reluctance to exchange the germplasm (itself nothing more than another form of information) that is the essential raw material on which plant breeding is founded.

Representatives of the corporations engaged in funding university research deny that such contracts have the feared effects. Testifying before Congress, Agrigenetics vice president James Chaney (1982:94) noted modestly that his company had agreements with "a number of institutions" and continued:

> It is our view that these agreements must not promote secrecy, impair the educational experience of students, diminish the role of the university, interfere with the choice of scientific issues addressed by the investigators or direct the energy of faculty or the resources of the university from their primary educational and research missions.

Yet, an Agrigenetics Corporation contract with Cornell University researchers for investigations into cytoplasmic male sterility in corn contains the following provisions:

1. No funding from any source but Agrigenetics to be used on the project[23]
2. Exclusive right of Agrigenetics to file for patents on results of the funded research
3. Restrictions on the dissemination of information provided to the researchers by Agrigenetics
4. Six-week publication delays to permit review of papers/speeches by Agrigenetics
5. Six-month publication delays to permit filing of patents by Agrigenetics
6. University forfeiture of royalties to any product/process that Agrigenetics is not permitted to maintain as a *trade secret*

This last provision is especially draconian in implication. The other elements of the contract restrict information flow in various ways, but trade secrets *stop* the flow entirely. Two faculty members of Cornell University, a public institution, have essentially been captured by Agrigenetics.

And there is evidence that a significant portion of the information flow does actually get stopped. Harvard's study reports that 41 percent of companies supporting biotechnology research in universities have derived at least one trade secret from the work they underwrite (Blumenthal et al. 1986:244, "Industrial Support"). In congressional testimony, Howard Schneiderman, Senior Vice President for Research and Development at Monsanto, stated,

> I am confident that we have not only safeguarded the academic freedom of Washington University, but we are enhancing it. We are as concerned as you that the academic enterprise be preserved. We are convinced that our contract not only preserves the goose that lays the golden eggs but will significantly increase its egg production for the public good. [Schneiderman 1982:20-1]

Yet a Washington University plant biologist with a Monsanto research contract reports that two of his findings, which he considers patentable, are being maintained by the company as trade secrets. The pious statements of corporate executives regarding their respect for the integrity of the university notwithstanding, there is real reason for concern for the preservation of collegial exchange of information.

Not all business interests are pleased with the penetration of plant science departments by private firms. Few companies possess resources on the scale needed to compete with Agrigenetics, Monsanto, and Sohio for the purchase of the services of university-based scientists. Independent seed companies, which depend on public researchers to supply them with the products of new technology, are particularly concerned (Pardee et al. 1981; Ingersoll 1983). Even Pioneer may be out of its league in this area, and President Thomas Urban has complained,

> I am concerned that the flow of new corporate money into the field [of plant science] is having a negative effect on universities... Free enterprise is a wonderful thing, but "hot stocks" probably do not benefit the world of serious basic research. [quoted in *New York Times* 1981:D2][24]

Biotechnology will raise competition in commercial plant improvement to a new plane. To compete at that plane successfully, a company will have to have a top research program or have access to university research, or perhaps it will have to have both. Insofar as access to university research can be purchased, large-scale capital has a decided advantage over competitive capital.

University administrators have not accepted this transformation of university/industry relationships with complete equanimity. They are clearly aware of the dangers of an enlarged corporate presence on the campus (U.S. House of Representatives 1981, 1982a). But, during a period in which federal support for the Ivory Tower is shrinking (a 38 percent reduction in real dollar value since 1968), the candle of private funding is proving as irresistible to the moths who inhabit the paneled offices of university administration buildings as to the professors in their labs.

Moreover, universities were poorly prepared to respond effectively to the corporate demand for intellectual labor power. Many institutions, especially the colleges of agriculture, lacked any formal guidelines for managing faculty interactions with industry (NASULGC 1983; Lower 1984:49). And if they did have them, they were frequently unenforceable. The establishment of the institutional frameworks detailed in Table 8.6 represents an attempt not only to facilitate inter-disciplinary work in biotechnology but also to systematize and control the flow of funds from, and the character of extramural linkages to, private industry. The appearance of biotechnology centers and institutes is actually as much a response to the need to rationalize the social division of labor as it is an effort to redefine the technical division of labor.

The emergence of the new biotechnology has been associated with a period of institutional flux in which the responsibilities of industry, state, and university vis-a-vis scientific research – what Stanford University president Donald Kennedy (1982) calls the "social sponsorship of innovation" - are being redrawn. The current social fluidity has opened space for the operation of interests that intend to transform the agricultural research sector. Their objective is to make the "Island Empire" (Mayer and Mayer 1974) of agricultural science a fully integrated part of the industrial capitalist mainland.

For more than a decade now the public agricultural research system has been the target of critics who question the quality of its work. Reports from the National Research Council (the so-called Pound Report, 1972b), the General Accounting Office (1977), and the Office of Technology Assessment (1981b) have bluntly indicted the USDA and the LGUs for their alleged parochialism, bureaucratic inefficiencies, and inability or unwillingness to support basic research of critical importance. In 1982, the Rockefeller Foundation and the White House Office of Science and Technology Policy (OSTP) jointly issued a report reiterating these charges and warning that unless steps were taken to improve matters, the nation would be unable to realize the potentials of the new biotechnologies for agricultural advance (Rockefeller Foundation 1982).[25]

The Winrock Report, as this document has come to be known,[26] is of special interest. The product of a workshop attended by fifteen elite decisionmakers from industry,[27] the federal government, prominent foundations, and the LGU community, the report provides a template for the reshaping of public agricultural research along lines more responsive to the changing needs of capital and the capitalist state.

In the ideal vision of the future expressed in the Winrock Report, the needed restructuring of public agricultural research would have three principal features. First, the highly decentralized institutional and financial structure of the ARS/LGU system would be streamlined and rationalized. Research effort would be concentrated in "centers of excellence" (i.e., the strongest institutions) rather than continuing to be dispersed among the myriad LGU, SAES, and ARS facilities. The creation of a competitive grant system would be a means for circumventing the "formula funding"[28] that had limited the access of institutions outside the land-grant system to research funds administered by the USDA.

Second, research would be redirected to emphasize basic science, and any real increases in funding would go to that end. Third, industry would be given more of an opportunity to determine the social division of labor in agricultural research effort:

> Private sector expertise should be fully utilized in efforts by the public sector to identify future research needs, estimate future demand for scientific and technical manpower, and define appropriate, comple-

> mentary roles and responsibilities for the various sectors and institutions involved in science for agriculture. [Rockefeller Foundation 1982:26]

Significantly, the report makes no effort to justify proposed changes on the basis of their benefits to farmers or consumers. Capital is implicitly recognized as the principal client of public agricultural research. The rationale behind the restructuring of agricultural research is essentially the enhancement of the capacity of a rationalized system to serve corporate interests effectively.

Over the last four years, corporations and their institutional allies have lobbied intensively for implementation of the principles embodied in the Winrock Report. *The New York Times* (1982) has called for creation of a "National Institutes of Agriculture" modeled on NIH. The National Agricultural Research and Extension Users Advisory Board (1983:iv) has warned that "without proper funding and necessary internal reforms, the traditional ARS/SAES system will become increasingly irrelevant to the type of technology-intensive agriculture that is likely to emerge in the decade ahead." With its *New Directions for Biosciences Research in Agriculture: High Reward Opportunities*, the NRC (1985) provided its own version of the Winrock Report and came to similar conclusions. Government officials worried about the competitive position of the United States in the world economy have added their voices to those calling for reform. Frank Press, president of the National Academy of Sciences, has warned that

> A nation with a weak base in plant biology hostages its future. It risks a serious disadvantage in world markets... Unfortunately, we have a growing list of industries that have done badly against competitive pressures from other nations: steel, textiles, consumer electronics and others. To keep agriculture from going down that sorry road, it is going to have to become even more efficient. And that means exploiting the powers of biotechnology. [quoted in Cochran 1985:1][29]

It is clear that these voices have been heard by USDA and LGU administrators. Both the Agricultural Research Service and the National Association of State Universities and Land-Grant Colleges are moving rapidly in the directions outlined by the Winrock Report (Agricultural Research Service 1983; Kinney 1983; NASULGC 1983). The ARS's latest five-year plan provides for a growing emphasis on basic research (ARS 1986). The USDA has established a Competitive Research Grants Program, and in financial year 1985 this fund funneled some $28.5 million to plant biotechnology work (*Diversity* 1984). In return for the enlarged flow of money into their laboratories, the institutions with strong programs in the new biotechnologies have been keeping their part of the bargain by moving away from applied work. Charles Hess (1986:18), dean of the College of Agriculture and Environmental Sciences at University of California-Davis, comments,

> When the private sector has the ability to develop ideas and concepts into marketable products, then the private sector should take the responsibility to do it rather than the university. In this way the university will continue to fulfill its role of conducting basic research and training graduate students but not be in competition with the private sector.

The restructuring now under way has not been uncontested. Some LGU administrators and faculty, unlike Hess, would rather fight than switch. Sylvan Wittwer (1985:16), Director Emeritus of the Michigan Agricultural Experiment Station, notes that the competitive grants program has "viciously been attacked by [some] experiment station directors, college deans, department chairmen."[30] To some extent, this old guard takes seriously the mission-oriented ideology of service to farm-level producers that has for so long been a characteristic feature of the land-grant community. Wholesale movement away from applied research represents an implicit abandonment of direct links to growers. But more important, opposition is generated by the dimming of prospects for their own institutions. Administrators at lower-rank LGUs correctly understand the competitive grants program to be a mechanism for diverting to other institutions the funds that might have come to them under formula funding arrangements. The program is, after all, competitive, and though every university scientist may have a chance to compete for the funds earmarked for biotechnology, the winners tend to be those from leading institutions with leading molecular biology and biochemistry programs. It is the realization that the current restructuring will accelerate the differentiation among LGUs (just as, at the farm level, some will gain and some will lose) that has been the material basis of faculty and administrator opposition to the current redefinition of the public role in agricultural research.

Support for the old guard of the LGUs has come from an unexpected quarter. Ever since Griliches' (1957) study of hybrid corn, agricultural economists have been interested in measuring the social rates of return to agricultural research investment. American public agricultural research expenditures for a variety of crops and time periods have been analyzed. Summarizing the findings of these studies, Vernon Ruttan (1980:531) comments,

> The rate of return studies, for both individual commodities or factors and for total research systems, suggest under-investment in agricultural research. The observed annual rates of return typically fall in the 30-60 percent range. It is hard to imagine very many investments in either private or public sector activity that would produce more favorable rates

> of return . . . the U.S. public sector agricultural research system appears to be relatively efficient in the allocation of research resources.

The returns-to-research literature directly contradicts the charges of bureaucratic inefficiency and low-quality work that have been directed against the ARS/LGU system. Economic analysis appears to show that public agricultural research is performing very well indeed. The decentralization that appears as lack of leadership and purpose in the Winrock Report emerges from Ruttan's analysis as an organizational structure that induces efficient resource allocation by its very consistency with the behavior of firms in a competitive market (Ruttan 1980:537).

In an article titled "An Unpersuasive Plea for Centralised Control of Agricultural Research," Nobel-laureate agricultural economist Theodore W. Schultz (1983:141) expressed his surprise at the conclusions reached in the Winrock Report. He wrote, "To appreciate [the report], one must have learned to enjoy the logic of Alice in Wonderland." It must be disconcerting for economists of the stature of Ruttan and Schultz to have what they regard as such clear evidence so completely ignored. But alas, it is not the logic of Alice in Wonderland that puts corporate interests at odds with their own neoclassical economic analysis, but the logic of accumulation and profit.[31]

The larger restructuring of public agricultural research that has been stimulated by the emergence of the new biotechnologies has important effects on the social division of labor characteristic of plant improvement. Owen J. Newlin (1986:44), president of the American Seed Trade Association, outlines the tasks toward which the ASTA has been trying to guide public breeders over the past several years:

> [W]e know that substantial public sector support of basic research is still needed. For example basic genetic studies are time-consuming and expensive . . . as is the development of new biotechnology techniques. Such basic studies are best suited to public research institutions since they do not have to depend on developing a saleable product . . . [I]f the seed industry has to dilute its applied product research and development dollars in order to conduct additional basic research, product improvements will be even slower, more expensive, and in my opinion, detrimental to U.S. agriculture and the general economy in the long run . . . Certainly the ability of genetic engineering to move individual genes offers mind-boggling potential for crop advances. However, the ability to move genes is not useful if we do not know which gene or genes control specific desired traits. Such mapping of genes is an example of basic research that would be best provided by our publicly-financed institutions to all qualified parties.

In his keynote address to the 1983 annual meeting of the American Society of Agronomy, former Secretary of Agriculture John Block demonstrated that he had been listening to such advice. He announced that

> Federal research is being phased out of conventional plant breeding programs where the private sector can meet these needs. That releases funding for basic research in plant genetics... Plans call for a larger share of the federal research dollar to be targeted at this basic research for increasing plant yields. In other words, it will be targeted to genetic engineering. The same developments would apply to the experiment stations. [quoted in Ryder 1984:810]

The "need" to emphasize biotechnology research becomes one more justification for levering public breeders away from the commodity-form.

The ARS has already abandoned the development of finished cultivars, and there is evidence that the states are following the federal lead in reducing applied breeding activities. A 1985 survey of public genetics and breeding programs in horticultural (floral, fruit, nut, and vegetable) crops projected a 50 percent reduction in the number of programs and a 39 percent reduction in personnel by 1990 (Brooks and Vest 1985). A similar survey of public agronomic (corn, soybean, wheat, etc.) crop programs was conducted in 1986 by the National Plant Genetic Resources Board. While the anticipated loss of programs and personnel is not as dramatic as it is for horticultural crops, the general trend is similar (*Diversity* 1986a:15). Agrigenetics' Robert H. Lawrence (Qualset et al. 1983:472) says a definite "shift in the balance of power" between public and private breeders is under way, and Pioneer Hi-Bred's Donald Duvick (1984:1) agrees that "The private sector is moving rapidly towards dominance." Ironically, in 1987, the centennial of the Hatch Act, which established the experiment stations, there will probably be more private than public breeders.

These trends do not sit well with all public breeders. The historic tension between the public and private sectors is finding overt expression as some ARS-SAES-LGU scientists resist pressures to back away from varietal release. Plant scientists critical of the current restructuring of public and private responsibilities argue that a variety of negative consequences may follow from the complete privatization of cultivar development (Vest 1984; Ryder 1984; Munger 1984; Childers 1986). Among their concerns are

1. The proliferation of lines that are genetically different in trivial ways but that are marketed as different varieties
2. Loss of satisfaction and prestige for public breeders as their work is circumscribed and they are unable to carry through the entire labor process of which they are capable (this is what the sociologist would call alienation)
3. Constraints on germplasm exchange as a result of proprietary considerations
4. Constraints on information exchange as a result of proprietary considerations

5. The complete elimination of new varietal development in crops with small seed markets and little profitability for private breeders
6. Declining emphasis on the local adaptation of cultivars
7. Concentration of the seed industry as small seed companies that depended on public varieties are forced out of business or absorbed by larger firms

One ARS breeder interviewed expressed his views fairly succinctly: "I resent like hell the idea that I was doing the wrong thing all these years."

Though some public plant scientists express concerns, these are not translated into a systematic critique. Public breeders have been living with the shifting division of labor for decades; it is only the rapidity with which it is now being changed that has stimulated overt resistance. Moreover, the opposition has a very narrow social base. It is concentrated among an older group of mostly horticultural breeders. There seems to be little support for the preservation of applied work among younger plant scientists, who prefer the challenges of biotechnology to the more mundane problems of conventional breeding. Arguing for what he calls "life in the slow lane," American Society for Horticultural Science president Edward Proebsting (1984) asks, "Do we want to be horticultural scientists or do we want to be plant physiologists or geneticists or some other discipline?" Younger plant scientists are not satisfied with lives in the slow lane. As James D. Watson phrased it on the thirtieth anniversary of his discovery of the structure of DNA, "If you are young, there is really no option but to be a molecular biologist" (quoted in Ryder 1984:809).

Conclusion: New generation, new division of labor

Pioneer's Duvick (1983:221) has suggested that a generational shift is needed to facilitate the integration of plant breeding and molecular biology.[32] And this may well be the way in which the transformed division of labor will settle into place. A large proportion of faculty in the fields of plant breeding, plant genetics, and agronomy is now approaching retirement age. These are the individuals who, as Qualset et al. (1983:485) discreetly put it, have not recently made "major programmatic changes in their research and teaching." These are also the professors who have been least sympathetic to the new biotechnologies and most resistant to halting variety release. In the next decade deans and experiment station directors have a major opportunity to redirect research efforts through the hiring of new faculty. According to University of Wisconsin Associate Dean Robert Hougas (1983:127), administrators should use retirement and vacancies to "reduce the emphasis on varietal development in order to devote increased resources to more fundamental plant breeding research."

The multi-disciplinary work entailed by a productive synthesis of genetic engineering, *in vitro* horticulture, and whole-plant manipulation also implies the need to restructure graduate training programs. There is widespread recognition of the need for plant breeders to acquire at least a familiarity with the new techniques (Qualset et al. 1983; Kalton 1984; NRC 1984). Duvick (1983:221) suggests establishing a molecular biology group in every agronomy department. The Curriculum Committee of Iowa State's Agronomy Department has considered the problem. They concluded that for their graduate students to achieve acquaintance with biotechnological techniques, six courses would need to be added to the graduate program. To achieve proficiency in the new technologies would require a total of fourteen (Hallauer 1984). The question becomes, What in the existing curriculum could be dropped?

For this question, private breeders have a ready answer. As early as 1969 they were arguing that practical breeding was something that could be taught just as well in industry as in the university (Buker 1969). In fact, this is true. The techniques of conventional breeding are both "easy in application and productive in results" (Qualset et al. 1983:482; see also Simmonds 1979). The proposition that training in applied breeding should be the responsibility of private enterprise is now being seriously voiced (Duvick 1982b; Johnson 1983; Pioneer Hi-Bred 1983). Should such an arrangement become an accepted part of the public-private division of labor, industry would, by forfeit, as it were, have achieved its goal of moving public agencies out of varietal development.

There is now developing a confluence of factors that appears to be setting the conditions for a significant recharacterization of the allocation of tasks between public and private breeding. It seems likely that graduate programs in plant breeding and agronomy will move toward the incorporation of biotechnological expertise in training. The retirement of the older cohort of breeders will remove the core of resistance to this restructuring and to disengagement from varietal release. At the same time, industry is willing and technically able to take over variety development. The shift will not be sudden, but plant breeding as it has long been known is ripe for privatization.[33] Public agencies will move increasingly toward basic research in which biotechnology will be used to evaluate and improve germplasm in what is known as the pre-breeding phase of plant improvement. The task of the public sector will be to provide private enterprise with improved raw materials, but it will be capital that determines how these materials will be combined and what form the product, the commodity, takes in the market.

9

Directions for deployment

> We've invented fire. The sky's the limit.
>
> Waclaw Szybalski,
> University of Wisconsin
> (in J. A. Miller 1985)

> Having biotechnology may be *necessary* for success, but it does not necessarily *guarantee* success.
>
> Arthur Klausner,
> *Bio/Technology* (1986)

What forms will the products of the new plant biotechnologies take? Ciba-Geigy's Mary-Dell Chilton has written that "Biotechnology is a completely new approach to solving old problems" (quoted in Rossman 1984). But qualitatively different tools do not necessarily imply qualitatively different sorts of solutions to those problems. Will we explore the possibilities of *all* the alternative solutions biotechnology offers, or will the social context in which the new techniques are being developed and deployed foreclose certain options? After all, what is considered a "problem" is as much a social as a technical determination. Biotechnology may simply be a new vehicle in which to drive down familiar roads.

Heading for hybridization

One of the most familiar of these roads is hybridization. A great deal of effort is being expended in attempts to use biotechnology to develop hybrids in crops that have proved intractable to hybridization by conventional breeding methods (Pramik 1982; Edwards 1983; Carlson et al. 1984; Klausner 1984; Sink 1984; Freifeld 1985; Orton 1985). Agrigenetics' David Padwa (1983:11) explained: "Biological proprietorships do not need a treaty organization and hybrid plants provide a form of economic protection that is actually more effective than the patent system. We should logically expect modern biology to give us new methods of generating and creating hybrids." The motivation behind hybrid research is less the prospect of realizing an enhanced yield than it is the prospect of achieving a more complete commodification of the seed. In fact, independent of hybrid research, some companies are trying to develop genetic mechanisms for the induction of biological sterility in specified generations of seed (Orton 1985). For his part, Phillips (1983:459) hopes that hybridization will be rapidly achieved in numerous crops, if only so that the one-half of private plant breeding expenditures that he estimates now go to that purpose can be directed to characteristics of agronomic importance.

That plant research in the private sector should be geared to uncoupling farmers from the autonomous reproduction of seed is not surprising. The annual rate of plantback – planting by farmers of bin-run seed saved from the previous year's harvest – in the United States runs about 60 percent in wheat, 40 percent in soybeans, 70 percent in oats, 50 percent in barley, and 50 percent in cotton (Freifeld 1985:219). Should hybridization – or some other mechanism for creating "economic sterility" - be achieved in these crops, annual seed markets could be greatly enlarged. Moreover, an absolute increase in seed sales would be accompanied by a higher rate of profit, for profit margins on hybrids run as high as 60 percent, compared with the 15-20 percent common with non-hybrid seed (Freifeld 1985:222). Explains L. William Teweles & Company's George Kidd (quoted in *Business Week* 1984b:70), "Hybrids are the driving force; everybody wants the profitability of Pioneer [Hi-Bred, Inc.]."

Biotechnology is also being turned to improving the efficiency of hybridization for crops in which hybrids have already been developed. For example, anther culture appears to be a way in which homozygous lines can be rapidly generated for evaluation as an inbred parent (Qualset et al. 1983; Duvick 1984; Carlson et al. 1984). Many labs are also using tissue culture and in vitro screening of corn cells to search for a T-type male sterile cytoplasm resistant to the corn-blight pathogen *Helminthosporium maydis*. The practical objective of this research is, of course, to eliminate the labor required by manual detasseling of the female parent in hybrid seed-corn production.

Biotechnology and genetic vulnerability

Tissue culture of elite corn inbred lines for the purpose of recovering somaclonal variants is another popular line of investigation (Earle 1984; Gebhart 1984a). The generation of somaclonal variants from existing inbred lines has some interesting implications for the problems of genetic uniformity and genetic vulnerability. At the 1983 Plant Breeding Forum sponsored by Pioneer Hi-Bred, Inc. (1984:25), plant breeders agreed that there still "remains a closer relationship than there should be among the leading current varieties of each of our leading major crops." The new biotechnologies appear to provide useful tools for generating the variation needed to broaden the genetic base of American agriculture. According to Carlson et al. (1984), "The current axiom is that passage of plant cells through tissue culture results in increased genetic variability."

But though tissue culture can indeed generate variability in the laboratory, it may result in an increasing level of uniformity in the field. Tissue culture is regarded as a means of obtaining somaclonal variants from *within* existing elite lines. Moreover, not all genotypes within a species regenerate *in vitro* with equal ease (Lawrence 1983:20). There is necessarily a bias toward use of those lines that have proven commercial value and from which whole plants can be retrieved in culture. This seems to ensure a reworking of particular elite genotypes. Even though new characters would be generated in such lines, that variation would be embedded in an increasingly narrow –and possibly more vulnerable – genetic matrix.

The application of rDNA transfer to crop improvement may also result in a greater degree of genetic uniformity among cultivars. The NBF Calgene has succeeded in isolating a bacterial gene that, when transferred to a tobacco plant and successfully expressed, confers resistance to the herbicide glyphosate (Monsanto's "Roundup"). Now it might be said that Calgene has added variability to the tobacco gene pool. But if that gene is a commercial success and is incorporated into most tobacco cultivars, the result may be increased genetic uniformity in that crop. Recall that it was the broad distribution of a single genetic character that led to the corn-blight epidemic of 1970.[1] And Calgene is now seeking to transfer the "GlyphoTol" gene into cotton, corn, rapeseed, tomato, and loblolly pine. Though the capacity to move genetic material between species is a means for introducing additional variation, it is also a means for engineering genetic uniformity *across* species.

Another technique that will affect genetic vulnerability of crops is a process known as somatic embryogenesis. Tissue from a seed embryo of an elite commercial line can be induced *in vitro* to form millions of individual embryos that, when regenerated, are *identical copies of the plant from which the original embryonic tissue was taken*. The cultured embryos can be encapsulated in an aqueous organic gel and then coated with a biodegradable polymer to make "synthetic" seeds. Embryo encapsulation can be automated, and it appears that it will be cost-competitive with natural seed at least in certain vegetable and fruit crops. Various NBFs – including Plant Genetics, Agrigenetics, and DNA Plant Technology (DNA Plant Technology 1983; Gebhart 1984b) – are vigorously pursuing the commercial application of somatic embryogenesis. It may be that fields of genetically identical carrot or celery plants are an acceptable social risk given the relatively low volume and minor economic importance of such crops. Could the same be said if the cost of synthetic seeds was to become competitive in crops such as wheat, corn, soybean, or cotton?

Actually, we may face such questions first not in horticulture or agronomy but in forestry. Tree crops have not been amenable to the kind of genetic improvement achieved in vegetable and field crops, largely because it is difficult to breed organisms that take twenty years rather than three months to reach maturity. Now, tissue culture offers the possibility of cloning superior individual trees. While technical problems remain, the United States Forest Service and companies such as Weyerhaeuser and International Paper are looking to the mass production of genetically identical seedlings. Dr. Rex McCullough, Weyerhaeuser's director of biological sciences, says of his company's efforts to clone Douglas Fir: "We can propagate the same genes now. If we could isolate the genes that control, say, yield, you could insert high yield genes in any tree you wanted and then multiply them to infinity. We are very close" (quoted in Malcolm 1986:20). But the very feature that makes tissue culture so appealing in forest crops – the slow growth of trees – also brings problems. Should cloned, genetically uniform trees prove susceptible to a pathogen or pest, millions of acres of forest and years of production might be lost.

It appears that biotechnology may well be used in ways that exacerbate rather than diminish genetic uniformity and the concomitant problem of genetic vulnerability. There is evidence that who uses the new technologies will influence the extent to which the problem of genetic uniformity is taken into account in plant improvement. In a survey of plant breeders, Duvick (1982c) found that, depending on the crop involved, up to a third of public breeders believed that genetic vulnerability was a serious problem.[2] In contrast, *no private breeders* felt that genetic vulnerability was a serious problem for any of the crops with which they were working. Insofar as public breeding activities are subordinated to the needs of private firms in the emerging division of labor between the public and private sectors, American agriculture may remain, as it was in 1970, "impressively uniform genetically, and impressively vulnerable" (NRC 1972a:1).

The chemical connection

The new biotechnologies have been welcomed for their apparent potential for reducing the chemical intensivity of modern, industrial agricultural production. Sam Dryden, president of Agrigenetics, has predicted that "in two decades we won't be spraying crap on plants anymore ... In time the entire insecticide industry may be totally displaced by plant genetics" (quoted in

Doyle 1985:90). Will we see a Schumpeterian wave of creative destruction in which, Monsanto's advertising paean that "all life is chemical" notwithstanding, a new biorational agriculture rises on the ruins of the toxic waste dumps?

Certainly, the realization of such a scenario would require a substantial shift in historical trajectory, even in the allegedly "ecology-plus" (Kent 1986:25) seed sector. After all, the seed-chemical connection is a relationship of long standing. As early as the seventeenth century, salt water was used to treat wheat seed to reduce the incidence of burnt smut disease (Copeland 1976:251). And in the last decade, the seed has come to be recognized as the ideal vehicle for the delivery of agrichemicals to the field.

With the seed industry rapidly coming under the ownership of companies with substantial agrichemical interests, seeds and chemicals have come to be linked in proprietary packages. Funk Seed introduced eight safener-treated sorghum varieties. Safeners are chemicals applied to seeds that block herbicidal action. After treatment with a safener in a process it calls "Herbishield," Funk sorghum seed can safely receive pre-emergence applications of parent-firm Ciba-Geigy's herbicides to which it would normally succumb. DeKalb-Pfizer Genetics has ten such sorghum hybrids available, and Northrup King (Sandoz) is close to commercializing herbicide-coated alfalfa seed. The costs of developing new herbicides have been rising rapidly, and safeners represent a low-cost means of extending the life of existing products and pushing them into new markets. John Ellis, director of biological research at Ciba-Geigy's Agricultural Division, says of the seed safener Concep, "It helped our [Funk's] seed business move to the number one sorghum seed producer and, at the same time, sell more Dual and Bicep [herbicides]" (quoted in O'Brien 1985:30).

The new biotechnologies open important new possibilities for such seed-chemical packaging. Somatic embryogenesis, for example, provides the biological foundation for the production of synthetic seed. The biodegradable polymer shell that surrounds the seed embryo can be filled with fertilizers, pesticides, bacterial innoculants, and other chemicals. The NBF Plant Genetics has obtained a patent on "GEL-COAT," its somatic embryo encapsulation system. Both Plant Genetics and DNA Plant Technology have signed contracts with agrichemical firms interested in exploring the possibilities of marketing their products not just *with* the seed, but as *part of* the seed.[3]

With safening or encapsulation, the union of seed and chemical is still mechanical. But biotechnology introduces the possibility of making the union at the genetic level – the seed might be genetically programmed to respond to, perhaps to require, the application of particular chemical compounds. The president of Asgrow Seed Company (Upjohn) notes that "the speculation that a variety could be developed that is dependent on a chemical for successful use is definitely within the realm of possibility" (Studebaker 1982:27). As Table 9.1 indicates, much biotechnology research, especially in private firms, is now directed to marrying seeds and chemicals through the achievement of herbicide resistance in a wide variety of crops.

The potential profit in herbicide resistance is considerable. Glyphosate (Monsanto's Roundup) is, at $400 million annual sales, the world's largest-selling herbicide. However, it is nonselective – it kills anything green, crops as well as weeds – and its use in agriculture is limited. If glyphosate-resistant crop varieties were developed, the agricultural market for Roundup would expand from a few million acres to some 150 million (Benbrook and Moses 1986:58). It has been estimated that development of atrazine-resistant soybeans would increase sales of that herbicide by $93 million per year and that phenmedipham-resistant rapeseed would give the developer an 80 percent share of the Western European rapeseed market (Teweles 1983:521). Calgene claims it will have a herbicide-resistant cotton plant in farmers' fields by 1989 (Gebhart 1984a:25), and a Michigan State University researcher says, "If somebody doesn't have a herbicide-resistant potato plant within the next year or two, I'd be very surprised" (quoted in NRC 1984:43).

A particularly attractive prospect for companies is the possibility of engineering crop resistance to new proprietary chemicals. The NBF Molecular Genetics has used tissue culture to select a corn line resistant to American Cyanamid's recently developed class of imidazolinone herbicides. American Cyanamid, which funded the research and has exclusive rights to the new plants, has sublicensed the germplasm to Pioneer Hi-Bred, which will develop and market herbicide-resistant hybrid seed-corn (Bishop 1985:27). And second place seedcorn producer DeKalb AgResearch has contracted with Calgene for the incorporation of the GlyphoTol gene into its corn lines. Proprietary chemicals will be increasingly linked with proprietary seeds. As one executive put it, "Genetics and chemicals together make the most long-term sense" (Donwen 1984:9).

Certainly, herbicide resistance makes sense for capital. It is less clear that society as a whole will enjoy net benefits. Herbicide applications account for 60 percent of the 500 million kilograms of pesticides used annually in the United States (Pimentel and Levitan 1986:86). The extensive use of herbicides has not been without costs. Some forty-one weed species now show resistance to herbicides (Sommers 1986:23). Of forty-five iatrogenically induced diseases of crop plants listed by Horsfall (1979), thirty-two were found to be caused by herbicides. Reduction in levels of soil organic matter, degradation and contamination of groundwater, human cancers, and general impoverishment of the ecosystem have all been associated with herbicide use (Hodges and Scofield 1983; Pimentel and Levitan 1986).

Table 9.1. *Companies and institutions working on herbicide resistance in plants*

Research by	Under contract to	Resistance to	Crop
Advanced Genetic Sci.		Experimental	Potatoes
Allelix		Atrazine	Rapeseed
ARCO (PCRI)	Heinz	Atrazine	Tomato
Biotechnica Intl.		Atrazine	Soybean
Calgene		Phenmedipham	
Calgene		Glyphosate	Cotton, corn
Calgene	Rhone-Poulenc	Bromoxynil	Sunflower
Calgene	Kemira-Oy	Glyphosate	Rapeseed
Calgene	Nestle	Atrazine	Soybean
Calgene	Campbell's	Glyphosate	Tomato
Calgene	DeKalb-Pfizer	Glyphosate	Corn
Calgene	Coker's Seed Co.	Glyphosate	Tobacco
Calgene	Phytogen	Glyphosate	Cotton
Calgene	U.S. Forest Service	Glyphosate	Loblolly
Du Pont		Chlorosulfuron	Tobacco
Du Pont		Sulfometuron	
International Paper			Douglas fir
Mobay (Bayer)		Metribuzin	Soybean
Molecular Genetics	American Cyanamid	Imidazolinone	Corn
Monsanto		Glyphosate	
Phyto-Dynamics		Trifluralin	Corn
Shell		Atrazine	Corn
Cornell Univ.		Triazines	Corn
Harvard Univ.		Atrazine	Soybean
Louisiana State Univ.		Glyphosate	
Michigan State Univ.		Atrazine	Soybean
Rutgers Univ.		Triazines	
Univ. of Alabama		Atrazine	
Univ. of California–Davis		Sulfometuron	Sunflower
Univ. of Guelph		Atrazine	Rapeseed
USDA-ARS		Metribuzin	Soybean
U.S. Forest Service		Glyphosate	Poplar
		Hexazinone	Jack pine

Source: Author's compilation from numerous sources.

The latest generation of herbicides avoids some of these problems. Glyphosate and the new sulfonylurea and imidazolinone herbicides now being developed are said to be biochemically active only in plants and therefore present little risk to other organisms. Moreover, they can be applied at very low rates and either break down rapidly after application or do not readily leach into groundwater. Some argue that, given these characteristics, the development of crop varieties resistant to the new classes of herbicides is environmentally desirable (Benbrook and Moses 1986:56).

But while the new compounds appear to be an improvement over many of the older herbicides, they do not alleviate all the problems associated with the chemical control of weeds. Their relative safety may well encourage more extensive and more liberal applications. After all, the sale of more herbicide is the companies' objective. Whatever the application rate, if the herbicide is effective it kills weeds. In conventional tillage systems, more effective weed control can mean increased erosion. In reduced tillage systems, which depend heavily on herbicides, heavier crop residues can exacerbate disease and insect infestation by providing better conditions for carry-over of pests and pathogens. Ironically, improved weed control can mean higher expenditures on insecticides and an acceleration of the pesticide treadmill (Pimentel and Levitan 1986). The widespread incorporation of herbicide-resistance genes – e.g., Calgene's GlyphoTol – could result in an unprecedented degree of crop genetic uniformity. And, of course, the new herbicides are not the only ones to which resistance is being sought (Table 9.1). There is also much interest in developing cultivars that will tolerate applications of established compounds. More intensive use of these chemicals will only deepen the environmental and human health problems with which the use of such herbicides has been associated.

Given the powerful position enjoyed by the agrichemical multinationals with regard to the seed industry and biotechnology, it would be naive to think that the plant-chemical connection will not exert a powerful influence on corporate plant breeding and plant genetic engineering goals. In spite of the promises of corporate proponents of biotechnology that the new techniques will soon be "genetically displacing various capital-intensive inputs such as chemicals" (Padwa 1983:11), the opposite may well be true.[4] One of the first applications of biotechnology to crop improvement, the development of herbicide-resistant plant varieties, can be expected to result in an *increase* in chemical usage.

There are, however, some countervailing tendencies. The agrichemical industry is keenly aware of the environmentalist critique. And as public opposition to the use of chemicals in agriculture has grown, there has been a gradual increase in the stringency of state and federal regulation of the development of agrichemicals and of the ways in which they may be used by farmers. A common complaint in the industry is the many years and great expense that must now be incurred to attain EPA and FDA approvals of new compounds. The new biotechnologies may provide a means of reducing these difficulties. Pests and pathogens are living organisms and thus are themselves subject to attack by other organisms such as bacteria, viruses, protozoa, and fungi. Genetic engineering might permit the development of "biorational" pesticides using the natural enemies of economically important pests. Because their active ingredient is "natural," such pesticides might be more easily approved and be subject to fewer use restrictions than chemical compounds.

Many company and university laboratories are directing at least some of their resources to the biorational approach. Some companies – such as the NBFs Ecogen, Mycogen, and Microbial Resources – are even specializing in the development of biological pest controls. Perhaps the most popular organism is *Bacillus thuringiensis* (B.t.), a bacterium that is lethal to the caterpillar stage of many insects. Monsanto has isolated the B.t. toxin gene and, in hopes of controlling cutworm in corn, has moved it into the bacterium *Pseudomonas fluorescens*, which colonizes corn roots. Pursuing a different strategy, both Agrigenetics and Rohm and Haas Company have successfully moved the B.t. toxin gene into tobacco plants, making the plants themselves resistant to B.t.-susceptible pests. By transferring genetic material from a tobacco mosaic virus to tobacco and tomato, cooperating researchers from Monsanto and Washington University have apparently succeeded in "vaccinating" the plants against tobacco mosaic disease. The capacity to selectively delete as well as add genes can also be important. The NBF Agracetus has removed from a tobacco[5] plant a gene that produces a substance necessary for the pathogen *Agrobacterium tumefaciens* to cause infection – thus providing the plant with resistance to crown gall disease. And from *Pseudomonas syringae*, Advanced Genetic Sciences has deleted a gene that causes the bacterium to act as a nucleus for the formation of ice crystals. Advanced Genetic Sciences hopes to use these "ice-minus" bacteria as a spray to prevent frost damage in crops.

From a social point of view, such advances seem very promising. The biorational approach to pest control appears to have major advantages over more chemical-intensive methods. Yet there are forces that may limit the scope of development of biological controls. And even the "natural" approach is not without its own unique problems.

The development of microbial pesticides is not entirely new. Formulations of B.t. have been available commercially for more than thirty years now. In addition to B.t., only one other biological agent (nuclear polyhedrosis virus) has been widely adopted for agricultural pest control in the United States, and these two products together account for a very small proportion of annual pesticide sales. By comparison, more than 45,000 chemical pesticides incorporating over 600 active ingredients have been developed (Shabecoff 1986:16). A consultant's study identified the factors that have precluded wider interest in and use of biological controls:

> The limitations are seen to generally include the amount of time it takes for damage to end or insects to die because these materials act on a biological method, and they have to be ingested or regenerated internally until they cause mortality or morbidity in the insect. The care required in timing applications, they have to be timed appropriately to catch the insect at a proper stage. They have limited control of secondary pests – they are very targeted, very specific. They have a lesser level of control or percentage kill than is thought to be desired, perhaps 80 to 85 percent of the insect population is affected rather than 90 to 95 percent, as with chemicals. Limitations exist on residuality or days of protection before reapplication, and they do have a narrow spectrum of control. It is like a rifle shot rather than a shotgun shot into a pest group. [Murphy 1982:105]

The consultant suggests that "it is in these areas of limitations that perhaps biotechnology may be able to assist in making them compete better with their chemical sisters" (Murphy 1982:104).

That is, biotechnology is to be used to make the biorational approach more competitive by rendering biological controls more *like* their chemical sisters. But the very characteristics identified as limitations to be redressed – carefully timed application, target specificity, low residuality – are those that make the biorational approach biologically rational. For example, the NBF Ecogen has reportedly increased the toxic effect of B.t. thirty times, thus extending its lethality to the cotton budworm and the cotton bollworm (Gebhart 1986:11). Are other, beneficial species affected? What might be the ecological impacts of achieving shotgun coverage with biological controls? Might not pests develop resistance to biological pesticides in the same way they have developed resistance to chemical pesticides? Might there not be a biological pesticide treadmill parallel to the chemical pesticide treadmill? Might the achievement of enhanced residuality result in persistence and even proliferation of the control organism? While biological pesticides appear to have enormous potential, it is a mistake to unequivocally equate "natural" or "biological" with "biorational."

Deliberate controls for deliberate release?

Genetically engineered organisms present a unique set of potential externalities. Whereas it is possible to isolate and clean up chemical spills, once released in a "genetic spill," new forms of life might be difficult to contain or eliminate. As living entities, they are capable of reproducing and exchanging genetic information with other life forms. Could the novel, chimeric organisms now being developed proliferate with unforeseen and possibly negative consequences?

Such concern was first raised within the scientific community itself. At the 1973 Gordon Research Conference on Nucleic Acids it became clear that a great many experiments involving the recombination of DNA were planned and that a number of these involved animal tumor viruses and other disease agents. A letter from the conference co-chairs to the NAS suggested creation of a study committee to examine the potential hazards of such work (Singer and Soll 1973). In a letter to *Science*, the eleven prominent molecular biologists appointed to the NAS committee, chaired by Paul Berg of Stanford University, noted "serious concern that some of these artificial recombinant DNA molecules could prove biologically hazardous" (Berg et al. 1974). The Berg letter proposed a moratorium on certain types of research pending further discussion of ways to assess and deal with potential biohazards.

The forum for such discussion was the Asilomar Conference Center, a haven by the sea in Pacific Grove, California, where in February 1975 assembled 140 of the world's top molecular biologists. Three days of presentations and deliberation culminated in a chaotic morning of heated and sometimes acrimonious debate as conference participants tried to agree upon levels of risk associated with various types of experiments and to decide whether or not restrictions on research were warranted. Despite the active resistance of a small but influential minority opposed to any restraints upon rDNA work,[6] the conference participants concluded that "it would be wise to exercise considerable caution in performing this research" (Berg et al. 1975:991). A "Statement of the Conference Proceedings" was adopted that proposed a set of guidelines specifying appropriate containment facilities for various types of research, and going so far as to suggest deferral of certain experiments. As James D. Watson (Watson and Tooze 1981:26) recalls, "Despite the confusion of the last session, many participants left Asilomar as exhilarated as they were exhausted . . . Having demonstrated their integrity, they naively believed that they would now be free of outside intervention, supervision, and bureaucracy." This was not to be the case.

A sense of social responsibility was certainly the most important factor leading molecular biologists along the road to Asilomar. It is nonetheless true that they hoped to preempt external regulation through self-regulation. But even before Asilomar, the NIH had decided to form a Recombinant DNA Advisory Committee (RAC) to draft guidelines for rDNA research. The final version of the Asilomar recommendations (Berg et al. 1975) provided the RAC with a template for its own task, and the NIH guidelines, issued in June 1976, generally reflected the conclusions reached at Asilomar. In addition to specifications on containment facilities, six types of experiments were explicitly prohibited:

1) the formation of rDNA derived from certain pathogenic organisms;
2) the formation of rDNA containing genes that make vertebrate toxins;
3) the use of the rDNA techniques to create certain plant pathogens;

> 4) transference of drug resistance traits to microorganisms that cause disease in humans, animals, or plants; 5) the deliberate release of any organism containing rDNA into the environment; and 6) experiments using more than 10 liters (l) of culture. [OTA 1981a:212]

Though not legally binding for any researchers not supported by federal money (i.e., private industry), the NIH rules constitute the regulatory framework under which biotechnology has been developed over the last decade.

Some participants in the Asilomar conference had also hoped that a display of scientific responsibility and self-restraint would defuse public concerns as to the safety of genetic engineering. But Asilomar and the subsequent release of the NIH guidelines drew additional attention to the question of hazard. Moreover, a vocal group of prominent biologists who were seriously concerned about the implications of the rapid development of the new biotechnologies continued to carry the issues into public forums. Some, such as Erwin Chargaff, Nobel laureate George Wald, Liebe Cavalieri, and Robert Sinsheimer called for a moratorium on all rDNA research or questioned society's capacity to deal adequately with so powerful a technology. Others, such as members of the Genetic Engineering Group of Science for the People (e.g., Jon Beckwith, Jonathan King, Frederick Ausubel, Ruth Hubbard), emphasized safety concerns and the need for enhanced public participation in policymaking.

The years 1976 and 1978 encompass the classic period of what Sheldon Krimsky (1982) has called "The Recombinant DNA Controversy."[7] In 1976, Harvard University's decision to construct a P3 (moderate risk) containment facility on campus led to divisions within the biology department and resulted in much-publicized hearings in the Cambridge City Council. Pressed by environmental organizations and public-interest groups, Senator Edward Kennedy held hearings on the implementation of the NIH guidelines and in 1977 proposed legislation that would make them mandatory for *all* researchers whether or not they were receiving federal funds (Kenney 1986:27).

Kennedy's bill never became law. By 1978, the balance of forces interested in rDNA was changing in an important way: Industry had begun to recognize the commercial possibilities of genetic engineering. To the lobbying efforts of such scientist-based groups as Friends of DNA were added the political clout and public-relations expertise of companies such as Eli Lilly, Monsanto, and Du Pont. With the proliferation of NBFs after 1980, biotechnology appeared to be one of the "sunrise industries" that could maintain America's competitive position and technological leadership in the world economy. Legislators might have been willing to regulate scientists to prevent hypothetical damage, but they were much more reluctant to delay the cornucopia of products that business claimed was just over the horizon. Safety concerns were submerged in the maelstrom of commercial excitement that accompanied the emergence of the biorevolution. In 1978, 1980, 1981, and 1983, the NIH guidelines were progressively relaxed (Watson and Tooze 1981; OTA 1981a). The 1978 revision was of particular importance, for it permitted the director of NIH, on advice of the RAC, to make case-by-case exemptions to the set of prohibited experiments. This effectively made the RAC the central regulatory body for biotechnology.

And safety has recently reemerged as an issue, though in a rather different form and context. The rDNA controversy of the 1970s focused on preventing accidental escape of laboratory organisms being used for research. But application of the new genetic techniques has been so rapid that companies have begun to request permission to field-test and even sell new biological products. In many cases the product is a living organism. Moreover, many of these novel organisms – e.g., live-virus vaccines, microbial pesticides, transgenic plants – are intended not for use in contained production facilities but for active introduction into the environment. Safety concerns no longer revolve around the question of *containment*, but around the implications of *deliberate release* into the biosphere. Just as the pivotal issue has changed, so the compositions of contending camps have shifted. Whereas the principals on both sides of the rDNA debate of the 1970s came from within the community of academic biologists, the 1980s debate finds industry ranged against ecologists and environmental activists.

There is no question that there will be regulation of deliberate release. Indeed, industry *wants* regulation. Rules provide a stable and predictable framework for business operations. Voluntary submission to regulation also helps legitimate the commercial development of biotechnology. Richard Godown, Executive Director of the Industrial Biotechnology Association, can boast that "We have an industry which has the unprecedented record of ASKING – from its inception – for federal regulation in order to assure the public, consumers, government officials and the media that its intentions are honorable and in the best interests of society" (Godown 1986:4). By embracing regulation, industry makes a virtue of necessity.

And while publicly welcoming regulation, industry has lobbied hard to prevent the promulgation of controls that it regards as restrictive. Industry spokespersons assert that there is no reason to expect that the deliberate release of recombinant organisms will cause problems qualitatively different from those associated with the development of new plant and animal varieties by conventional methods of breeding (Brill 1985; Hardy and Glass 1985). A tobacco plant with a *Salmonella* gene need not be treated any differently than any other tobacco plant. It follows that an adequate regulatory framework is provided by existing legislation (Schneiderman 1985; Godown 1986). Indeed, it is argued that restrictions more stringent than the NIH guidelines would be detrimental to the national interest. The twin spectres of the loss of American technical leadership to other nations and the inability to feed the hungry billions of the world are frequently cited as the projected consequences of overregulation of biotechnology.[8]

Legislators and government officials have not been the only targets of corporate lobbying efforts. Mindful of the lessons of the 1970s rDNA debates, companies committed to the commercialization of biotechnology have no intention of repeating the debacle suffered by the nuclear power industry. Anticipating opposition to deliberate release, they have engaged in a broad program of public relations designed to introduce biotechnology to the American people. By seizing the initiative, they hope to establish public confidence in the new genetic technologies and defuse criticism by setting and dominating the terrain of ideological struggle (Kleinman 1986).

Monsanto's efforts along this line have been particularly notable. The company has produced several educational films on biotechnology, printed a variety of informational pamphlets, funded a national survey of the attitudes of religious, environmental, and science policy leaders toward biotechnology (J. D. Miller 1985), and organized a traveling museum exhibit on genetic engineering. Additionally, in Columbus, Ohio, and Columbia, South Carolina, Monsanto moved from genetic to social engineering by testing a carefully integrated media and public-relations blitz. The program used television spots, newspaper advertisements, shopping mall exhibits, speeches by Monsanto executives, sessions with business and university leaders, and even appearances by astronaut Charles Walker, a McDonnell-Douglas engineer who had performed electrophoresis experiments on the space shuttle.

The central message of this media extravaganza is expressed in the title of Monsanto's pamphlet, "Genetic Engineering: A *Natural* Science" (Monsanto Company, n.d., emphasis added). Genetic engineering is depicted as natural and, by implication, as familiar and safe. A Monsanto newspaper advertisement adds another dimension. Below a picture of a corn plant growing in what appears to be either the moon or the Empty Quarter of Saudi Arabia is the following caption: "Will it take a miracle to solve the world's hunger problems? It might seem miraculous today for a plant to grow in an environment like this. But thanks to the science of biotechnology, in the future, it won't take a miracle." Monsanto attempts to convey the message that their recombinant organisms not only will be "natural" - and therefore intrinsically benign, not requiring regulation – but also will confer tremendous benefits upon society. Given the logic embodied in the advertisement, careful regulation of deliberate release appears not only needless but also, inasmuch as it might slow the movement of food to the hungry, downright irresponsible.

For the most part, academic biologists have not challenged such logic. Indeed, those who had always doubted the wisdom of regulation must find such logic congenial. For example, an excerpt from a recent letter to *Science* from a prominent biologist might serve very well as a caption to Monsanto's advertisement of the desert-dwelling corn plant:

> The argument is simple: (i) the known present and future benefits of genetic engineering are enormous; (ii) the hypothetical, inadvertent risks, if any, are balanced by the hypothetical inadvertent benefits; and, (iii) the overall cost of unnecessary regulation is very high. Thus, the balance sheet clearly shows that regulations are not justified at present and are against the best interests of society. [Szybalski 1985:115]

Few biologists would go so far, yet it is true that they have not been as vocal about the need for regulation in the 1980s as they were in the 1970s. This derives in large part from changed perceptions of levels of risk.[9] But now that they are deep into recombinant work, it may also reflect a natural reluctance to have their research delayed. Moreover, many molecular biologists, including some of those who were the strongest supporters of regulation, now have ties to industry.[10]

One cannot but agree with part (i) of the Szybalski quotation. The present and future benefits of genetic engineering do seem large, though certainly their magnitude is often exaggerated (the Monsanto desert corn advertisement is a case in point).[11] But while most biologists may now agree with part (ii), the same cannot be said of many ecologists.

There is no question that the introduction of organisms into new ecological niches can be very desirable. The United States has enjoyed enormous benefits as a result of the systematic transfer of crop species from other areas of the world. Yet in addition to maize and the cow, we also now have gypsy moths, kudzu vines, Dutch elm disease, Japanese beetles, carp, chestnut blight, and English sparrows. According to Dr. Elliot Norse (1986:173), public affairs director of the Ecological Society of America, ecologists "have learned an important lesson during more than a century of research: when novel organisms are introduced into the environment, the consequences can range from effectively nonexistent to very serious, very expensive and irreversible." There is no *a priori* reason to think that all of the products of genetic engineering will necessarily be environmentally benign. As Cornell microbiologist Martin Alexander (1983:7) observes,

> It is difficult to see why a manmade genetic change would necessarily behave any differently from those occurring spontaneously in nature . . . It is, thus, my view that alien organisms that are inadvertently or deliberately introduced into natural environments may survive, they may grow, they may find a susceptible host or other environment, and they may do harm.

Indeed, genetically engineered organisms might even be more likely to cause problems than naturally occurring mutations. Products of biotechnology such as microbial pesticides must be designed to out-compete other microorganisms if they are to persist long enough to be effective. Moreover, microorganisms are able to exchange genetic material with other microbes. Dr. Patrick Flanagan, a microbial ecologist and director of NSF's ecology program, notes that

> If we release bacteria with pesticide genes or toxin genes, the likelihood is those genes can move through the environment. The question of whether recombinant organisms can spread their genes throughout the soil and waters of the world, and possibly cause a problem, has not been answered. [quoted in Schneider 1986e:48]

Plants are less mobile and less subject to rapid mutation than microorganisms, but genes from a genetically engineered crop variety could be transferred to other lines by normal processes of sexual crossing. And because many weeds cross freely with their domesticated relatives, the deployment of herbicide-resistant plants might conceivably result in the development of herbicide-resistant weeds (Colwell et al. 1985; Keeler 1985).

The ecologists who construct sobering scenarios of environmental externalities are the first to admit that it is very difficult to assign likelihoods to the possibility of ecological harm associated with the deliberate release of genetically engineered organisms. Predictive ecology is a young and relatively undeveloped discipline. But while the probability of damage for any one release is probably low, the environmental consequences of a low-probability genetic spill may be very high (Alexander 1983; Norse 1986). Testifying before Congress, Martin Alexander (1983:7-8) concluded that

> Although genetically engineered species have not been subjected to all the risk factors I have cited, we have never made meaningful tests of the probabilities of any of these risks. Therefore it seems foolhardy to make dogmatic statements as to whether there will or will not be a detrimental effect. The prudent course of action is to establish the risk factors and simultaneously develop a regulatory procedure to assess the survival, growth, and deleterious effects. In this way we may gain the benefits of genetic engineering while not being exposed to the likely hazards from the misuse of the technology.

Ecologists are not calling for a moratorium on research, but for a measure of caution and for the development and careful application of formal protocols for assessment of risk.

Yet there is little enthusiasm for restraint among microbiologists and molecular biologists, to whom the alleged risks of deliberate release seem mere hypothesis and conjecture (Kolata 1985; Brill 1985; Szybalski 1985).

Deficiencies in the base of ecological knowledge are taken not as evidence of a need to improve understanding of the interactions within biological communities but as evidence of the futility of environmental risk assessment. In an editorial reminiscent of Donald Jones' dismissal of the possibility of developing synthetic rather than hybrid corn lines, the editor of *Bio/Technology* asserted that satisfying the demands of ecologists would "require near infinite knowledge, and would require near infinite amounts of money" (McCormick 1986:1045). Even some of those actually responsible for the task of regulating genetic engineering share *Bio/Technology*'s perspective. RAC member Susan Gottesman remarked, "Let me make the plea that we do not ask every possible question before we do the first [field] test" (quoted in Kolata 1985:35). And David Kingsbury, a NSF assistant director and advisor on regulatory policy for biotechnology regulation, has observed, "If we approach biotechnology as if it's dangerous until we prove it's not, we'll never prove it's not, and we'll never go anywhere" (quoted in Maranto 1986:57).

Only a handful of RAC-approved instances of deliberate release have taken place, and all of those in the last year. This is due less to the testimony of ecologists than to legal challenges initiated by public-interest groups. In September of 1982, University of California-Berkeley scientist Steven Lindow applied to the RAC for permission to field-test *Pseudomonas syringae* bacteria whose ice-nucleating capacity had been genetically deleted. This would have been the first deliberate release of a recombinant organism. However, in September, several environmental groups – most prominently Jeremy Rifkin's Foundation on Economic Trends – filed suit against the NIH claiming that approval of the experiment had violated the National Environmental Policy Act (NEPA) by failing to require either an environmental assessment or a more complex environmental impact assessment. In May 1984, Federal District Court Judge John Sirica found in favor of the environmental groups and issued a preliminary injunction barring approval of all deliberate releases.

The chill cast over the biotechnology community by Sirica's decision deepened as the Foundation on Economic Trends sued, in July 1984, to extend the ban on field testing to private as well as federally funded researchers. With plans for both university and commercial experiments going on hold, the Foundation on Economic Trends filed suit again in October, this time to challenge USDA experiments. In the crucible of legal challenge, it became clear that the regulatory structure for genetic engineering in the United States was not as well constructed as many had thought. Faced with the prospect of extensive litigation, commercial and academic interests clamored for creation of a coherent regulatory policy that would break the logjam of blocked experiments and, in the words of the Industrial Biotechnology Association's Alan Goldhammer, "make it that much harder for someone to make a suit stick" (Hoppe 1986:29).

Concerned by the clouds gathering over what it regarded as one of America's sunrise industries, the White House OSTP issued a "Proposal for a Coordinated Framework for Regulation of Biotechnology" on the last day of 1984. Though it was the subject of much criticism, the proposal recognized the limitations (really, the impossibility) of exclusive reliance on the RAC and addressed the need to coordinate the activities of the EPA, the FDA, and the USDA for the regulation of genetic engineering. In February 1985, a federal court of appeals lifted the injunction against deliberate release, though it ruled that such experiments cannot be approved without an environmental assessment. In June, another court found that private companies do not require NIH sanction for field tests. Despite failure to produce a final version of the Coordinated Framework, RAC approval of Advanced Genetic Sciences' (AGS) request to field-test ice-minus *Pseudomonas* seemed to signal that regulatory procedures were getting back on track.

But at a January 1986 meeting of the Monterey County Board, AGS found that although it had jumped through the requisite EPA and RAC regulatory hoops, it had neglected to consult with residents of the proposed test areas and was faced with substantial local opposition (Sun 1986b). Then, in February, it was revealed that AGS had already performed a deliberate release of the bacteria a full year earlier. As part of the safety studies required by the EPA for approval of release, the company had injected the microbes into fruit trees atop its Oakland, California, headquarters. John Bedbrook, AGS research director, contended that "Since the trees were inoculated in a nonaerosol manner, and the bacteria were contained in the woody plant tissues, the conditions of test were under physical containment and the experiments did not constitute an environmental release of the organisms" (quoted in Pramik and Sterling 1986:10). Disagreeing, the EPA revoked the release permission and fined the company $20,000, the maximum permissible under federal law.

Then, only days after the March 25 transmission of a General Accounting Office (GAO) report titled "Biotechnology: Agriculture's Regulatory System Needs Clarification" (GAO 1986), the USDA admitted that over the past year it had approved the testing and sale of a live, genetically engineered viral vaccine for the immunization of pigs produced by Biologics Corporation. While technically in violation of no rules, the approvals were given without being brought before the USDA's own biotechnology review committee, without consultation with other federal agencies, and without notification of agencies in the states where the vaccine was tested and sold (Schneider

1986a). With *The New York Times* (1986a:22) editorializing against what it saw as "A Novel Strain of Recklessness" represented by these two incidents, the USDA suspended Biologics' license for the vaccine.

Stung by criticism and fearing a resurgence of regulatory fervor, industry moved to limit the damage (Crawford 1986b). Given his position as executive director of the Industrial Biotechnology Association, a remarkable article by Richard Godown deserves special attention. Godown's theme is given in his title: "The Real Question Is Overregulation." He asks the reader to "Keep the [AGS] Frostban injection into rooftop trees in perspective. Critics have taken a shoot-from-the-hip, narrow view of procedural compliance." The controversy over the Biologics vaccine is attributed to inter-divisional jealousies within the USDA. *The New York Times* is taken to task for "shouting 'fire' in the theater" with its "A Novel Strain of Recklessness" editorial. The reader is reminded that biotechnology will be "beneficial to millions of people around the world" and that "the industry is not deceitful." Finally, Godown warns us that "If the regulatory process becomes too cumbersome, time consuming, costly, and uncertain, economics will drive the industry elsewhere."

And it has. In November of 1986, the Wistar Institute admitted that it had field-tested a live rabies vaccine without the knowledge or approval of either American or Argentine authorities. Because Argentina has no rules governing biotechnology, explained Wistar's director, Dr. Hilary Kropowski, "It was not my business to bring this [experiment] to the Argentine government" (quoted in *The New York Times* 1986b:22). Even more telling is the response of the NSF's Dr. David Kingsbury, a central figure in the development of the proposed Coordinated Framework. He observed, "This is a very important product, a rabies vaccine. Wistar must have felt that the [American] regulatory framework was too stringent. We may be overregulating and pushing companies to test their products overseas" (quoted in Schneider 1986c:9). The lesson Kingsbury draws from the Wistar case is not that Argentina's regulations are too weak, but that those in the United States are too strong.

Actually, deliberate release is proceeding in the United States even now. Having obtained approval from the RAC, the USDA, and NIH, the NBF Agracetus quietly planted recombinant tobacco plants on a test plot near Madison, Wisconsin, in June 1986.[12] And both Rohm and Haas and Ciba-Geigy have received permission to field-test transgenic tobacco plants (Rigl 1986). But in the wake of the negative publicity surrounding the AGS and Biologics incidents, the EPA has adopted a more cautious approach to the approval of deliberate release. Monsanto's application for an experiment involving an outdoor test of soil bacteria engineered to express the B.t. toxin gene was rejected by the agency. Of the seventeen preliminary safety studies conducted by Monsanto, thirteen were found to be inadequate or inconclusive. Acknowledging the need for caution, EPA assistant administrator Dr. John Moore wrote the company that "it is in the best interest of Monsanto and the E.P.A. that the general public develop a feeling of trust and confidence and that all decisions to permit experiments of this sort be based on expert evaluation of reliable data" (quoted in Schneider 1986e:108).

If the EPA is willing to exercise restraint, the White House has been counseling relaxation of regulatory restrictions. In a June 1986 revision of its Proposal for a Coordinated Framework for Regulation of Biotechnology, the OSTP proposed new rules that would effectively exempt many recombinant experiments from any special regulatory oversight (*Federal Register* 1986). The RAC voted to adopt the new guidelines (Schmeck 1986), but the redesigned Framework has been heavily criticized (Crawford 1986c; Gibb 1986; Schneider 1986d). Ecologists – within and without EPA – are especially upset. The committee that drafted the revision of the Framework never contacted the Ecological Society of America for advice, and consequently, according to the society's Elliot A. Norse (1986:173, 177), "the Framework clearly reflects the absence of our input. The tilt toward minimizing safeguards begins with the first page... The web of interactions among species outside the ideal world of laboratory glassware is vastly more complex than the Framework implies."

Concurrent with publication of the OSTP Framework, the principal regulatory agencies – EPA, FDA, USDA – issued their own policy statements. Not all the approaches were procedurally or scientifically consistent, and the Coordinated Framework was revealed to be less than coordinated. The USDA has since abandoned its proposed rules, and the Framework is likely to be rewritten again. The regulatory tangle has yet to be unraveled, and with congressional and public interest in regulation increasing, the nature of controls on deliberate release of recombinant organisms is yet an object of struggle.

Plants, products, processes, and patents

In 1930, Paul C. Stark advised the American Seed Trade Association's Plant Patent Committee to drop their efforts to have sexually reproducing species included in the proposed Plant Patent Act. He suggested that it was best to let the establishment of patent rights to asexually reproducing species set a principle that new plant forms could be considered patentable items (ASTA 1930:66). Just over half a century later, the United States Supreme Court issued a decision in *Diamond v. Chakrabarty* that appeared to bring all products of plant breeding under the standard utility patent statute.

Because living organisms are central to production processes in biotechnology – and often are the products of those production processes – the provisions of the juridical framework under which ownership rights to living organisms can be established are matters of great concern to private companies interested in commercializing the new genetics. Of course, living organisms have long been ownable in the sense that a cow may be one's property. But, with few exceptions, living organisms were not held to come under the purview of the utility patent act that provides property rights to inventors of "any new and useful process, machine, manufacture, or composition of matter, or any new and useful improvement thereof" (35 U.S.C. 101).[13] In the brave new world of biotechnology, such an interpretation posed difficulties for industry, and General Electric decided to challenge U.S. Patent and Trademark Office (PTO) doctrine by appealing the PTO's rejection of an application for a patent on an oil-degrading microorganism developed by General Electric scientist Ananda Chakrabarty.

The case was eventually heard by the Supreme Court. In its 1980 decision in *Chakrabarty*, the court determined that "a live, human made micro-organism is patentable subject matter" (U.S. Supreme Court 1982:303). The court held that whether the invention in question is animate or inanimate has no bearing on its patentability as long as it meets the criteria of novelty, utility, and non-obviousness, and as long as it is a product not of nature but of human manufacture. Living organisms were declared patentable, and, of course, plants are living organisms.

In view of the implications of *Chakrabarty*, the Plant Variety Protection Office took the precaution of appending to its list of abandoned applications the warning that "Varieties published in this list may possibly be protected under the Patent Act." Yet there was no immediate rush to patent varieties of corn, wheat, and other sexually reproducing species. This reticence did not reflect a lack of interest among private plant breeders in extending property rights to new areas. Indeed, Agrigenetics' David Padwa (1982:102) told the Battelle Memorial Institute Conference on Genetic Engineering,

> I remind everybody here that it may be of marginal commercial utility to develop something that isn't proprietary. You may get a Nobel Prize, a Kettering Award and a whole bunch of other honors, but if you're trying to make a return for your shareholders, you've got to have a way of protecting your product and it's not as simple as putting a marker or fingerprint gene in or something.

That there was no rush to patent plants – or parts of plants – reflected uncertainty as to the legal implications of the Supreme Court's decision.

The source of this uncertainty was the existence of the Plant Patent Act and the PVPA. In *Chakrabarty*, the PTO argued that enactment of those laws showed that Congress did not consider living organisms generally to be patentable subject matter and was making special dispensation in providing such protection for plants. Though the Supreme Court did not agree,[14] the status of the 1930 and 1970 acts in relation to the utility patent statutes was problematic. The potentially overlapping protection provided by the different laws raised substantive and procedural difficulties that could be resolved only in litigation, a prospect that pleased no one, except perhaps lawyers (Neagley et al. 1984:10; Williams 1984:19).

These uncertainties were exacerbated by initial indecision within the PTO as to the application of *Chakrabarty* to plants. In the wake of the Supreme Court's decision, a number of product patents claiming plant germplasm of various types did in fact issue (Table 9.2). However, breeders soon found their patent applications for anything but hybrids rejected on the grounds that Congress had expressly articulated separate property-rights policies for non-hybrid plants. The PTO explained that

> In the absence of judicial guidance, the Patent and Trademark Office has for the present adopted a practice based on the legal principle of "preemption." Any subject matter protectable under either the plant patent law or the Plant Variety Protection Act is preempted by that law and cannot be protected under the general patent law. [quoted in Bent 1985]

In September of 1985, the PTO received judicial guidance. In its decision in *Ex parte Hibberd*, the United States Board of Patent Appeals and Interferences overturned a half century of federal patent policy. Molecular Genetics (a Minneapolis NBF) scientist Kenneth Hibberd and his co-inventors were granted patents on the tissue culture, seed, and whole plant of a corn line selected from tissue culture. In approving the patents, the Board of Appeals rejected the PTO's contention that the Plant Patent Act and the PVPA in any way preempted protection under utility patent legislation. The effect of this decision is to permit breeders to choose among the several statutes for the best form of protection.

For several reasons, utility patents are likely to be preferred to PVP certificates and to plant patents. At $300 per application, PTO fees are substantially less than those levied by the Plant Variety Protection Office ($2,000 per PVP application). Moreover, applicants get more for their money. The PVPA and the Plant Patent Act permit only a single claim for the new plant variety as an indivisible whole. Utility patents may encompass claims not only to multiple varieties but also to the individual components of those varieties: DNA sequences, genes, cells, tissue cultures, seed, and specific plant parts, as well as the entire plant. The *Hibberd* application, for example, included over 260 separate claims. The ability to make multiple claims

Table 9.2. *Plant breeding patents*

Assignee	Date	Process
Research Corporation	1956	Hybrid corn seed production
Teweles Seed Co.	1971	Hybrid alfalfa seed production
Univ. of Illinois	1973	Hybrid corn seed production
Univ. of Illinois	1975	Hybrid corn seed production[a]
Pfizer Inc.	1975	Hybrid soybean seed production
Kent Feeds, Inc.	1977	Production of hybrid alfalfa seed
Research Corporation	1977	Hybrid corn seed production
Pfizer Inc.	1978	Hybrid soybean seed production
Research Corporation	1979	Hybrid wheat seed production[a]
Weyerhaeuser Co.	1980	Embryogenesis of gymnosperms
David & Sons, Inc.	1981	Sunflower pollination[a]
USDA	1982	Hybrid rice seed production[a]
Agrigenetics	1982	Hybrid seed production
DeKalb AgResearch	1983	Semi-dwarfism for hybrid corn seed
Agrigenetics	1983	Hybrid cabbage seed production
Red River Commodities	1983	Hybrid sunflower seed production[a]
International Paper	1983	Clonal propagation of gymnosperms[a]
Sandoz	1983	Selection of stringless beans[a]
Pioneer Hi-Bred	1983	Wheat breeding[a]
Univ. of California	1983	CMV virus DNA as vector system[a]
Agrigenetics	1984	Mutant selection in cereal crops
Molecular Genetics	1985	Tryptophan-overproducing corn mutant[a]
Du Pont	?	Selection for herbicide resistance
ARCO	Pending[b]	Regeneration of plants from protoplasts
Calgene	Pending[b]	Promoter system for Ti vector
DNA Plant Technology	Pending[b]	Generation of somaclonal variation
IPRI	Pending[b]	Vector system for monocots
Phytogen	Pending[b]	Cotton plant regeneration from callus
Univ. of Illinois	Pending[b]	DNA transfer by pollen vector

[a]Includes claims on products of patented process.
[b]Patent may already have been issued.

Sources: Author's compilation from numerous sources.

significantly broadens the protection afforded the invention. It also permits the licensing of particular components – e.g., a gene for herbicide tolerance – for use by third parties. Because genetic engineering in plants is geared to transformations at the cellular and molecular levels, utility patents provide a significant advantage over PVP certificates, which can provide property rights only in the whole organism.

This is not to say that the protection conferred by a utility patent is complete. In return for the right to exclude others from use of the invention, the inventor must provide sufficient information on the invention "to enable any person skilled in the art... to make or use the same" (35 U.S.C. 112). For living organisms, this disclosure requirement must be satisfied by depositing a sample organism in an approved repository with unrestricted access. In *Hibberd*, a deposit of seed was mandated, and this practice will likely become the norm. Competitors therefore have access to the patented genetic material, and because research *with* (but not *on*) a patented product is considered fair use (Lesser 1986:10), the deposited seed (or tissue) may be used in the production of new plant types.

Use of the utility patent statutes will not, therefore, necessarily eliminate the problem of cosmetic breeding. Indeed, it may be easier to trivially alter a DNA sequence in patented genetic material to produce a "new" gene than it is to develop a "new" plant variety under PVPA. However, unlike the PVPA, the utility patent law does require that in addition to novelty, an invention must be non-obvious. Some "minimum distance" between like products will be required. But what degree of similarity will constitute infringement is now unclear and will need to be determined in the courts. In anticipation of such litigation, much attention is being given to development of techniques for the fingerprinting of genetic material. Scanning electron micrographs of leaf hair and pore patterns, two-dimensional gel electrophoresis, and restriction-fragment length polymorphisms (RFLPs in the vernacular) are being examined for their utility in defending property rights in plant genetic material (Orton 1985; USDA 1986).

But the real importance of *Hibberd* is not the new legal context it provides for corporate competition. The most profound impact of the Board of Patent Appeals decision will be felt at another level of competition entirely: that between farmers and the seed industry. With *Hibberd*, a juridical framework is now in place that may allow the seed industry to realize one of its longest-held and most cherished goals: to bring all farmers in all crops into the seed market every year.

Unlike the PVPA, the utility patent statute does not include a farmer-exclusion clause. Farmers are no more exempt from the legal obligation to respect the property rights of developers of patented seed than are their corporate competitors. Legal precedent is that the purchase of a patented product brings with it the right to *use* the product, but not the right to *make* it. Applied to seed, this principle implies that a farmer purchasing patented seed would have the right to use (to grow) the seed, but *not* the right to make the seed (to save and replant). Though it will certainly be tested in court, the farmer who saves and replants seed of a patented plant variety will be in violation of the law (Neagley et al. 1984; Lesser 1986).

The seed industry is keenly aware of the opportunities that are being presented by changing legal structures. In an article in *Seedsmen's Digest* published just prior to the *Hibberd* decision, the secretary general of the International Seed Trade Federation, Dr. Hans Leenders, wrote,

> [I]n most countries the farmer may do on his own farm whatever he chooses, i.e., among other things produce young plants for a commercial production of the produce without the authorization of the breeder, the seed industry will have to fight hard for a better kind of protection... Plant reproductive material has increasingly become a technical product in which much money has been invested. Even though it has been a tradition in most countries that a farmer can save seed from his own crop, it is under the changing circumstances not equitable that farmers can use this seed and grow a *commercial* crop out of it without payment of a royalty... I expect, however, that the seed trade in many countries will increasingly be prepared to raise this issue, which is rather based on tradition with grandma's varieties than on common sense. [Leenders 1986:9]

With *Hibberd*, the seed industry has the legal means to bring the farmer to see what Leenders claims is, in any case, only common sense.

Enforcing seed patent rights among farmers will be no small task. Yet companies are not without resources or experience in the matter. They have already had sixteen years of practice enforcing the PVPA, and many of the tactics employed in detecting and proving violations of the PVPA should also prove useful in policing patent infringements. Cooperatives, very large operations, and farmers who advertise sale of seed to their neighbors can be monitored. ASTA consultant William Lesser (1986:13) comments, "With the widespread [seed] dealer network (many dealers are actually neighbor-farmers) supplying data on the identification of likely infringers seems a plausible task." A letter from a seed company attorney has generally been sufficient to stop suspected violations of the PVPA by farmers, and the same would probably be true of unauthorized propagation of patented seed. Selective prosecutions would also have a useful demonstration effect. If effective methods of genetic fingerprinting are developed, proof of violations will be quite simple. Enforcement of property rights in patented seed is a practical proposition and even at modest levels may provide substantial returns to seed firms by reducing plant-back of saved seed.

The elimination of plant back will be gradual. In enforcing their patent rights, seed companies will have to avoid unnecessarily antagonizing the farmer, who is, paradoxically, both competitor and customer. Replacement of non-patented cultivars by patented cultivars will be facilitated to the extent that public sector breeding continues to move away from production of finished varieties. The demise of public varietal breeding might eventually solve the problem of identifying patent infringement. If only private breeders are developing commercial lines, and if those lines are patented, then at some point all seed will be patented, and anyone saving and replanting seed will by definition be in violation of the law. Without competition from plant-back seed or publicly developed varieties, the increasingly oligopolistic seed sector might become very profitable indeed.

Unhappily, public breeders have had little to say about seed patenting. They have been preoccupied less with the commodification of the end product of research – the seed – than with an emerging commodification of the research process itself – with the very techniques of plant improvement. It is *process* patents rather than *product* patents that are now causing concern among public plant scientists.

Patents on breeding techniques have long been anathema to public breeders, because their purpose is to exclude others from access to the means to perform research. What is called into question is not ownership of the product but the right to "do science." It was reaction to the violation of this norm that was behind the unprecedented measures taken by the American Society of Agronomy in publicly censuring Donald F. Jones and Paul C. Mangelsdorf when, in 1956, they applied for a patent on Jones' fertility-restorer system that made double-cross hybrid production without detasseling possible (see Chapter 5). More recently, with the advent of new biotechnological techniques and the proprietary concerns introduced by their use by industry, there has been a rapid growth in the number of plant breeding patent applications that include process claims (Table 9.2).

One of the more controversial of these is United States patent no. 4,326,358, issued in 1982 and assigned by the inventors to Agrigenetics Research Associates Limited. The patent's simple descriptive title – "Hybrids" - gave little indication of the importance of the claims made in the body of the application. Yet the descriptive title was most appropriate in its breadth, for patent 4,326,358 potentially established property rights over an extremely broad area. The patent made fourteen separate claims, but in its essentials it gave Agrigenetics Research Associates rights to the process of using clonally propagated parental lines to develop new hybrid plant varieties.

If it could hold up against challenge in court, patent 4,326,358 would give Agrigenetics Research Associates an enormous degree of control over an important cutting edge of the field of plant breeding. Anyone using clonally propagated breeding lines would have to come to Agrigenetics for licensing of the technology. Ownership of so fundamental a patent could confer a considerable commercial advantage on the company. Trumpeting what it called a "major technological break-through," Agrigenetics, in press releases reported in the June 1982 *Seed World* and the July 1982 *Seedsmen's Digest*, announced that it would commence licensing negotiations with interested parties immediately and would "aggressively defend its patent position, energetically use discovery procedures, and actively pursue any entity infringing on its proprietary patent rights." Both the seed industry and the plant science community were put on notice that in the brave new world of biology, science is the handmaiden of business, and that at the core of business are property rights.

A response was not long in coming, although it was not the response Agrigenetics wanted to hear. In a letter published in the August 26, 1982, issue of *Nature*, N. L. Innes (1982), chairman of the British Association of Plant Breeders, accused Agrigenetics of arrogating to itself "rights over techniques that have been part of the stock-in-trade of plant breeders for some considerable time and have already been used commercially." Innes went on to assert that the patented principles and techniques were known and practiced before the patent was filed, that the particular combination of techniques detailed in the patent had been recognized, and that the entire approach was obvious to anyone with ordinary skill in plant breeding. He concluded by bemoaning what he regards as a bald-faced attempt to "restrict the use of techniques and combinations of techniques that are common currency among plant breeders worldwide." Essentially, Innes declares that the emperor has no clothes, that Agrigenetics' break-through was established practice, and that, by implication, the patent should never have been issued. In a very unusual action at its meeting of October 1982, the board of the European Association for Research on Plant Breeding reviewed the issue and concluded that "none of the techniques in the patent are new; some of them have actually been applied for ages... From the scientific point of view it would be greatly deplored if certain breeding techniques should be patented and consequently their application not be allowed in the production of new varieties" (Lamberts and Sneep 1984:2).

In the United States, sentiment among publicly and privately employed plant breeders seems consistent. There is virtually unanimous agreement in the plant science community that the Agrigenetics patent is, as an Asgrow Seed Company executive put it, "not worth the paper it is printed on."[15] It appears that no one but Agrigenetics and the Patent Office believes that the patent is legitimate. Plant breeders are able to cite a whole series of instances of prior art. Agrigenetics, whatever it might say publicly, has implicitly confirmed the questionable nature of its patent by its failure to pursue gross and publicly recognized infringements by both public and private breeders.

In fact, the patent appears to have been sought even though executives of Agrigenetics realized its tenuous legitimacy. As yet, there are few actual products that have resulted from application of the new biotechnologies to plant breeding. In order to raise the financing needed to perform research, plant biotechnology NBFs such as Agrigenetics have had to depend as much (and often more) upon outside funding as on product sales. There is widespread opinion among both public and private breeders that the patent was pursued in large measure for its publicity and fund-raising value rather than for its actual scientific legitimacy or commercial potential. In the absence of products, patents become the visible sign of a biotechnology company's vitality. It is true that Agrigenetics was seeking secondary financing and that it issued a prospectus in December 1983 in preparation for an issue of common stock (the public offering was subsequently aborted, and Agrigenetics was purchased by Lubrizol in late 1984). In its prospectus, the company emphasized that its newly patented breeding "technology will permit additional rapid product introductions in the future." Agrigenetics was trading on the appearance of patent 4,326,358, though not its substance. As Levins and Lewontin (1985:201-2) write, "Once the scientific report becomes a commodity, it is also subject to two other features of the business world: the stagecoach can be hijacked and the beer can be watered, that is, scientific commodities may be stolen or debased."

There is also the problem of the effect of patent 4,326,358 on information exchange and the willingness of researchers to pursue particular lines of inquiry. Prominent scientists at well-known institutions indicate that they are continuing to use the patented technique. At the margins, it may be that the threat of litigation has been enough to constrain research in small seed companies or small universities. The case has contributed to the trend toward narrowing the types of information scientists are willing to exchange and to a growing awareness of the commercial implications of the work accomplished in universities and other publicly funded institutions. One breeder in a private company comments that as a result of the patent, "I suspect the public side will be a lot more careful, it seems already obvious they are not so willing to divulge their information. They say 'wait until it's published.' "

Restrictions on the flow of information are becoming increasingly evident. At a seminar on vectors for crop improvement, Dr. Johannes de Wet of the University of Illinois Genetic Engineering Center described his use of a new pollen vector in corn. According to *Biotechnology Newswatch* (1984b:3, emphasis added), "De Wet also hinted, in response to a question from the floor, that he has successfully transformed tomatoes and sorghum with this technique. *But upon advice of patent counsel, declined to elaborate.*"[16] De Wet's process is now patented (see Table 9.2) and licensed exclusively to Agrogene Plant Science (Klausner 1984:775). Such reticence applies to the exchange of germplasm as well as to conduct in meetings. In 1982, Winston Brill (1982:8), then of the University of Wisconsin, now of Agracetus, told a congressional committee that publicly funded researchers, "realizing the potential for financial gain to themselves or their institutions distribute materials only after they have decided that there is no value or after patents have been filed and/or agreements made." What *Bio/Technology* Contributing Editor Bernard Dixon (1986:1046) has called "creeping concealment" is becoming ubiquitous.

Biotechnology and plant genetic resources

The emergence of the new biotechnologies adds another level of complexity to the cluster of issues associated with international patterns of control over and access to plant genetic resources. Germplasm is the fundamental resource of the new genetic technologies. The commercial excitement and patent controversies attending the biorevolution have focused attention on the very questions that the FAO found to be most contentious: value and property rights in germplasm.

Techniques such as protoplast fusion and rDNA transfer offer means of circumventing the barriers of sexual compatibility that have long established the natural parameters within which plant breeders have had to work. In theory, the gene pool for any one organism has suddenly expanded beyond the boundaries of its species and now encompasses *all* living organisms. According to Calgene's Raymond Valentine, the plant biotechnologist's motto is "any gene from any organism into plants" (quoted in California Agriculture Lands Project 1982:13). Germplasm that has been of little or no economic importance may now be useful. As Agracetus' Winston Brill observes: "We are now entering an age in which genetic wealth, especially in tropical areas such as rainforests, until now a relatively inaccessible trust fund, is becoming a currency with high immediate value" (quoted in Myers 1983:218).

There is hope among environmentalists and others concerned with the processes of species extinction and genetic erosion that biotechnology's capacity to render any organism potentially useful – and therefore potentially valuable – will encourage the conservation of genetic resources (Murray 1982; Myers 1983; Plucknett et al. 1983; Wolf 1985; Lesser 1986). Following the same logic, FAO delegates from Third World nations might also find a rationale supporting their arguments that existing patterns of plant germplasm exchange are inequitable. And might the apparent enhancement in the value of genetic resources make the advanced capitalist nations more responsive to such concerns?

But at least one of the techniques of biotechnology may actually tend to devalue genetic resources. As we have seen, plants that have been regenerated through tissue culture often exhibit somaclonal variation. Genetic variability can be generated *in vitro*. The vice president of the NBF Calgene has observed that such methods "release us from our dependence on nature for providing new genetic material" (Al Adamson, quoted in California Agricultural Lands Project 1982:13). He might well have added that they could also ultimately release the advanced industrial nations from their dependence on Third World genes in plant improvement.

However, there is reason to believe that such a view is seriously flawed. It is true that superior genotypes are recoverable from tissue cultures. But *in vitro* selection techniques are still in their infancy and even when more fully developed will be subject to significant limitations (Carlson et al. 1984). For example, there is no way to screen cells in a petri dish for traits that are not expressed at the cellular level. Relatively simple characters controlled by single genes (e.g., disease resistance, herbicide tolerance) may be successfully selected *in vitro*, but more genetically and physiologically complex traits expressed in whole plants (e.g., stress tolerance, yield, maturity, standability, plant architecture) will more efficiently be engineered using other methods (Ammirato et al. 1984; Carlson et al. 1984; Fraley et al. 1986).

In fact, biotechnology should increase rather than reduce the need for and utility of exotic plant genetic resources. The variability that breeders need already exists in nature; it simply has not been easily accessible. A major bottleneck has been our inability to efficiently characterize even the germplasm deposited in gene banks, much less uncollected exotics. New techniques of gene mapping will make it much easier to identify useful sequences of DNA. Says Thomas Lovejoy of the World Wildlife Fund, "Natural species are the library from which genetic engineers can work. Genetic engineers don't make new genes; they rearrange existing genes" (quoted in Eckholm 1986:20).

And there will be a continuing need for access to plant genetic resources in source areas of diversity. Biotechnology may provide the breeder with a greater degree of variability with which to work, but will also provide the means to engineer a higher degree of genetic uniformity in the field. Even now the life of a commercial variety is only three to nine years (Plucknett and Smith 1986:42), and the increasing instability of yield in American cereal production has been associated with genetic uniformity of cultivars (Hazell 1984). Breeders will need to use a broader variety of genetic material if they are to keep up with the accelerating varietal relay race and address the potential problems associated with the narrowing genetic base of commercial cultivars. Also, because most of the crops now grown in the United States originated elsewhere, they have been isolated from many of the diseases and pests indigenous to their areas of origin. As those pests and pathogens find their way to the United States, we will need to look to the source areas of diversity for needed resistance (Harlan 1980).[17]

Even given the materials already collected and stored in the gene banks of the West, the prospect of a reduction or stoppage of the flow of plant germplasm from the periphery is a sobering prospect for public and private scientists alike. Dr. J. P. Kendrick (quoted in Mooney 1979:3) of the University of California warns that

> If we had to rely only on the genetic resources now available in the United States for the genes and gene recombinants needed to minimize genetic vulnerability of all crops into the future, we would soon experience losses equal to or greater than those caused by southern corn leaf blight several years ago – at a rapidly accelerating rate across the entire crop spectrum.

And Dr. Donald Duvick, Pioneer Hi-Bred's Director of Research, says, "I can't conceive of plant breeding without an international network [of seed exchange]" (quoted in Paul 1984). The developed capitalist nations have a continuing need for Third World germplasm.

This is now being recognized in the wider agricultural community and, significantly, in Washington. In December 1985, Representative James Weaver proposed to the House Agriculture Committee that $250 million be appropriated for a comprehensive strengthening of the National Plant Germplasm System, including the expansion of collection activities. That same month, genetic conservation bills were introduced in both the House and the Senate that directed the United States Agency for International Development (AID) to earmark $10 million per year for activities dealing with genetic conservation and tropical deforestation. AID prepared a Biological Diversity Action Plan designed to incorporate genetic resource issues into all its development programs. In 1986, the Council on Agricultural Science and Technology issued a report on "Plant Germplasm Preservation and Utilization in U.S. Agriculture," and recommended further plant exploration and the conversion of exotic germplasm as priorities. More recently, the congressional Office of Technology Assessment completed a study of biological diversity, and the Board on Agriculture of the National Academy of Sciences is now embarking on a two-year, $2-million blue-ribbon study of genetic resource issues.[18] A statement by the newly appointed head of the National Plant Germplasm System, Dr. Henry Shands, provides some insight into the objective of all this activity:

> The U.S., moreso than most nations of the world, is deficient in having centers of origin of important food and fiber crops. In fact, the U.S. has no major economic crop native to its lands. *One goal of this nation is to become self-sufficient in germplasm. Attainment of this goal presents a major task in the acquisition and characterization of germplasm.* [Shands 1986, emphasis added]

The United Sates continues to require plant genetic resources from the Third World, and it is preparing to strengthen its capacity to collect those resources. The question for the developing nations is, On what terms will access to genetic resources be granted?

If biotechnology does not release the advanced industrial nations from their genetic dependence on the South, neither does the socioeconomic manner of its development mitigate the contradictions that have given rise to the FAO Undertaking on Plant Genetic Resources. In the advanced capitalist nations, the range of phenomena and objects that can be considered private property is being rapidly extended (viz., *Chakrabarty* and *Hibberd* in the United States). Japan has followed the American lead and granted a patent on a plant, and the Federal Republic of Germany's Patent Office is now considering a similar case (*Genetic Engineering and Biotechnology Monitor* 1985:39-40). Efforts to compel Third World nations to honor such patents cannot but reinforce perceptions of the asymmetries in the treatment of commercial cultivars as private property and other types of plant germplasm as common heritage.[19]

The internationalization of the seed industry and its global merger with the agrichemicals sector continue apace. A truly transnational structure is emerging as joint ventures are established between companies with locations on different continents. For example, Celanese (USA) has concluded a cooperative agreement with Lafarge Coppee (France) for research in vegetables, the NBF Phyto-Dynamics (USA) has initiated joint research with British Petroleum (UK), and SeedTec is undertaking sunflower investigations with Rhone-Poulenc (France). Bill Ward (1986:110), publisher of *Seedsmen's Digest*, explains the rationale behind such unions:

> Joining with a foreign seed firm to establish an axis of strength that will develop multiple advantages of climate, markets, and expertise is a growing consideration. This far surpasses any boundary lines previously experienced. Management that occurs on a worldwide level will better withstand uncontrolled negative environments or political barbs.

The seed trade is achieving a truly global reach.

This global reach is being constructed so as to encompass Third World markets. The NBF Crop Genetics International is already shipping sugar cane seedlings produced by tissue culture. DNA Plant Technology and United Fruit have established a joint venture to clone elite oil palms for sale in Africa and Southeast Asia. Cloned oil palms are already on Unilever test plots in Malaysia. Rohm and Haas Company and Japan's Sumitomo Chemical Company have announced formation of a joint research program for the development of hybrid rice seed for Japanese and Southeast Asian farmers. And Plant Genetics has licensed its "GEL-COAT" synthetic seed technology to Japan's Kirin Brewery Company for product development in Asia and Oceania.

Third World nations are asked to supply plant genetic resources – the raw material of the new genetic technologies – as common heritage. In return, they are offered the opportunity to purchase the products of biotechnology or to supply the labor for the production of those products. Third World nations need to consider how the right to buy improved, genetically engineered (perhaps hybrid, perhaps patented) seed or synthetic seed (with its chemical additives) balances against the gift of their genetic heritage to the advanced industrial nations. This is not the entire equation. Development of industrial plant tissue culture techniques means that alienation of plant germplasm may result in direct damage to the donor nation's economy.

It is one of the ironies of underdevelopment that the advanced capitalist nations are now exporting wheat, corn, and rice to the very nations in which those crops originated. Wherever climatic conditions have been sufficiently similar to permit transfer, the developed nations of the North have adopted the crops of what is now the Third World. But though they could be moved laterally across the globe, many tropical crops could not be grown in the United States or Europe. Third World nations have thus retained a unique capacity to produce certain tropical plant products. The central position of such crops as sugar, coffee, and cocoa in the economies of many less developed countries (LDCs) is well known. But the Third World also supplies the developed nations with a wide variety of plant-derived flavoring agents, scents, spices, dyes, chemicals, and drugs. Though prices for these primary commodities are notoriously unstable, the export of plant products is a vital component of the economies of most LDCs. The new biotechnologies threaten to undermine and perhaps to eliminate these markets.

We have seen that plant tissue culture can be used to regenerate a whole plant from a single cell. Plant cells can also be treated in such a way that they do not differentiate into the complex structures characteristic of a whole organism, but are organized to produce specialized, economically useful by-products called secondary metabolites (Japanese Association for Plant Tissue Culture 1982; OTA 1984:179). Industrial plant tissue culture offers the developed nations the possibility of replacing imports of tropical plant products with domestic production. It permits effective appropriation of a new class of Third World germplasm: that of plants that will not grow in the temperate climes of the Northern Hemisphere but whose cells can be grown in stainless-steel fermentation vats anywhere.

Current techniques of industrial plant tissue culture are not as sophisticated as those available for microbial fermentation. Moreover, few plant cell lines have undergone rigorous selection for productivity in culture, the genetic mechanisms that control production of secondary metabolites are not well understood, and cultured plant cells tend to be genetically unstable (Tudge 1984:25). At present, only plant products worth $250-500 per kil-

Table 9.3. *Industrial plant tissue-culture research*

Plant	Secondary product	Currently produced in	Research by
Cacao	Cocoa butter (food)	West Africa, Latin America	Hershey (USA), Nestle (Swiss), Cornell Univ.
Chili pepper	Capsaicin (flavor)	Latin America	Univ. of Edinburgh/ Prutech (UK)
Cinchona	Quinine (drug)	Latin America, Asia	Plant Sciences Ltd. (UK)
Coffea	Coffee (food)	Asia, Africa, Latin America	Bio-Foods/Fluor (USA)
Jasminum	Jasmine (scent)	Asia	Fermenich (Swiss)/ DNA Plant Tech. (USA)
Lithospermum	Shikonin (dye)	East Asia	Mitsui (Japan)
Pyrethrum	Pyrethrin (insecticide)	East Africa Latin America	Univ. of Minnesota, Biotec (Belgium)
Saffron	Crocin (dye), saffronin (flavor)	Asia	Univ. of Edinburgh/ Prutech (UK)
Sapota	Chicle (gum)	Central America	Lotte (Japan)
Thaumatococcus	Thaumatin (sweetener)	West Africa	Beatrice/Ingene (USA), Monsanto/ DNA Plant Tech. (USA), Tate & Lyle (UK)
Vanilla	Vanillin (flavor)	Africa, Asia, Caribbean	Univ. of Delaware/ David Michael & Co. (USA)

Sources: Author's compilation from numerous sources.

ogram could economically be produced by tissue culture. However, the success enjoyed by Japan's Mitsui Petrochemicals Industries with the plant dye shikonin has demonstrated the practical efficacy of industrial plant tissue culture. More efficient techniques such as fluidized-bed and immobilized-cell systems are being rapidly developed, and there seems no reason to believe that production costs will not be substantially reduced in the near future.

Table 9.3 reflects the substantial interest that private companies have already shown in the possibilities of industrial tissue culture. Note that the range of secondary products for which production processes are being developed is not limited to high-value, low-volume products like drugs and scents. Research on industrial production of bulk commodity crops such as coffee and cocoa is also under way. And plant germplasm is just as much the raw material of the industrial tissue culturist as it is of the plant genetic engineer or the plant breeder. Has Ghana supplied Hershey with the cocoa germplasm that, multiplied in a fermentation vat, will ultimately displace Ghana's cocoa industry?

In conjunction with the principle of common heritage, the new biotechnologies may permit the advanced capitalist nations to displace not just existing LDC production but also *potential* areas of production. Global biotic diversity is increasingly being recognized as encompassing, as the *New Scientist* (1982:158) phrased it, "an Eldorado of plants that cure, feed and fuel." The United States' National Cancer Institute intends to make full use of this natural medicine chest. Over the next five years it will collect over 20,000 plants from tropical rainforests and assess them for their therapeutic potential (Klausner 1986).[20] Insofar as it may be possible to produce any new drugs isolated from the collected plant material microbially or through tissue culture, the nation donating the material will have lost a potential market for its plant products. The free donation of genetic resources as common heritage can hardly be termed costless for the LDCs.

There seems little reason to anticipate a rapprochement between the Third World and the advanced industrial nations regarding the issues surrounding use and exchange of plant germplasm. The advent of biotechnology may well exacerbate rather than mitigate the emerging conflict. As the representative of Trinidad and Tobago put it during the debate at FAO in 1983:

> We have entered into a new era of technology, with gene splicing and bio-engineering and so on, and, it would seem, we are on the threshold of a new kind of power play. [FAO 1983c:6]

Conclusion: Logical extensions

Writing of the *Chakrabarty* decision, Bruce Collins of the American Patent Law Association observed that it would have

> been news had it gone the other way. Then it would have been a complete subversion of the intent of the patent law, *whereas this is just the logical extension of it.* [quoted in Sylvester and Klotz l983:118, emphasis added]

The emerging social impacts of the new genetic technologies in the plant sector are, substantially, logical extensions of historically established processes. The logic is that of the capitalist mode of production: the concentration and centralization of capital in the seed industry, the commodification of the seed, the decline of the petty commodity producer, the struggle over the control of the state apparatus, and the continued appropriation of the plant genetic resources of the Third World.

Dr. James Delouche, of the Mississippi Seed Technology Laboratory, makes essentially the same point, though from a conventional perspective. Writing of what he terms the coming "sea change" in plant production, he asserts:

> Of this I'm sure: *agriculture in the future will be much more "scientific" than it is now*; there will be fewer farmers (or production units) and their inputs will be supplied by a relatively small number of high-tech companies; varieties of major crops, for example, will be almost wholly developed by private companies. Thus, there will be a great shrinkage in the number of seed companies, most of which are presently producing seed of publicly developed varieties. Seed certification and seed control procedures will have to adjust to the new emerging situation, and their importance will inevitably diminish. The spillover from these and other changes in agriculture will also affect the roles of the agricultural extension services and the agricultural experiment stations. [Delouche 1983:8]

To the extent that public scientists – like Delouche – hold such a view of the future, it is likely to become a self-fulfilling prophecy.

10

Conclusion

Let me give you my "take-home" message straight away. Clausewitz is supposed to have said that, "War is too important to be left to the generals," and the message relevant to this discussion is "Research priorities are too important to be left in the hands of research directors and other kinds of management types" ... Priorities, to some very large extent, ought to be generated from within rather than from the top down.

J. Eugene Fox,
ARCO Plant Cell Research Institute
(Qualset et al. 1983)

As plant breeding, *per se*, is a wholly benign technology, any enhancement of it must be welcomed as being in the public good, no matter who does it; this statement is, I think, true, though contradictory of silly arguments heard in recent years to the effect that any commercial involvement in plant breeding is in some sense wicked, destructive of genetic resources, and socially discriminatory.

Norman W. Simmonds,
Edinburgh School of Agriculture (1983)

I dwell in Possibility –
A fairer House than Prose –
More numerous of Windows –
Superior – for Doors –

Emily Dickinson

Let me give my "take-home" message straight away as well. Atlantic Richfield's J. Eugene Fox has paraphrased Clausewitz by saying that "Research priorities are too important to be left in the hands of research directors and other kinds of management types" (quoted in Qualset et al. 1983:475). Scientists, he asserts, should be the final arbiters of what sort of research is to be pursued, and how. My message is that *research priorities – in plant breeding and improvement no less than in any other field – are too important to be left to research directors, management types, or scientists.* The public has a right to demand not just accountability from the scientific community but also a voice in determining the goals and purposes to which science and technology are directed.

Simmonds' comments point to the fallacy inherent in the expectation that the denizens of the "Republic of Science" are capable of objective assessment and regulation of their own activities. The scientistic perspective that perceives plant breeding as "wholly benign" can hardly be expected to see the need for any limits to the "freedom" of the scientific enterprise. I have tried to show in this book that in addition to it undoubted positive achievements, plant breeding has a darker side that has been little acknowledged. I have also tried to show that, contrary to Simmonds again, it makes a great deal of difference who does plant breeding. Whatever objectives private plant breeders may have, their overriding goal is to produce a new variety that can be sold at a profit. That the new variety may not be coterminous with a social optimum is not the product of "wickedness," but is a consequence of the imperative of profitability in a capitalist system.

In contrast, state-supported agricultural science – the land-grant complex – is not dependent on profitability for its reproduction. Public breeding programs have historically been relatively autonomous from, and even directly competitive with, private enterprise. By releasing finished plant varieties, public breeders have exerted a substantial amount of direct influence over market structure and product character. It is true that, since 1930, this relative autonomy has been increasingly circumscribed as capital has sought, often successfully, to subordinate public science to its own purposes. But there yet remains a residual of autonomy, an institutional space that may be turned to the generation of plant improvement strategies that are not constructed around the overriding criterion of profit. We are now poised on the threshold of what may well be a new era of productivity growth in agriculture. Biotechnology holds both promise and perils. It is crucially important that we preserve, and if possible expand, our capacity to generate alternatives as we explore the applications of these new technologies.

So the second half of my take-home message is this: *We must not allow our options to be foreclosed by ceding to capital the exclusive power to determine how biotechnology is developed and deployed.* We must dwell in the full range of technical possibility and not be limited to corporate prose. The land-grant complex is an established institutional vehicle through which at least a modicum of public control can be exerted over technological innovation. Not only can options be generated but, through the release of finished plant varieties, these options can be introduced into the marketplace. Public plant breeding programs must be strengthened and broadened. They should move quickly to incorporate the new biotechnologies. But in doing so, they must not abandon applied breeding and the direct link to the market that is the source of their unique potential for the realization of broadly public, rather than narrowly private, purpose.

Bridges to the island empire

James O'Connor (1975:51) has observed that what is unique about American capitalism is "the difficulty that it had in building firm foundations, not the ease with which it took hold and developed after these foundations were constructed." This was particularly true for the agricultural sector, which presents a wide variety of barriers to the penetration of capital. In 1974, Jean Mayer and Andre Mayer (1974:87) wrote that "Intellectually and institutionally, agriculture has been and remains an island – a vast, wealthy, powerful island, an island empire if you will, but an island nevertheless." Though Mayer and Mayer clearly did not have my theoretical perspective in mind when they coined their metaphor, I think it serves nicely as an image of the historic separation of the agricultural sector from the capitalist mainland. This book has really been about the manner in which capitalism's engineers have tried to bridge those straits.

The difficulty of this engineering feat is nowhere clearer than with regard to plant breeding and seed production. The development of a capitalist seed industry in the United States has historically faced two sets of obstacles, one biological and the other institutional. First, the very reproducibility of the seed tended to undermine the prospects for its commodification. Farmers were – and for many crop species still are – the commercial seed company's principal competitors. In the face of constraints on the willingness of private enterprise to invest in plant improvement, the state assumed responsibility for this crucial activity. Out of early global germplasm collection activities ultimately grew the enormous edifice of the federal Agricultural Research Service and the experiment station and land-grant complex. The state itself then became an institutional obstacle to the penetration of plant breeding by capital.

But since the 1930s, capital has increasingly been able to build bridges to the island empire. Its enhanced engineering capabilities are closely tied to the rapid, if somewhat belated, development of the biological sciences during the last 50 years. In this book I have tried to show how the historical development of scientific understanding and technical capacity in agricultural plant science has interacted with the social processes of commodification, the elaboration of an institutional division of labor, and genetic resource transfer in an increasingly interdependent world economy.

Commodification: primitive accumulation and the propertied laborer

The application of science to the problem of commodifying the seed has been a most effective means of eroding the barriers constraining the penetration of plant improvement by capital. The hybridization of corn broke the unity of seed as grain and seed as means of production. In doing so, it opened the space for capital accumulation that private industry needed in order to set down firm roots in plant breeding and commercial seed production. Hybridization has been assiduously pursued in other species and is now one of the principal objectives of those who would apply the new biotechnologies to the production of new commodities. I have argued that agricultural research of this sort can be regarded as part of the continuing process of primitive accumulation, for it functions to uncouple agricultural producers from the autonomous reconstitution of their own means of production. This is a contemporary instance of a broader process the outlines of which Marx was able to discern in the mid-nineteenth century: "Agriculture no longer finds the natural conditions of its own production within itself, naturally arisen, spontaneous, and ready to hand, but these exist as a separate industry from it" (Marx 1973:527).

As we have seen, it was neither chance nor an immanent and ineluctable technical logic that produced the development of hybrid corn. In the 1920s there were several possible paths to corn improvement. At least one of these, population improvement, may well have been as productive as hybridization. That hybridization was the route that was pursued was determined not by strictly scientific considerations but by the provision of funding incentives and the manner in which political power was wielded within the Department of Agriculture. Similar patterns are evident in regard to exploration of the potentials of the new biotechnologies.

Indeed, what is striking is the extent to which scientific objectives and outcomes in agricultural plant science have been and are now being shaped by forces originating in the larger political economy. Advances in genetic knowledge in the 1920s and the 1970s clearly opened new historical possibilities. Yet these advances in the forces of production did not contain specific characters that unilaterally determined the direction of technical change. Rather, existing relations of production molded the manner in which the new technologies developed. Hybrid corn, rather than improved open-pollinated varieties, emerged in the 1930s. And today, tissue culture is being used for inbred development rather than the creation of a homozygous commercial cultivar whose seed can be saved and replanted. As Marx and Engels (1970:94) put it, "The conditions under which definite productive forces can be applied are the conditions of rule of a definite class of society." But these conditions are themselves in flux, not least in response to the potential of new technical possibilities. Hybrid corn galvanized extensive changes in social relations, and the new biotechnologies are now stimulating an even more comprehensive social transformation. The model of change that emerges from this analysis is fundamentally dialectical – the forces and relations of production are mutually conditioning.

The progressive elaboration of a legal framework in which plant germplasm has become increasingly appropriable as private property has constituted a second historical vector of commodification along which capital has penetrated plant breeding. The "social" solution to commodification of the seed represented by the Plant Patent Act, the Plant Variety Protection Act, and the *Chakrabarty* and *Hibberd* decisions complements the technical solutions provided by the achievements of science. The legal and technical capacities are now in place that will permit capital to realize the apotheosis of the seed as a commodity-form. It is now possible to use a patented breeding (or genetic engineering) process to produce a patentable plant variety. Embryos of the patented (and possibly hybrid) seed can, in another patented process, be encapsulated in a biodegradable polymer that may also contain patented agrichemicals or biological pesticides (which are also patentable, of course).

And in the new technical milieu represented by biotechnology, the completion of the commodification of the seed provides a vehicle for capital to gain an unprecedented degree of control over the shape of the crop production process as a whole. Dramatic though it may be, we must not become completely preoccupied with the chemical connection alone. Having gained access to the instructions coded in DNA at the molecular level, a "package" approach to cultivar development potentially encompasses any or all plant traits relevant to the production process. As crop varieties are increasingly "programmed," they will require sophisticated monitoring and management packages if their productive potential is to be realized. Genetically engineered seed should reinforce the trend to on-farm use of electronic data evaluation. Microcomputers and other information-processing equipment may well follow new plant varieties as closely as the mechanical corn picker followed hybrid seedcorn.

Up to the present, farmers have retained a substantial degree of control over the on-farm production process and over the allocation of their own labor. They have selected individual inputs and technologies "cafeteria style" (Agricultural Research Service 1983:38) and assembled them as they saw fit. But the "package approach" to input linkage and marketing that is facilitated by biotechnological advances accelerates the erosion of farmers' managerial prerogatives already begun by such practices as contract farming. Just as hybrid corn stripped the seed reproduction function away from the farmer, so the genetically engineered seed will strip away parts of the managerial function.

Dr. Klaus Saegebarth, Du Pont's Director of Agrichemicals Research, sees "the breadth of Du Pont's line of crop protection chemicals as literally representing a Du Pont Crop Management System" (*Farm Chemicals* 1982a:21). Monsanto's Howard Schneiderman (NOVA 1985) has a similar vision:

> The farmer would provide his labor and his land and Monsanto could provide him with the system which would be seeds, chemicals, and perhaps microorganisms.

As variety-specific "packages" become increasingly sophisticated and complex, farmers will in effect also be determining the shape of their own labor process at the same time they select the cultivar they intend to plant. Or they may not even select the variety themselves. For example, the contracts that food processors now establish with farmers frequently specify the variety to be grown, and, in an arrangement known as "bailment," the processor supplies the seed and retains title to both the seed and the crop that grows from it (Pfeffer 1982:77).

The new plant biotechnologies should also parallel the introduction of hybrid corn in their effects on the technological treadmill on which farmers run. Nearly all the agricultural applications of the new genetic techniques that are contemplated are output-enhancing. Biotechnology should set the stage for a particularly rapid series of cycles on the technological treadmill. Biotechnology can be expected to increase the reliance of farmers on purchased inputs even as it accelerates the process of differentiation among farms and facilitates further concentration of operations. The myth of the yeoman-farmer as a skilled craftsperson may persist, but the reality may be the "propertied laborer," given instructions by a computer that monitors the progress and needs of a crop grown from genetically programmed seed provided by a corporation to which the farmer is contractually bound and that already owns the crop in the field.

Division of labor: w(h)ither public varieties?

As private involvement in plant breeding has grown, seed companies have recognized public breeding activities as an institutional barrier to further capital accumulation. Evenson and Evenson (1983:215) have written of agricultural research that "Public systems thus have reason for concern at the prospect of more private activity and more competition. All public institutions have reason to fear competition because of the basic nature of bureaucracies." In fact, this book has shown that quite the opposite is true. It is private firms engaged in plant breeding that have regarded their *public* competitors with fear.

A principal theme of the development of plant breeding and the seed industry in the United States has been the persistent effort of private enterprise to circumscribe the activities of scientists in the experiment stations and land-grant universities. The division of labor between the public and private sectors has pivoted around the question of varietal release: Should public breeders develop lines appropriate for use as commercial cultivars, or should this be the exclusive prerogative of private industry? Historically, private enterprise has sought to move public breeders away from the commodity form and into more basic sorts of research that do not directly challenge it in the marketplace.

The emergence of biotechnology has stimulated the seed industry to redouble its efforts to emasculate the public breeder. As a result, the Agricultural Research Service has made it a matter of policy that varietal release should be discontinued, and a parallel tendency is apparent in the experiment stations and land-grant universities. Public acquiescence to the demands of capital is unfortunate and short-sighted. There are numerous good reasons why the development and release of finished plant varieties by public agencies should be continued.

One reason has to do with satisfaction of the scientist. Plant breeding is, after all, a labor process, and decoupling varietal release from other breeding activities is effectively a decomposition of that labor process. Public breeders understandably lack enthusiasm for carrying a task 90 percent of the way to completion and then turning the job over to a private breeder who will fine-tune the germplasm and receive credit for the new variety. There is enough alienated labor in this country already, and there ought to be a better justification than the "right" of private enterprise to make a profit before we add to these numbers.

Moreover, there are technical efficiencies associated with the maintenance of institutional continuity between germplasm enhancement and varietal release. The full utility of breeding lines cannot be explored unless they are combined with material with which they might constitute a finished variety. To limit germplasm enhancement to non-commercial combinations is to introduce inefficiencies into the breeding pipeline. Also, the problems that arise in applied work often direct attention to the need for work in more basic areas. To force research to conform to arbitrary distinctions is to limit the serendipitous and synergistic interactions of the various segments of the continuum of scientific investigation. Finally, one of the responsibilities of public plant breeders is the development of new breeding methodologies and the training of students. How can these duties be discharged if public scientists and their students are not engaged in a full range of plant improvement techniques? As one breeder put it, "If you are not producing corn hybrids [i.e., finished cultivars] how can you claim you are training corn breeders?"

Preservation of an active public role in varietal development in the United States is also important to the LDCs. The land-grant universities now train many students from the Third World and provide many other sorts of direct assistance. Should applied breeding be discontinued in public institutions, such expertise might be available to developing nations only from private companies, and at a price. Not just seed, but the knowledge of how to produce that seed, would become a commodity for those most in need of it but least able to afford it.

An additional argument in favor of public varietal release is its unique utility as a restraint on the activities of private industry. By producing finished varieties, the public sector exerts an important degree of *direct* regulatory control over capital. The crucial point is that a public institution is giving shape to the commodity-form and is directly articulated to the market. The best check on the potential for oligopolistic or anti-competitive pricing practices is the availability of quality, publicly developed varieties at reasonable prices. And without public varietal release, the majority of seed companies in the United States would quickly disappear, and the crop improvement associations, which now provide a mechanism for moving new varieties from public laboratories through farmer-growers to other farmers, would collapse. Cultivar development by public breeders has the effect of maintaining a more competitive industry structure than would otherwise be the case.

Most important, we need continued public presence in varietal release in order to maintain our options in plant improvement. A wide variety of useful characters has been incorporated into varieties by private breeders only after public breeders have proved their commercial utility in public cultivars. Without public varietal release programs there may simply be no new cultivars developed for certain minor species or for geographic areas with special needs but little market potential. Unfettered by the need to turn a profit, public breeders can also turn their attention to real alternatives to conventional varieties. For example, multi-lines – mixed varieties consisting of two or more cultivars incorporating different genes coding for resistance to the same disease – have been developed by public rather than private breeders. The lack of interest shown by private breeders in multi-lines has less to do with their efficacy than with the lack of uniformity that has made them difficult to protect under the PVPA.

The foregoing points take on added weight in the context of the emergence of the new plant biotechnologies. The tremendously powerful new genetic techniques offer an enormous range of practical applications. But which of the myriad technical possibilities will be selected? In an advertisement in a recent issue of *New Farm*, Northrup King (a seed company subsidiary of Sandoz) boasts an alfalfa variety that "grows like a weed." Genetic engineering has the potential to make this posturing literally true. Could we not identify the characteristics that permit, say, a dandelion to thrive and to persist so tenaciously? Could we not transfer the genes that code for those characteristics to an alfalfa plant – or a corn or soybean plant? If we could design crops literally to grow like weeds – to enhance their competitive advantage – might that not allow us to reduce or eliminate herbicide applications? Of course, this is not what is now being done. Rather, much effort is being put into moving bacterial genes into plants to allow them to tolerate herbicides so that more of these chemicals can be used. Both a literal and a metaphorical approach to making crops grow like weeds can be pursued with biotechnology. How is the choice made, and who makes it?

There is no question that biotechnology holds unprecedented promise for plant improvement. But we cannot rely upon private industry to explore the full range of technological possibilities. The research programs of private firms are necessarily limited by the inescapable parameters of profitability and the need to protect their own interests. How can private companies afford to investigate the utility of polygenically controlled traits in the dandelion when a simple, single gene transfer from a bacterium is available? Is industry likely to use the perenniality of *teosinte* to breed a perennial corn and thereby eliminate half of the American seed market? Will companies produce genetically diverse multi-lines or composites that cannot be effectively protected under patent law or breeders' rights legislation?

In the contemporary political economy, there is a clear tendency for private enterprise to determine not only what plant improvement research gets done but also who does it. The subordination of public research to private ends has the effect of foreclosing the options available to society as a whole. This is something we cannot allow to happen, especially as we face the brave new world of genetic engineering. In varietal development the public sector provides a unique counterbalance to the prospect of corporate "laisser innover" in plant breeding. To permit this measure of public control to be further eroded, or even eliminated, would be unconscionable.

Germplasm transfer: seeds and sovereignty[1]

Plant genetic resources are as much the raw material of the genetic engineer as they have been of the conventional plant breeder. And as the coming decades witness the elaboration of a new regime of production based on the manipulation of genetic information, germplasm will become an increasingly central resource. The nations of the Third World have legitimate grounds for demanding to have all types of plant germplasm treated similarly; certainly they are justified in their pursuit of common heritage. But given the contemporary structure of the world economy, the material consequences of the decommodification of all plant germplasm – if that were achievable – might actually work to the detriment of the nations of the South.

Application of the principle of common heritage to all plant germplasm would actually result in only minor alterations in existing patterns of plant genetic resource use and exchange. Paradoxically, when applied to one sector alone, common heritage may exacerbate rather than reduce inequality. This is the case with common heritage in plant genetic resources. Equality of access to plant genetic resources does not imply a necessary equality in the distribution of benefits accruing to that access. Given the genetic vulnerability of their high-performing agricultures, the advanced capitalist nations have a greater need to utilize plant genetic diversity than do the countries of the Third World. They also have a much greater financial and scientific capacity to do so. Formal institutionalization of common heritage might simply legitimate the differential abilities of North and South to appropriate, utilize, and benefit from plant genetic resources. Implementation of the principle of common heritage not only would allow the advanced capitalist nations to "mine" plant genetic resources with increasing intensity but also would preclude donor nations from realizing any return benefit – financial or in-kind – from the extraction of the genetic information contained within their borders. Given the substantial economic benefits accruing to the use of plant genetic materials in crop improvement, these forgone rents could be very large indeed.

True, under a regime of common heritage, the South would gain access to genetic materials that it previously has been unable to obtain. But is access to advanced breeding lines and other elite germplasm developed by commercial seed firms in the industrial North actually a benefit? Such lines are developed for use in industrialized, capital-intensive, energy-intensive agricultural production systems and will in many cases not be appropriate to the needs of the bulk of Third World producers of the South. For example, does the Sudan really want access to a Funk Seed Company sorghum line that, through recombinant DNA transfer, has had a bacterial gene added that provides resistance to a proprietary herbicide produced by Funk's corporate parent, the transnational agrichemical giant Ciba-Geigy Corporation? Access to the elite lines developed in the advanced capitalist nations might simply reinforce processes of social differentiation among peasant producers, facilitate the global elaboration of factor markets, accelerate environmental degradation, and deepen relations of technological dependence between North and South.

In any case, it is doubtful that the advanced capitalist nations would agree to extensions in the application of common heritage. The elite and breeders' lines of private-sector seed companies are now private property, and capital intends to do all it can to ensure that they remain so. There is no indication that the advanced capitalist nations are willing to begin dismantling the institutional arrangements that confer proprietary rights to genetic information. Indeed, current developments are bearing in the opposite direction. Given the uncertain unity of the South on the issue, and the tenacity with which capitalist interests are likely to defend the sanctity of private property, the prospects for actually achieving common heritage status for all types of plant germplasm are not bright.

On balance, then, given the contemporary geopolitical realities, pursuit of common heritage is not a strategy that is likely to enhance the possibilities for improving the lives of most of the world's people. The South's demands may be legitimate, but, given geopolitical realities, they are also misplaced. It makes little sense to permit access to a vast storehouse of plant genetic diversity in exchange for access to genetically narrow lines of great technological sophistication but dubious utility. The real problem for the South is not acquiring access to the elite lines of the North but establishing control over and realizing some benefit from the appropriation and utilization of its own resources. This requires a political strategy other than common heritage.

Third World nations have little to gain from quixotic pursuit of common heritage in plant genetic resources. But they have a great deal to gain through international acceptance of the principle that plant genetic resources constitute a form of national property. Establishment of this principle would provide the basis for an international framework through which Third World nations could be compensated for the appropriation and use of their plant genetic information. Codification of the status of plant genetic resources as national property has a clear basis in international law. Moreover, while capitalist interests are unalterably opposed to decommodifying their breeding lines, there are indications that they would be willing to provide compensation for use of plant genetic resources.[2] Even given the volume of materials already stored in their gene banks, the advanced capitalist nations still require fresh infusions of Third World germplasm. On the whole, the recognition of exchange-value in land races will prove more palatable for private companies than continued conflict and possible restrictions on the flow of what is, for them, an essential raw material.

A national-property initiative is by no means an ideal solution to the plant germplasm controversy. A principal problem with establishing a compensatory framework for plant genetic resources is that they are distributed unequally within the Third World as well as between North and South. With material recognition of the value embodied in plant germplasm, Third World nations might be tempted to charge each other, as well as the advanced industrial nations, for use of plant genetic information. The extent to which such problems can be avoided will depend much upon the manner in which

compensation mechanisms are structured. Bilateral agreements may tend to produce a market for plant genetic information. A market-oriented approach may isolate Third World nations and press them into roles as competing suppliers. Multilateral approaches that build upon the Third World's current willingness to confront the issue of plant genetic resources as a North-South issue are to be preferred. The FAO's International Undertaking on Plant Genetic Resources provides a useful institutional framework for preserving Third World unity on the matter.

The FAO might replace the principle of common heritage with that of national sovereignty, and then specify a legal and institutional structure for managing the exchange of and compensation for plant genetic materials. The Undertaking could mandate the creation of an "International Plant Gene Fund," managed by the FAO, that would support plant genetic conservation, construction of gene banks, and the training of plant breeders in the Third World. Money would be paid into this fund by advanced industrial nations, which would in turn have the right to access global collections of plant genetic materials collected and stored under the auspices of the FAO. The size of these payments could be determined by considering a number of factors, such as size of national seed industry, value of national agricultural production, and frequency and size of drafts upon the FAO's global "gene bank." While they provide no fully adequate model for managing the exchange of plant genetic resources, various existing international arrangements, from commodity agreements to the Law-of-the-Sea treaty, provide evidence that multilateral arrangements can be constructed.

Structuring some such compensatory framework will require much negotiation and compromise, both within the Third World and between the developing nations and the industrialized countries. But by pursuing a multilateral approach based on the principle of national sovereignty, it may be possible to recognize plant genetic resources as social rather than private property and to preserve the principle of free exchange within the developing world. Such an arrangement also has the advantage of placing the determination of compensation in a political rather than market setting. Avoidance of the market mechanism mitigates the centripetal forces that tend to separate competing suppliers of a good and avoids implicit acceptance of the necessity of markets for structuring access to resources.

Common heritage may be intuitively appealing, but, even if achieved, would not necessarily bring material advantage to Third World nations. Recognition of national sovereignty and the creation of compensatory mechanisms, on the other hand, would help redress a significant asymmetry in the economic relationship between the advanced capitalist nations and the less developed countries.

An epigram for an epilogue

Preservation of a vital and autonomous public presence in the full range of plant improvement activities will not be easy. Indeed, this book has shown that current trajectories are bearing in another direction. Capital is currently ascendant in plant improvement. As Agrigenetics' Robert Lawrence has pointed out, the balance of power is changing rapidly. Significantly, it is the generation of public breeders now close to retirement that is most committed to the development of finished varieties. Will the new generation of public breeders – or plant genetic engineers – acquiesce in their own emasculation? Or will agricultural science finally generate a cohort of internal critics, as physics and molecular biology have done before it? Plant scientists would do well to ponder these words of Henry Wallace (1961:31):

> Scientific understanding is our joy. Economic and political understanding is our duty.

11

Still the seed: plant biotechnology in the twenty-first century

Traditionally, we shrink from permitting small, authoritarian minorities to dictate our social agenda, including what kinds of research are permissible and which technologies and products should be available in the marketplace.

Henry I. Miller, Hoover Institution, and Gregory Conko,
Competitive Enterprise Institute (2001)

I view my job of public breeder as being public. Therefore, I am opposed to the commercialization of anything a public breeder creates. . . . I'm not necessarily against the science used to create GMO's, but there is potential for trainwrecks. What concerns me more are the practitioners, not the practice.

Stephen Jones, Washington State University wheat breeder,
quoted in Schubert 2001

The last – and perhaps the most important – of the risks that I want to examine is the effect of GM foods on sustainability. The single biggest concern in the developing world may be that millions of poor farmers will become dependent on a dozen or so multinationals for their future livelihoods.

Gordon Conway, president, the Rockefeller Foundation (1999)

Farmers' Rights was a fundamentally flawed argument that had been proposed by some who feared that to confront the robber who was already in the house might be to court conflict and disaster. A more discreet course, they thought, might be to "*negotiate*" terms which would permit him to proceed with his plunder but, at the same time, work out some sort of a "*just*" settlement that might placate his victims. In short, those defending plunder's victims armed themselves with the weapons of the enemy – the recognition of property rights, however legitimately or illegitimately that property had been acquired.

Erna Bennett (2002)

The reissue of *First the Seed* has given me the opportunity to add a chapter to the book. My history of plant breeding has, so far, taken its narrative and analytical structure from the interaction of scientific development with three themes of political economy: progressive commodification, the changing division of labor between public and private science, and asymmetries in patterns of seed commerce and exchange between North and South. These themes will still undergird this new chapter, but the focus now will be on the recent emergence of hopeful prospects for significant redirection of the long-established historical trajectories that the book has so far described.

Much has changed since the initial publication of *First the Seed.* "Biotechnology" is now a familiar term and transgenic crops – now commonly referred to as GM ("genetically modified") crops or GMOs (genetically modified organisms) – are planted on some 145 million acres worldwide, and their by-products are found in some 70 percent of processed foods sold in the U.S. The possible negative effects of GM corn on both Monarch butterflies and human consumers of tacos have been front-page news. "Terminator Technology" designed to genetically sterilize seed saved by farmers has come to symbolize corporate greed and lack of concern for the social and environmental impacts of genetic engineering. Protesters dressed as Monarchs have become a common sight not just at rallies against corporate "Gene Giants,"[1] but at demonstrations opposing the trade and property regimes being installed by the World Trade Organization (WTO). In 2003 the United States filed a formal grievance with the WTO, hoping to overturn European Union (EU) regulations that since 1998 have effectively barred the importation of most GM grain by EU member nations. Agricultural biotechnology has moved to center stage in the most salient struggles of contemporary political economy.

How, then, should I resume the story after a hiatus of sixteen eventful years? I have chosen to reopen analysis through the explication of four quotes which I think exemplify some of the principal continuities and some of the ways circumstances have changed between 1988 and 2004. From these quotes I will draw out key themes, to which I will return in this chapter as I review the events of the last sixteen years and as I assess the prospects for achieving broader social control over plant biotechnology in the coming century.

The first quote, from conservative analysts Henry Miller and Gregory Conko, is a clear and succinct statement of the central conclusion that I hoped readers would take away from the first edition of *First the Seed,* and that I still hope they will take away from this version of the book. I quite agree: a small, authoritarian minority ought not to dictate what kinds of research are permissible and which technologies and products should be available in the marketplace. Yet this is precisely what is happening in the plant biotechnology sector as corporate capital has, since 1988, significantly extended and consolidated its capacity to shape how the new genetic technologies are used, who uses them, and for what they are used.

However, those corporations are emphatically *not* the "small, authoritarian minority" that Miller and Conko had in mind when writing their essay. Those who actually concern Miller and Conko are what they term a "consortium of radical groups" that "proselytize for illogical and stultifying regulation or outright bans on product testing and commercialization" of GMOs. That Miller and Conko should feel the need to write their polemic at all is indicative of a critically important distinction between conditions in 1988 and those in 2004: corporate biotechnologists must now deal with substantial opposition from civil society. Sixteen years ago, concerns about the manner in which biotechnology was being developed were limited to a small number of academics and a few far-sighted advocacy groups. Today, opposition to GMOs and GMO food is robust, globally distributed, increasingly well organized, and grounded in a broad range of issues. Further, it is also occurring in the context of a diffuse but powerful social movement that is resisting what has come to be understood as the project of corporate "globalization." Indeed, because of the material and symbolic potency of its constitutive elements and associations (e.g., "Frankenfood," "Golden Rice," "Terminator Technology," "biopollution," "biopiracy," "Gene Giants") agricultural biotechnology has become a key locus in this larger struggle.

The second quote, from Stephen Jones, illustrates a critical emergent dimension of the social composition of the opposition to corporate plant biotechnology. Jones is a wheat breeder at Washington State University who has resisted heavy pressures to incorporate proprietary genes from Monsanto and BASF into public wheat varieties. He explicitly notes the "public" character of his job and unambiguously embraces a commitment to service that rejects the patenting of plants and the private appropriation of the labor of public breeders. While the degree of Jones's defiance is unusual, his stance is emblematic of a developing undercurrent of widespread dissatisfaction among public scientists who are finding their appointed place in the division of labor in plant biotechnology to be more confining than they had anticipated.

In the concluding lines of the previous chapter of this book, I asked in 1988, "Will the new generation of plant breeders – or plant genetic engineers – acquiesce in their own emasculation? Or will agricultural science finally generate a cohort of internal critics?" I am pleased to report that in 2004 it appears this latter possibility is developing, and discontent is now beginning to coalesce into organized, collective action directed at reestablishing the vitality of a public plant science committed to public service. This is an exceedingly hopeful turn of events. The forms of civil society opposition that so exercise Miller and Conko *could* themselves be sufficient to alter the trajectories on which the development of biotechnology is now proceeding. The likeliness of such a redirection would be materially improved were those social elements to be allied with a set of public scientists willing to speak to both technical and social concerns with the uniquely potent authority with which the scientific enterprise is now imbued.

However promising, prospects for such an alliance are now constrained by the rather different viewpoints held by scientists and activists concerning biotechnology itself. Jones's quote is especially instructive in this regard. He acknowledges the health and environmental risks – the "potential for train-wrecks" – associated with the use of genetic engineering. But he is careful to affirm that he is "not necessarily against the science used to create GMO's." He goes on to make a point of absolutely fundamental importance: "What concerns me more are the practitioners, not the practice." That is, he believes that the critical determinants of the impacts of biotechnology have to do more with who uses the tool than with the nature of the tool itself. Correctly, I think, Jones privileges *social* relations over *technical* relations.

In the third quote, Gordon Conway, president of the Rockefeller Foundation, suggests that further interrogation of the "practice" itself will also be useful. The passage is taken from a remarkable speech given in 1999 in which he outlined a variety of serious risks associated with the activities of multinationals involved in biotechnology research generally, and those of the Monsanto Corporation in particular. There are three points that I would like readers to take from his quote. First, chief among the risks Conway sees is not a technical problem but a social one: the deployment of biotechnology in such a way as to create a dependent relationship between farmers and a small set of companies. Corporate concentration in agricultural biotechnology has now proceeded so far that its menace is of concern even to individuals and organizations that, like the Rockefeller Foundation, are committed to maintenance of a liberal form of capitalism. The opposition to corporate control of the means of reproduction[2] now has a very broad base.

The second point is that, as Conway's explicit use of the term signifies, "sustainability" is now a rhetorical and ideological touchstone in contemporary constructions of what used to be called simply "development." There are, of course, many versions of "sustainability." Having achieved canonization as a kind of cultural shorthand for "the green and good," the term is deployed by all sorts of organizations and actors who want to access the word's discursive potency but whose goals and interests are not necessarily compatible. Still, the very ubiquity of the word and its use in this context by Conway is an indicator that environmental and social struggles have opened up spaces and possibilities for positions and initiatives – organic, participatory, democratic, agroecological, alternative – that heretofore were consigned to the margins but are now engaged as legitimate.

The third point is that, for Conway, biotechnology is an essential tool for realizing what he has called a "Doubly Green Revolution, one that is as successful in productivity terms as the old Green Revolution, yet is environmentally friendly and equitable" (1999). He suggests that when removed from its corporate integument, biotechnology might well be used in the service of socially and environmentally sustainable objectives. For many in the activist community "sustainable biotechnology" is an oxymoron. But what is referred to generically as "biotechnology" is actually a sizeable and expanding constellation of knowledges and technologies with different characteristics, capabilities, and uses. Biotechnology need not necessarily mean "genetic engineering" in the sense of interspecific gene transfer. Certainly, the techniques that fit under the rubric of "biotechnology" are uniquely powerful. Their development and deployment warrant special care. But blanket condemnation of the tools may, in fact, limit the capacity to understand and engage complex systems that will be the foundation of the truly agro*ecological* science that needs to be developed. The challenge is to create the social conditions in which that expression of technics is possible.

In the final quote, that longest-serving veteran of the "seed wars," Erna Bennet, reminds us that changing social relations is no simple task. "Farmers' Rights" was to have conferred on peasants and indigenous peoples a moral and material recognition of the utility and value of the labor they had expended, and continue to expend, in the development and preservation of crop genetic diversity. But to the extent that Farmers' Rights mimics "breeder's rights" in its linkage to the market, suggests Bennett, it is not the alternative it was intended to be but a shadow of the very system it had been intended to challenge. Rather than insulating farmers and indigenous peoples from a predatory and expansionist capitalism, might it actually institutionalize a framework for their continued – but legitimized – expropriation? And thus we come full circle to the fundamental questions with which this book began. Are market-based arrangements for the exchange of genetic resources inevitably, as Bennett and Marx agree, "the tools of the enemy"? To what degree must the social formation be changed so that plant biotechnology, or any form of institutionalized knowledge production, can be used justly and sustainably?

Sixteen years ago, I did a pretty good job of defining the trajectories on which the plant sciences were moving. Today, I find reason to believe that events are giving rise to a conjuncture that might allow some quite different directions to be taken. In the pages that follow, I will note the incredible concentration of economic and scientific power that has been accumulated by a few companies. I will show that this very concentration of power, and its often-egregious expression, has given rise to a vital and broad-based social opposition. I will suggest that there are reasonable prospects for recruiting public plant scientists to that movement, and good strategic reasons for doing so.

The race to cash in on the genetic code: 1988–2004

If there has been one constant over the past sixteen years, it has been the remarkable persistence with which a narrowing range of companies has striven mightily, and with increasing urgency, to somehow squeeze profits from a set of technologies that have always seemed pregnant with commercial promise but whose returns have always been a step away. The gestation period for profits in plant biotechnology turned out to be considerably longer than its corporate and academic proponents had anticipated during the 1980s. It was not until 1994 that Calgene's tomato, the inelegantly named 'Flavr Savr,' became the first transgenic plant to reach the market. It was rapidly withdrawn not as a result of popular protest against the technology, but because its "antisense" gene for delayed ripening didn't actually give the tomatoes any advantage in the field or in the market.

The next commercial release of GMO seed came in 1996 with regulatory approval of several corn varieties incorporating Bt toxins. These were followed shortly by "Roundup Ready" soy, cotton, and canola varieties containing bacterial DNA that provides resistance to the herbicide glyphosate (Monsanto's "Roundup"). The diffusion of those varieties has been exceedingly rapid. Between 1996 and 2003, the acreage planted to GM crops worldwide soared from less than 2 million acres to nearly 167 million acres (James 2003). Since high fructose corn syrup and soy, cotton, and canola oils are ingredients in a high proportion of processed foods, nearly every resident of the United States has consumed foodstuffs containing GMOs or their products. The rapid rate of adoption, unprecedented for an agricultural technology, has been touted as *prima facie* evidence of the desirability and benefits of GM crops, at least from the farmer's point of view. Similarly, the apparently unproblematic presence of GM components in many processed foods over several years in the United States has been taken as evidence of the safety of GM foods and their acceptability to American consumers.

Of course, statistics regarding the commercial deployment of GM crops can be made to tell different sorts of stories. Fully 98 percent of the acres planted to GM varieties in 2003 were located in only five countries: the United States (63 percent), Argentina (21 percent), Canada (6 percent), China (4 percent), and Brazil (4 percent). Moreover, only four crops accounted for virtually 100 percent of the area planted to GM varieties: soybeans (62 percent), corn (21 percent), cotton (12 percent), and canola (5 percent). And in all those plants on all those acres, only two GM traits were actually being expressed: herbicide tolerance (73 percent), Bt insecticidal action (18 percent), or a "stacked" combination of both herbicide tolerance and Bt action (8 percent) (James 2003). Thirty years after Cohen and Boyer produced the first biological chimera, farmers in a narrow range of countries plant a narrow range of GM seeds which incorporate only two agronomically relevant GM traits.

The rather constricted character of commercial agrobiotechnology reflects the narrowing range of companies that are now involved in that undertaking. What first drew me to the topic of this book was the wave of acquisitions of seed companies by agrichemical firms that swept the industry during the 1970s (see Table 6.3). That process has continued over the last sixteen years and has even accelerated as the roster of desirable firms has been reduced. With one $7.7 billion outlay in 1999, DuPont became the world's largest and leading seed company by purchasing Pioneer Hi-bred. Monsanto has been particularly active over the last decade with acquisitions of over a dozen companies including Asgrow, Holden's Foundation Seed, and DeKalb Genetics. Concentration has advanced to the point that even economists and federal regulators are concerned about the effects of growing monopoly power in the seed sector (Hayenga 1998; Oehmke and Wolf 2002). In 1999 the U.S. Justice Department went so far as to press Monsanto to withdraw its proposed merger with the leading cotton seed firm, Delta & Pine Land Co. Regulators' fears of anticompetitive behavior may well be justified. Recently, news reports have appeared suggesting that from 1995 to 1999 Monsanto executives sought, with varying success, to persuade other seed companies to fix the prices of GMO seeds at artificially high levels (Barboza 2004).

The "new biotechnology firms" (NBFs) that emerged in the 1980s have also been swept into the consolidation movement. With no product base and tenuous funding, the NBFs have been absorbed by multinationals. Agrigenetics and Phytogen were acquired by Dow, and Plant Genetic Systems has been bought by Aventis. Calgene and Agracetus are now units of Monsanto. Outside public agricultural research systems, there is little in the way of either agrobiotechnology research capability or commercial seed production capacity that is not now controlled by a small set of very large firms.

Given the historical trajectories described in previous chapters of this book, consolidation among seed companies and NBFs is not unexpected. What is surprising is the degree of cannibalism and recombination that has taken place since 1988 at the very top of the corporate biotech food chain. Compare the multinationals listed in Table 8.3 with those in Table 11.1. The companies that in the 1980s were leading actors at the intersection of biotechnology and seeds are, at first sight, apparently absent from the list of today's dominant "Gene Giants." Several firms listed in Table 8.3 have indeed reduced their seed or biotechnology holdings. Most, however, are still there but appear under new names as a result of having been involved in mergers and acquisitions. Syngenta was created in 2000 by a merger of the agrichemical and seeds units of AstraZeneca and Novartis, Novartis itself being the product of a 1996 merger between Ciba-Geigy and Sandoz. Aventis, acquired by Bayer in 2003, is the result of a 1999 merger involving Hoechst and Rhone Poulenc. In 2000 Monsanto merged with Pharmacia, but in 2002 it was spun off as an independent company.

This flux in industrial structure has several sources. One is simply the general character of this moment of global political economy, which is witnessing an incredible pulse of capital concentration across all sectors. In biotechnology, this general trend is reinforced by uncertainties associated

Table 11.1. *Top five gene giants 2003 (with mergers and acquisitions)*

Gene Giant	Mergers and Acquisitions
Syngenta	Novartis, AstraZeneca, Ciba Geigy, Sandoz, Imperial Chemical Industries
Monsanto	Asgrow, Calgene, DeKalb Genetics, Agracetus, Holden's Foundation Seeds
Bayer	Aventis, Rhône-Poulenc, AgrEvo, Hoechst, Schering
Dupont	Pioneer Hi-bred
Dow	Cargill North America

Sources: Author's compilation from numerous sources.

with how new genetic technologies may facilitate new connections and synergies across related production sectors. DNA has been revealed to be a common substrate for the pharmaceutical, medical diagnostic, chemical, veterinary, agricultural, and food industries. Firms specializing in those areas have explicitly tried to remake themselves into "life sciences companies" by acquiring – and sometimes divesting – holdings across multiple sectors. This process occurs unevenly and even chaotically as firms try to position themselves to take strategic advantage of rapidly changing technical possibilities, which are themselves far from assuming a determinate shape.

A further problem for companies involved in biotechnology generally has been the difficulty of developing marketable products. This problem has been particularly acute for firms with substantial commitments to agricultural biotechnology. From its inception, of course, commercial expectations for biotechnology have been high. Indeed, they consistently have been highly exaggerated. However misleading, over the last sixteen years, this public relations strategy has been eminently successful in generating an investment stream of absolutely massive proportions. During the 1990s, however, investors and Wall Street began to wonder when a reciprocal stream of revenue might begin to flow. Periodically, the business press has erupted with articles along the lines of the cover story of the September 26, 1994, issue of *Business Week:* "Biotechnology, why hasn't it paid off?" The need to maintain investment levels and stock prices has fostered enormous pressure on biotechnology companies to develop and deploy products as quickly as possible. The headline of a 1999 article in the Money & Business section of *The New York Times* accurately summarized the resulting dynamic as "The race to cash in on the genetic code" (Fisher 1999:BU1).

Despite this race to cash in, agricultural biotechnology is not yet paying off in the way that corporate biotechnologists had expected or hoped. In large measure, this is because life is much more complicated than either they or their academic counterparts had anticipated. At the beginning of chapter eight, I cited a 1984 statement from Ciba-Geigy's (now Syngenta by way of Novartis) Mary-Dell Chilton: "The solutions are coming very fast now. In three years we'll be able to do anything that our imaginations will get us to" (quoted in Rossman 1984:84). Certainly, a great deal has been imagined in regard to desirable traits for genetically engineered plants: not just herbicide tolerance, but insect resistance, disease resistance, drought tolerance, frost tolerance, enhanced photosynthesis, nitrogen fixation, delayed ripening, improved flavor, and the metabolism of a whole raft of desirable biochemical products. Few of these ambitions have been realized. As of 1999, fifty-three transgenic crops had been approved for unregulated release in the United States, but half of those are herbicide tolerant or Bt insecticidal varieties of corn, soy, cotton and canola (Shoemaker 2001:18). All the other approved GM varieties – e.g., virus resistant squash and papaya – are planted, if at all, on a negligible number of acres.

So few of the anticipated or imagined commercial applications of genetic engineering have been developed because identifying what genetic sequences ought to be moved, getting them moved, and getting them to function as intended in their new locations are not simple matters. The term "genetic *engineering*" implies a level of precision and routinization that is far from being achieved. In principle, it is indeed possible to move any gene out of any organism and into any other organism. In practice, it is very difficult to determine what genes, combinations of genes, or gene fragments are of interest or utility.

Relatively few single genes are associated with a single character, and traits of economic or agronomic importance are mostly multigenic. The "molecular ecology" of how genetic material (e.g., genes, alleles, gene fragments, promoters, DNA and RNA sequences, allegedly "junk" DNA) interacts and functions is very poorly understood. The set of robust techniques for gene transfer is fairly narrow, relying on brute force (e.g., the "gene gun" which literally shoots gold-covered pieces of DNA into a target cell) or on a few widely used plasmid or viral DNA vectors (e.g., *A. tumefaciens*). These transfer mechanisms can move genetic constructs between species, but their action is unpredictable in regard to the incidence of successful transformation and random as to the location at which insertion occurs. Even when transformation is achieved, the positioning of the transgene in a novel genomic environment may "silence" or alter its expression, or the transgene itself may disrupt or change the operation of its host chromosome in unexpected ways. In *Science,* Harvard molecular biologist William Gelbart (1998:659) has described himself and his colleagues as "functional illiterates" when it comes to reading the genome. No wonder actual products of agricultural biotechnology have been slow in coming.

The recalcitrance of life to easy manipulation, combined with pressures to provide tangible returns on investments, have given the race to cash in on the genetic code some interesting features. The need to bring products to market has reinforced the impetus to acquire seed companies which not only are necessary to any commercialization of crop genetic engineering but also carry the appeal of already having established product lines. It was the failure of the NBFs to develop marketable products that made them vulnerable to takeover. And the need to commit very large sums of money to research over many years is one of the major factors that has led to the consolidation which has left us with the handful of Gene Giants (see Table 11.1) that are willing to stay the course and continue the race to make biotechnology pay off.

Being big and having deep pockets has allowed a narrowing range of firms to stay in the race, but this has also made the stakes higher. In order to defray the high cost of bringing a refractory biology to heel and to allay the concerns of investors, companies focused an enormous effort on what was technically feasible in the short run in crops with large markets. What was easiest and most potentially profitable was the incorporation of herbicide resistance and Bt toxin into commodity crops such as corn, soy, cotton, and canola. Although both herbicide resistance and Bt toxicity are single gene traits, engineering even these "simpler" genetic constructs took a decade and entailed the expenditure of large financial and scientific resources under circumstances of heated inter-firm competition. As corporate executives struggled through this process toward commercialization, they were disinclined to accept any delay in introducing their products into the market. At Monsanto, the message was "We do not have the luxury of doing this the right way. We are going to do this the way that gets it done the quickest, because our entire future depends on the success of this program" (Charles 2001:67).

This urgency has manifested itself most clearly in the approach of the industry to regulation in the United States. Corporate lobbyists, with assistance from prominent academic biologists and NSF bureaucrats, had engineered promulgation of the Coordinated Framework for Regulation of Biotechnology in 1985. The Framework mandated no new regulations for biotechnology and determined that release, use, and consumption of GMOs would be governed by existing laws and agencies. This principle of "substantial equivalence"[3] between GM and non-GM crops and foods, reaffirmed formally by the FDA in 1992 (Food and Drug Administration 1992), has been the cornerstone of policy ever since. With few exceptions, the companies have been relentless in defense of substantial equivalence, fighting off multiple initiatives for more stringent health and environmental safety testing, and for labeling of foods containing GMOs or their products.[4] According to Henry

Miller, the FDA's point man on regulation of biotechnology from 1979 to 1994, "In this area, the U.S. government agencies have done exactly what agribusiness has asked them to do and told them to do" (Eichenwald et al. 2001). This corporate insistence on doing it their way surely facilitated the earliest possible introduction of GM crops, but the fragmented, transparently thin, and sometimes illogical oversight structure that it engendered has also fueled an emerging popular concern with corporate involvement in agricultural biotechnology (Pollan 1998).

Pressed to bring products to market as quickly as possible, the characteristics of the technologies the companies had available for deployment did not lend themselves to easy or enthusiastic embrace by the public. In 1993, after intensely pushing its propaganda for a decade, Monsanto received FDA approval to begin selling its recombinant bovine growth hormone (rBGH) to a dairy sector already glutted with milk. Consumers saw neither lower prices nor an increase in milk quality. The introduction of rBGH did, however, galvanize farm and environmental groups to active opposition, which in turn led to broad public dissemination of possible problematic effects of the hormone on the health of both people and cows (DuPuis 2000). Bans on the use of rBGH in the European Union (1994) and Canada (1999) highlighted the different ways scientific data can be interpreted. The foundering of the 'Flavr Savr' tomato shortly after its debut in 1994 – despite Calgene's advertised promise of "a bite of summer every time . . . exceptional taste . . . whether it's summer, spring, winter or fall" – did not enhance the stature of biotechnology nor the believability of corporate claims.

Into this uncertain environment, in 1996 and 1997, were pushed Roundup Ready soybeans and Bt corn. As with rBGH, these GM crop varieties carried no direct nor tangible benefit for consumers. Insect resistance and herbicide tolerance are *input* traits; they affect the agronomic performance of the varieties but contribute little to the utility or quality of the end product. Apart from some hypothetical and trivial reduction in the cost of food, GM corn and soybeans carry no intrinsic appeal for consumers. GM varieties could, however, be discursively constructed to carry various sorts of risks to community, human, and environmental health. As a growing variety of activist organizations worked to publicize those alleged hazards, a number of events transpired which tended to reinforce negative associations with crop biotechnology for the U.S. public.

In 1999 a Cornell University study (Losey et al. 1999) suggesting that pollen from Bt corn varieties could pose a threat to the Monarch butterfly became front-page news (Yoon 1999). In October came the WTO protests, and the cameras trained on the "Battle in Seattle" brought images of demonstrators dressed as Monarchs into many American homes. That year, too, Monsanto was forced to pledge not to commercialize its "Terminator" seed technology after global criticism of the ethical implications of engineering a "sterile spring" for farmers, especially those in the developing world. In 2000 Aventis's StarLink Bt corn, a type not then approved for human consumption and destined for animal feed, was found to have been inadvertently mixed with food-grade corn that had been shipped to a number of millers. Extensive news coverage of the subsequent recall of Taco Bell taco shells and Aventis's $100 million buy-back of the entire StarLink crop underlined for the public the weakness of regulatory mechanisms and the uncertainties regarding the human health effects of GMOs (Pollack 2000a).

In 2001 it became clear that not only was Bt corn commonly cross-pollinating with non-GMO corn varieties in the United States, but that it had also somehow crossed with the landraces of campesinos living in the mountains of Mexico, lending weight to charges that such "biopollution" will inevitably accompany the use of GMO crops (Quist and Chapela 2001). While the companies touted the potential of Bt corn for displacing harmful synthetic pesticides, their arguments that Roundup Ready crops would reduce herbicide use seemed much less plausible given that the point of herbicide resistant varieties is to make it easier to spray. The prospect of more widespread use of one particular herbicide and one particular insecticidal toxin also brought attention to the possibility of accelerating pesticide resistance in both weeds and insects.

If they were not given much material reason to welcome the advent of GM crops, American consumers attentive to news reports did have a variety of opportunities to be exposed to the problems, real and potential, associated with the rapid introduction of transgenic corn and soybeans. A growing number of activist organizations opposed to GM crops and foods also employed a variety of public relations channels to reinforce and extend those negative associations.[5] Awareness of GMOs is now fairly widespread. A national poll commissioned by the Pew Initiative on Food and Biotechnology found in 2001 that 44 percent of Americans say they have heard some or a great deal about genetically modified foods (Toner 2002). Whatever the quality of the information on which they are basing their opinions, it is clear from the many polls conducted over the past decade that consumers are uncertain about the value and safety of GM crops and overwhelmingly support labeling of foods derived from them.[6] Mandatory labeling is exactly what agribusiness wants to avoid, believing that marking their product in such a way will be taken as an implicit warning of risk or lower quality. Attempts by activists to mobilize what both they and the companies regard as a very large but latent popular opposition have had some success. In a preemptive move to protect its french fry market, McDonald's single-handedly sank Monsanto's market for its 'New Leaf Bt-potato' by quietly informing suppliers that they wanted no GM potatoes. Several other companies, including Gerber (ironically, a

Table 11.2. *Public controversies over biotechnology*

Monsanto introduces rBGH (1993)
Calgene introduces Flavr Savr tomato (1994)
European Union bans rBGH (1994)
Brazil nut allergen in GM soybean (1995)
Mad cow disease (1996)
Dolly the sheep is cloned (1996)
Novartis (now Syngenta) deal with UC-Berkeley (1998)
Celera announces it will beat Human Genome Project (1998)
USDA withdraws proposal for GMs in organic definition (1998)
European Union implements de facto moratorium on GM crops (1998)
Canada bans rBGH (1999)
Cornell study of Monarch butterfly and Bt corn (1999)
Monsanto no-"Terminator" statement (1999)
Death of Jesse Gelsinger in gene therapy (1999)
Turning Point Project advertisements in *The New York Times* (1999)
WTO meeting protests in Seattle (1999)
Celera/Human Genome Project announce sequencing of human genome (2000)
Aventis's Starlink Bt corn contaminates food-grade corn (September 2000)
GM corn bio-pollution in Mexico (2001)
Percy Schmeiser/Monsanto patent infringement court decision (2001)
ProdiGene biopharm escape (2002)
Oregon GM labeling initiative (2002)
Cancer in French children shown related to gene therapy (2003)

Sources: Author's compilation from numerous sources.

Novartis subsidiary) and Heinz, have declared that they will not use GM ingredients in their products. Public awareness of the issues has been maintained by substantial press and magazine coverage of the continuous stream of difficulties generated by industry in its rush to cash in on biotechnology (Table 11.2).

If the U.S. populace has remained predominantly skeptical of rather than hostile to GM corn and soybeans, European consumers were galvanized into extensive, active resistance by the appearance of these crops on their shores (Margaronis 1999). For complex historical and cultural reasons, the transformation of food is a more highly charged topic in Europe than it is in the United States (Joly and Lemarie 1998), and critical perspectives on capital, the way technology is used in relation to the environment, and social justice are more widely held there. Reflecting the incorporation of the "Precautionary Principle" into many EU regulatory frameworks, mandatory labeling of foods containing GMO-derived ingredients was put in place in

1997. Although by 1996 several varieties of GM crops had been approved for production and use by the EU, it was the prospect of large quantities of American-grown Roundup Ready corn and soybeans entering the food chain that exercised European activists. The outbreak of Mad Cow Disease in 1996 and the contamination of Belgian animal feed with dioxins also supplied Europe's consumers with timely object lessons in the unanticipated perils associated with the industrialization of the food supply and the willingness of private and public authorities to dissimulate and misinform.

Opposition to "Frankenfoods" blossomed across Europe. Demonstrators destroyed GM crop test plots in Britain. French farmer José Bové became the icon and inspiration of the movement with his trashing of a McDonald's restaurant and his uncompromising call to direct action (Daley 1999; McNeil 2000). A heavy-handed and culturally misplaced publicity campaign by Monsanto only exacerbated the resistance (Charles 2001; Alvarez 2003). With labeling required, it became impossible to sell any food containing GM ingredients in Europe. Further, widespread popular protest led the EU nations to enact, in 1998, a moratorium on approval of new GM crops until such time as a policy could be formulated that would provide not only for labeling but also for a system permitting effective traceability of GM components in food.

The StarLink incident in the U.S. stimulated testing of grain shipments and foods in many countries. "Adventitious" contamination by GMOs unapproved for human consumption was discovered in cornmeal and potatoes in Japan and in tortilla chips in Britain and Denmark (Pollack 2000b). In the wake of these incidents and with the EU's example, many countries – including major importers of U.S. corn and soy such as China and Japan – have imposed a wide variety of labeling and safety verification procedures on GMOs, which have complicated trade and seriously threatened U.S. commodity markets (Kahn 2002).[7]

European governments are not necessarily hostile to genetic engineering, but they would prefer to follow a precautionary approach to introducing GMOs into the food system and to contain popular opposition by clearly addressing food safety and environmental concerns. In June 2002 the EU proposed a new set of rules calling for labeling of any food containing more than 0.5 percent GM per ingredient and imposing strict arrangements to permit the tracing of all GM crops and foods back to sources of production. Distressed by the precipitous decline in exports of corn and soy since promulgation of the 1998 moratorium and frustrated by European commitment to the anathema of labeling, the U.S. is now trying to force open the EU market to its GMOs through a complaint brought under WTO dispute settlement rules. Joined in its suit by fellow GM-exporting nations Canada and Argentina, the U.S. argues that the EU's restrictions constitute an illegal trade barrier. This pressure tactic has met with some success. In 2004 the EU approved strains of GM corn for both human (a sweet corn containing Bt) and animal consumption (a Bt feed corn). Although they can be imported, neither variety can yet be actually grown in the EU (Becker 2004). These actions have failed to mollify the U.S., which continues to regard strict labeling and traceability requirements as non-tariff barriers to trade.

With nations of the developing world, the U.S. has added another tactic to direct pressure. It has been sending GMO corn varieties as food aid to countries in Asia and southern Africa, using pressing human need as a means of eroding trade barriers and silencing opposition. It appears that the U.S. objective in including GM grain in shipments for famine relief is as much to maintain pressure on the Europeans as it is to persuade reluctant and uncertain nations of the South to accept GMOs. When controversy erupted over the propriety of using food aid in so apparently instrumental a manner, President Bush himself commented in a speech that "European governments should join, not hinder, the great cause of ending hunger in Africa" by themselves acquiescing in the production and consumption of GMOs (Sanger 2003:A3).

If attitudes to GMOs vary widely among consumers and among governments, the same also holds true for farmers. Acceptance of GM crops in the United States has been widespread with 32 percent of corn, 71 percent of soybean, and over 70 percent of cotton acreage being planted to Bt and herbicide resistant varieties in 2002 (Seely 2002:A1). Given the rapid pace of adoption, it is ironic that studies comparing yields and cost of production of GM and non-GM varieties have found no unambiguous advantages for farmers in their use of the biotech seeds. This appears to be because the companies are charging high "technology fees" as a premium above seed costs, because the cost of Roundup and competing herbicides has been falling, because variable pest pressures do not necessarily reach levels where treatment is economically warranted, and because the technology has not always functioned as well as promised or expected (Myerson 1997; Duffy 2001). Bt varieties are used as insurance against the prospect of insect infestations, which do not always become serious. The appeal of Roundup Ready soybean varieties, despite a clear "yield drag" of 6–7 percent compared to conventional varieties, is located in how they simplify management decisions – don't think twice, just spray – for overburdened farmers having to deal with large amounts of land (Benbrook 1999). The technology licensing agreements that must often be signed as a condition of sale for GM varieties can be, as we will see, a source of considerable resentment. Whatever they think of genetic engineering itself, farmers growing crops in which GM varieties are close to commercialization (e.g., wheat) are debating the wisdom of proceeding with the development of varieties that might be excluded from important export markets.

Predisposed to try any technology that promises to lower their average costs of production, a good many conventional farmers in North America find GM varieties a moderately useful tool in their ongoing struggle to stay in business. In contrast, organic farmers have found Roundup Ready and Bt crops to be direct threats to their livelihoods. One of the most interesting features of the contemporary American food system is the rapid growth over the last decade of organic farming. Though still quite small as a proportion of total production (on the order of 1 percent of sales), it is the most rapidly growing sector of the food industry at around 20 percent per year. This rapid increase is itself a significant indicator of dissatisfaction with the conventional food and fiber system and of the willingness of certain segments of the population to embrace concrete manifestations of more sustainable production. When the USDA moved in 1998 to formally establish a set of rules defining and regulating organic standards, it included GMOs in its definition. The resulting outcry from organic producers and consumers forced the USDA to withdraw its proposal. In the final version of the standards approved in 2003, no product may be certified organic if it contains GM ingredients or was produced using GMOs.

Widespread contamination of organic corn and soybean crops by pollen from neighboring fields of GMO varieties now threatens the ability of organic producers to market their crops as organic (and, for that matter, the ability of conventional producers of non-GM crops to sell to markets where GMOs are banned). Not only are individual producers faced with the loss of premium prices for their products, but the possibility of any organic production at all is called into question. Who is liable for losses to adventitious contamination is not yet clear, though a recent decision in an Illinois class action suit includes provisions for paying farmers whose crops were contaminated by StarLink pollen (Garden City Group 2003). Several additional suits have been filed by organic farmers, and the courts will certainly be some time in working matters out. The prospects and parameters for "coexistence" of GMO and organic crops are now a major topic of discussion and negotiation among producers, industry, and government organizations (Iowa State University 2001; GRAIN 2004).

More than a quarter of the GMO crops now planted around the world are being grown in developing nations of the geopolitical South. Most of this acreage is in soybeans in Argentina, Brazil, and China, though India has a small but significant planting of BollReady Bt cotton. The demand for Roundup Ready soy is even stronger in Argentina than it is in the United States, with adoption rates approaching 100 percent. The biotechnology industry must also have been pleased by the substantial illicit sales of GM varieties from Argentine farmers to those in neighboring Brazil, where GM varieties were yet to be approved. Under pressure from its commercial farmers, the Brazilian government reversed its position and legalized planting of GM soy in September of 2003 (Smith 2003). Although royalties are not being paid on the portion of the Argentine crop sold as seed in Brazil, the companies can point to the active black market as clear evidence of desire by Third World farmers for GM seed (Barboza 2003).[8] Unapproved GM corn, soy, and cotton are now reportedly being grown in Thailand and Pakistan, and surely elsewhere.

It is not clear, however, how well such varieties perform or whether they are as useful to small farmers as to large commercial producers. There is considerable disagreement, for example, over whether the adoption of Bt varieties by some small cotton producers in India has been a success or a disaster (compare Satheesh 2003 with Qaim and Zilberman 2003). Throughout Asia, small and subsistence farmers, and indigenous peoples, have been the backbone of scores of movements and initiatives organized not simply to oppose GMOs, but to engage the broader issues of corporate power, social inequity, and sovereignty, with which the question of seed supplies and new agricultural technology is entwined (Food First 2003). Developing nation governments are deeply conflicted. While they are tempted to pursue what is alleged to be the cutting edge of agricultural development and a route to serious participation in world markets, they are also mistrustful of the aggressive tactics with which seed companies and the U.S. government are pushing GMOs and are leery of reprising, in a Gene Revolution, the problems that had accompanied the earlier Green Revolution.

But, at the turn of the millennium, the much-anticipated Gene Revolution had not yet appeared. After more than a decade of racing hard to cash in on the genetic code, the Gene Giants had been able to bring to market only two GMO traits in only four crops. The financial costs had been enormous. Ultimately, the cost in terms of misspent cultural and political capital in fighting regulations and labeling may have been even higher. The haste of the companies in trying to generate revenue had led them to focus their efforts on products with inherent liabilities and to push harder and faster than civil society was prepared to accept them. At the turn of the millennium, they had succeeded in unsettling many Americans and outraging many Europeans, peasant farmers, and indigenous peoples. Through their own actions, the Gene Giants had done much to catalyze the emergence and growth of a global opposition to the technology on which they had bet their corporate futures.[9]

In the wake of the Monarch controversy, activist organizations in the U.S. conceived a bold public relations move that they hoped would arouse, among Americans, the same level of concern for and opposition to biotechnology that Europe was witnessing. Some sixty groups formed the optimistically named Turning Point Project and sponsored a series of five full-page advertisements critical of genetic engineering in prominent newspapers, including *The New York Times.* The companies understood that they would need to launch a counteroffensive and, in April of 2000, they created the Council for Biotechnology Information (CBI), which promptly launched a $50 million campaign for the hearts and minds of North Americans (Barboza 2000a). The effort has involved what CBI refers to as "advertorials" on television, in newspapers, and in magazines such as *National Geographic, Gourmet,* and *Natural History.*[10]

The campaign's inaugural advertisements in *The New York Times* clearly illustrate the themes that CBI intends the public to absorb. The March 16, 2000, issue of the paper carried a photograph of an Asian mother holding a child who is using chopsticks to eat. The headline reads: "Biotechnology researchers call it 'golden' rice. For the color. For the opportunity." A second advertisement, published two months later, featured a North American farmer and the caption: "Biotechnology gives her a better way to protect her crop – and her children's planet." The copy goes on to elaborate: the GM variety she uses "lets her use fewer chemicals" and is kin to other GM crops that are "helping provide ways for developing countries to better feed a growing population." A third and similar advertisement in the series, featuring GM soybeans, makes an additional claim: "Extensive testing by scientists shows foods derived from plant biotechnology are as safe to eat as traditional foods." The messages are very simple: GM crops meet the needs of the poor and hungry, they do not harm the environment, and they are safe for you to eat.

The plausibility of such claims was enhanced by the June 27, 2000, announcement that the human genome had been sequenced and by the global attention that the prospect of developing 'Golden Rice' attracted a few months later. The "successful deciphering" (Wade 2000a:A1) of the genome was celebrated by the press as another critical milestone in "learning the language in which God created life," as President Clinton so sententiously put it. Shortly thereafter, 'Golden Rice' made the July 31, 2000, cover of *Time* with the headline "This rice could save a million kids a year." As has so frequently been the case with biotechnology, these claims were exaggerated. The "deciphering" was really more of a first draft in which even the number of genes – much less their function and operation – remains unknown and is yet to be determined. Similarly, although 'Golden Rice' does indeed use daffodil transgenes to manufacture beta-carotene, it remains a hypothetical technology, the practical efficacy of which is highly debatable, even assuming that a biotechnological solution is preferable to the simple provision of carrots or tomatoes or leafy greens – other sources of beta-carotene – to vitamin A-deficient populations of hungry people.[11] For the biotech companies, the chief value of the genome decoding announcement was that it reinforced the faith of investors in the scientific progress of biology. And the virtue of 'Golden Rice' lay in the fact that it was "*not* the moral equivalent of Roundup Ready beans" (Nash 2000:41).

That is, the new genomics and 'Golden Rice' give biotechnology proponents something concrete to point to when an old Green Revolution warhorse, like economist Vernon Ruttan (1999:58), begins to grumble about the "irrational exuberance" of the gene jockeys and asks them to demonstrate "measurable impacts of biotechnology on either human health or agriculture." Of course, the Gene Giants would love to be able to satisfy Ruttan with something more than promises. What they want more than anything else is to actually have on offer of something besides the "first generation" of biotechnology products represented by herbicide resistance and Bt corn. A few "second generation" products such as a high oil corn variety, cereal varieties fortified with various amino acids, and some pest and disease resistant cultivars are close to release. But none of these are qualitatively superior to existing technologies, nor do they have appreciable consumer appeal or the kind of ethical cachet carried by a 'Golden Rice.'

However, the sequencing of the human genome – like those of a variety of bacteria, a roundworm, the mouse, and the plant *Arabidopsis thaliana* in previous years – does indeed represent a critical milestone on the route that must be taken to additional products. Now that the DNA sequences of organisms are being mapped, attention is being turned from the structure of genomes to their *function.* The emerging field of *functional* genomics is uncovering the ways genes work with each other and with their protein products to shape how an organism develops and operates. Researchers anticipate moving beyond the limitations of single gene transformation to the development of generic tools for manipulating cell circuitry and comprehensively redesigning organisms from the inside out (Lander 1996). The tools of proteomics, bioinformatics, combinatorial genetics, and directed molecular evolution are said to promise a "third generation" of agricultural biotechnology products.

This "third generation" will feature multigenic reworking of crops to create "Golden Rice" analogs of interest to the wealthy and, ironically, the often-overfed consumers of the North. Research is under way on "cancer fighting tomatoes" and "oils with reduced levels of saturated fats" to help prevent heart disease (Council for Biotechnology Infromation 2003). A great effort is being devoted to "pharming," the production of high value pharmaceutical, and biochemical, substances in crop plants.[12] According to a review in *Science* (Enriquez 1998:926), products featuring these "output traits" will blur the distinctions between product sectors:

> Soon medical prescriptions may be personalized to our genotype, along with specific nutraceutical foods. Some vaccines will be delivered through foods such as raw potatoes or bananas. . . . These new products may be delivered through your health management organization, a merger of supermarket and pharmacy, or perhaps even through a series of national health club chains.

Enriquez knows readers have heard the hype before, for he adds, "Genomics is not the biotechnology of the 1980s which promised much and delivered little." According to Nobel laureate and DNA elucidator James D. Watson, "The pace of discovery is going unbelievably fast" (Wade 2003b:D6). So the race to cash in continues, with the technical promise of genomics now serving to justify staying the lengthened course.

Slowing down?

Putting brakes on those racing to cash in has been a consistent objective of those troubled by the directions that biotechnology research has taken ever since Herbert Boyer founded Genentech. Currently, the principal goal of the Organic Consumers Association's Food Agenda 2000 campaign is: "A Global Moratorium on all Genetically Engineered Foods and Crops. Because these products have not been proven safe for human health and the environment, they must be taken off the market" (Cummins 2000:4). In many public lectures and workshops and in classes over the last decade, I myself have made similar representations regarding a moratorium (see Massey 2001); though my rationale for such an approach encompasses more than potential hazards to human and environmental health. Such views are in stark contrast to suggestions by corporate executives that "sustainable agriculture is possible only with biotechnology and imaginative chemistry" (Schneiderman and Carpenter 1990:472) and that their participation in the race to cash in is simply an expression of "planetary patriotism."

Since publication of *First the Seed* in 1988, there has been an enormous outpouring of analysis and commentary concerning the development and commercialization of biotechnology. A great number of books treating agricultural biotechnology has been produced, ranging from a paean to genetic engineering in agriculture (e.g., McHughen 2000) to an avowed "self-defense guide for consumers" (Cummins and Lilliston 2000).[13] Burgeoning interest in the scientific and business potentials of genetics has not only drawn the focused attention of existing organizations, it has also spawned new professional and business associations (e.g., Biotechnology Industry Organization, Council for Biotechnology Information), new journals and newsletters (e.g., *Nature Biotechnology, AgBioForum, Biotechnology and Development Monitor*), and new NGOs (e.g., Genetic Resources Action International, Indigenous Peoples Council on Biocolonialism).

Simultaneously, the emergence and growth of the internet has revolutionized the provision and retrieval of various biotechnology data. Information pertaining to nearly any facet of biotechnology is quickly and easily accessed in a number of formats from many different web sites hosted by a wide range of individuals and organizations (from the FDA to the American Seed Trade Association, to Mothers for Natural Law, to the Third World Network, to the National Science Foundation). The sheer quantity of available biotechnology data is impressive, but it also varies widely in quality. Ultimately, despite the huge amount that has been written over the past sixteen years, what I find most striking about biotechnology, and compelling for my own position, is not what we know but *what we do not know.*

In order to allay popular concern and to preclude what they regard as unpleasantly restrictive regulation, the biotechnology industry and its academic allies have consistently claimed that the genetic engineering of crops is not substantially different from conventional plant breeding. This principle of "substantial equivalence" has been widely applied both to the procedures of genetic engineering, and to its products, and it finds wide expression in corporate propaganda and the policies of U.S. regulatory agencies. Of course, it is true enough that humans have been using sexual recombination to manipulate the genomes of other species for millennia. But, in a very concrete biological sense, the incorporation of jellyfish genes into a monkey (Vogel 2001) is qualitatively different from the breeding of corn. In the latter case, sexual recombination entails the movement of "clusters of functionally linked genes, primarily between similar chromosomes, and includes the relevant promoters, regulatory sequences, and associated genes involved in the coordinated expression of the plant" (Altieri 2001:16).

In contrast, University of Wisconsin plant pathologist Robert Goodman (2002) likens the random insertion of transgenes by genetic engineering techniques to "throwing a grenade into the genome." Harvard's eminent geneticist Richard Lewontin (2001) describes the potential consequences: "When DNA is inserted into the genome of a recipient by engineering methods, it may pop into the recipient's DNA anywhere, including in the middle of some other gene's regulatory element. The result will be a gene that is no longer under normal control." Lack of normal control could result in the production of new substances or in unexpected changes in the way the organism functions or interacts with other organisms and the environment. Lewontin concludes that "the process of genetic engineering has a unique ability to produce deleterious effects and . . . this justifies the view that all varieties produced by recombinant DNA technology need to be specially scrutinized and tested."

I find this persuasive. If we don't know with considerable confidence where transgenes are going, or what they will do when they arrive, it seems that we ought to make sure that we learn more before proceeding at speed. Advances in genomics are surely exciting, but they also reveal how much we do not yet know, and how we have sometimes been mistaken in what we thought we did know. Recently, geneticists have been surprised by how few genes comprise the human genome, and now they are having to rethink gene interactions and the functions of RNA and "junk" DNA in accounting for the complexity apparent in the human organism (Wade 2003a; Pollack 2003c). Indeed, the classical concept of the "gene" is itself regarded by some as an outmoded construct whose persistence is a barrier to an effective understanding of the manifold intricacies of the ways components of the genome interact (Gelbart 1998).

Fundamental uncertainties at the procedural root of genetic engineering are reflected in additional uncertainties at other levels. Critics have raised a variety of questions regarding the human health and environmental impacts of the use and consumption of GMOs. Might the bacterial disease vectors used to transfer DNA facilitate unanticipated transfers into additional organisms? Might the instability of transgenes cause plants to produce novel toxic substances? Might widespread use of antibiotic marker genes that are used to monitor transformation events contribute to the existing problem of antibiotic resistance in bacteria? Might novel transgenic proteins in GM foods cause allergies in some people? Might herbicide resistance engineered into crop plants be transferred to weedy relatives, or even unrelated species? Might Bt toxins affect non-target species? Might widespread deployment of Bt and herbicide tolerant varieties accelerate the development of Bt and herbicide resistance in insects and weeds? Such questions have occasioned extensive discussion and debate among scientists, as well as between the biotechnology companies and their activist critics.

I will not recap those debates here.[14] What seems clear to me, however, is that while there is no consensus, a significant number of scientists unbeholden to the companies and without close ties to activist groups have found good reason to question the headlong rush to deployment of GM crops (Lewis et al. 1997; Ervin et al. 2000; Benbrook 2003; Kapuscinski et al. 2003; Pollack 2004a). They regard the questions raised by critics as plausible and worthy of response. What they frequently conclude, however, is that, given the current level of knowledge, it is not now possible to adequately answer those concerns. In some cases this is because studies simply have not been done, and in other cases because, given available knowledge and methods, they *cannot* yet be done (Snow and Palma 1997; National Research Council 2000; Wolfenbarger and Phifer 2000; General Accounting Office 2002; Nestle 2003). The capacity to develop appropriate tools for answering outstanding questions is also constrained by asymmetries among disciplines. While the development of a robust and powerful (but largely reductionist) molecular biology has been underwritten by public and private monies, the field of ecology is in its infancy and has been badly undernourished financially.

Given current patterns of resource allocation and the passion for genomics, an understanding of the molecular ecology of the genome is likely to be developed before we have any effective understanding of the higher-level hierarchical systems (e.g., farm fields, ecosystems, food chains, human stomachs) into which transgenic organisms are deployed. How can the potential impact of Bt crops on the environment be assessed if we don't even know what lives in the soil (99 percent of soil microorganisms are uncharacterized), much less know how to effectively model the interactions that occur there? In a recent assessment of the future of plant breeding, Ronnie Coffman suggests that "Genomics is likely to take plant breeding in directions unforeseen and unfathomable" (1998:3). If that is so, and if we would like to live in a world that is as safe, equitable, and sustainable as we can make it, then surely we ought to proceed judiciously while doing our best to redress our knowledge deficits so as to illuminate the choice of paths we might take.

Moreover, there is emerging a good deal of evidence that there are indeed some real pitfalls on the little known and apparently unstable path of genetic engineering. Cloned animals are exhibiting severe developmental defects that seem to derive from random errors in individual genes (Kolata 2001). A transgenic variety of the *Arabidopsis* plant was found to be some twenty times more likely to outcross to wild-types than conventional plants (Bergelson et al. 1998), and strains of GM crops including sunflower, canola, and sugar beet have been found to exhibit similar characteristics. A variety of Monsanto's GM soy was found to contain extra fragments of DNA that the company's scientists were not aware of transferring. Australian researchers added a mouse gene to a mousepox virus and accidentally created a mouse-killing pathogen (Finkel 2001).[15] Most recently, the FDA suspended twenty-seven gene therapy trials after the therapeutic gene inserted into patients' cells landed randomly on or near oncogenes, inadvertently turning them on and giving three boys leukemia (Pollack 2003b). Both the British Medical Association (1999) and the American Medical Association (2000) have called for an end to the use of antibiotic resistance marker genes in genetic engineering because of the possibility that such genes could be passed on to bacteria affecting humans. The AMA considered a variety of the potential risks associated with genetic engineering, found them to be small but plausible, and called for a broad plan to study the issue.

While it is true, as the Council for Biotechnology Information (n.d.) is at pains to point out, that "there hasn't been a single documented case of an illness cause by biotech foods," the apparent instability of transgenes, the potential for unintended exchange of engineered DNA, and the appearance of unanticipated and inexplicable effects argues for caution. I find compelling Lewontin's (2001) argument that genetic engineering is "unusually likely to produce unpredictable results," and I believe that a moratorium on the further release of GMOs is justified until our understanding of their biological and ecological effects has been substantially improved. We simply need to know more.

That there is an unusual and biologically intrinsic *potential* for problems associated with genetic engineering does not mean that those problems must necessarily be manifested. The techniques of biotechnology are tools created by human beings who make choices about how the tools are used as well as about how they are designed. Of course, neither scientists nor engineers make technology any way they like, just as people make history, but not in any way they please. Nature presents certain constraints, but which elements of nature are explored, how nature is interpreted, and how nature is shaped to serve human desire are matters of human choice. I believe that biotechnology in general – and genetic engineering in particular – might be used in safe, socially progressive, and sustainable ways, *if* social circumstances allow them to be developed in a manner appropriate to those goals. Ultimately, my own interest in slowing down the "race to cash in on the genetic code" has more to do with the social dimensions and meanings of the "race to cash in" than it does with the technical circumstance in which the genetic code is involved. Ecological and social sustainability will follow principally from the social arrangements we construct, not from the technologies we create. To imagine otherwise is to succumb to a species of technological determinism. A focus on technical questions can be a deflection from the difficult but necessary task of social transformation.

So, while I do not believe that we now possess the scientific knowledge to use the new genetic technologies safely, I am even more concerned that we do not now have social institutions in place to see that they are used properly and well. Henry Miller and Gregory Conko (2002) are quite right: a small, authoritarian minority is now dictating what kinds of research are permissible and which technologies and products should be available in the marketplace. The powerful tools of biotechnology are now being wielded largely by a narrow set of corporations which claim to want to use them to eliminate hunger, protect the environment, and cure disease, but which in fact simply want to use them as quickly as they can to make money just as fast as possible. Achievement of that narrower goal is predicated upon three social processes that have run as themes throughout this book and whose operation are as salient today as they were sixteen years ago: the extension of the commodity form, the shaping of the public/private division of labor, and the securing of access to that essential raw material, DNA.

Commodities and commodification

In *The Communist Manifesto,* Marx and Engels claimed that what they called "the bourgeois epoch" was characterized by "everlasting uncertainty and agitation." The last sixteen years have witnessed an enormous amount of scientific and social restructuring as the possibilities and limitations of genetics and genomics and proteomics have unfolded and as companies large and small have merged, and divested, and invested, and struggled with an awakening global civil society. A great deal of analytic effort has been expended in trying to interpret the fluid and often contradictory course of events involving biotechnology. In this book, I have argued that the commodity form is an underlying and constitutive regularity which shapes and limits the particular forms taken by the episodic and often chaotic expressions of a developing capitalism. To extend the reach of the commodity form is to extend the reach of capitalism. No matter how immense it may already be, the very essence of capitalism is the enlargement of the collection of commodities by which we are already surrounded.

The most familiar way in which the self-expansion of capital occurs is through accumulation, the reinvestment of profits in additional rounds of production. But for accumulation to occur there must be commodities to be sold, and the availability of products is precisely what the biotechnology companies have had difficulty achieving. An emphasis on getting something – anything – into the marketplace has led to work on products such as champagne, fish, and grass that glow in the dark (Riordan 1999; Barboza 2000b; Pollack 2004b). Although these are properly regarded as frivolous, they are being seriously explored by certain companies because they are technically feasible and because, while their overall social utility may be virtually nil, some people would be willing to pay for them. The biotechnology companies' need for income – accumulation *now* – has shaped their release of products in ways that have significant consequences for the environment, farmers, and society as a whole.

Bacillus thuringiensis has long been a useful tool for vegetable growers who have not wanted to use synthetic pesticides for the control of certain caterpillars. Insects did not develop resistance to this natural pesticide because it was used sparingly in widely dispersed fields, accounting for a relatively small total acreage. When the Bt toxin gene is incorporated into crop plants, however, the toxin is expressed in every cell of the plant and that plant makes the toxin available throughout an entire growing season. The prospect of planting millions of acres with Bt corn and other Bt crops generated concern among entomologists and ecologists whose experience with the "pesticide treadmill" led them to anticipate widespread development of insect resistance to Bt toxins if they were indiscriminately introduced. Writing in *Science*, McGaughey and Whalon (1992:1455) bluntly declared that what was at stake was preservation of the efficacy of "the most scientifically, environmentally, and sociologically acceptable pest suppression tools of this century and possibly the next."

But the companies' need for revenue confounded efforts to design a considered, socially rational approach to the introduction of Bt corn and cotton.

After extensive negotiations, the EPA was at least able to require that farmers using Bt crops plant a portion of their fields in "refuges" of non-Bt varieties so that resistant insects might mate with susceptible partners surviving among the conventional plants. Refuge requirements are poorly publicized by seed companies, not actively enforced by the EPA, and widely ignored by farmers (Pollack 2003d).[16] If the refuge strategy is a sham, Monsanto is developing its own solution to the anticipated emergence of resistant insects. The company's vice president for regulatory affairs told Michael Pollan (1998:50), "there are a thousand other Bt's out there. . . . We can handle this problem with new products" – that is, more commodities.

If the experience with herbicide resistance is any indication, the appearance of Bt resistant insects should not be too distant. Contrary to corporate claims that adoption of herbicide resistant varieties would reduce the need to spray, it is now clear that farmers growing Roundup Ready soybeans are using two to five times more herbicide (in pounds applied per acre) than those using other weed management systems (Benbrook 1999:2). It is not now unusual to find Roundup Ready corn and Roundup Ready soy being grown in rotation. One result is a doubling of the use of glyphosate since the introduction of GM varieties in 1996, another is the appearance of Roundup resistant weeds (Pollack 2003a). The gene for Roundup resistance now is embedded in some 70 percent of the soybeans, 65 percent of the cotton, 55 percent of the canola, and 10 percent of the corn grown in the United States. It will likely soon be available in wheat, alfalfa, and turf grass. This is an impressive level of genetic uniformity. The short run interests of farmers and biotechnology companies converge to produce a situation in which the new technical possibilities are used not to seek truly innovative and sustainable solutions to production problems, but to patch and reinforce a system whose characteristic attributes – monoculture, chemical intensity, genetic uniformity – are widely regarded as unsustainable.

If current modes of deployment associated with crop genetic engineering reproduce existing problematics, they also introduce the novel dimension of "biopollution." The adequacy of containment of GMOs has been a principal concern of biotech watchdog groups since the 1980s, and it was a coalition of activist organizations acting on its own initiative, not a government regulatory agency or company lab, that uncovered the contamination of foodstuffs by unapproved StarLink corn in September of 2000. The immediate corporate response was to try to discredit the testing facility, Genetic ID, and to attempt to engineer an ex-post approval for Aventis's controversial variety (after all, people were already eating it). These efforts failed.

In the ensuing months, additional testing confirmed what many suspected or covertly knew. Transgenic DNA was all over the place where it wasn't supposed to be and where people didn't want it. Some of the contamination was the result of physical commingling of grain in harvest, shipping, and processing. But what was most disturbing was evidence that much of the contamination was the result of pollen flows from field to field, that it was occurring in self pollinated crops like soybeans as well as open-pollinated crops such as corn and canola, and that government recommended buffer zones intended to isolate GM from non-GM crops were grossly inadequate.

By the harvest season of 2000, organic producers realized that they potentially faced the wholesale loss of markets. In a letter to fellow organic certifying agencies, Farm Verified Organic (2000) laid out what was at stake:

> We are facing a crisis in organic production, an invasion that threatens the very existence of organic production as we know it. The problem is the contamination of organic products by Genetically Modified Organisms. . . . In fact, the problem of GMO pollution has become so pervasive that, in the major corn producing areas of the United States, it may not be possible to purchase seed or to grow corn that meets current organic standards.

Early in 2001, the *Wall Street Journal* tested twenty foods that were labeled "GM-free." Sixteen were found to contain transgenic DNA. In Europe, despite the moratorium and labeling requirements, ten out of one hundred products supposedly containing no GM ingredients tested positively. The managing director of a leading U.S. natural foods miller said frankly, "There's no such thing as certified GMO-free" (Callahan and Kilman 2001).

In 2002 consumers found that their own health might be threatened by biopollution associated with the development of "third generation" biotechnology products. In two incidents, the Texas company ProdiGene failed to completely destroy plants in test plots containing corn engineered to produce pharmaceutical substances, subsequently contaminating soybean grain and cross pollinating with corn plants in a nearby field.[17] The contaminated plants did not enter the food supply, but the company was fined $3 million (Pollack 2002), and the prospect of inadvertent leakage of crops containing vaccines or antibiotics or birth control substances was alarming for many. Calling such research "an unacceptable risk to the integrity of the food supply" (Clapp 2002), the president of the National Food Processors Association joined food manufacturers and grocery organizations in asking that biopharming not be undertaken in food crops. The biotechnology industry refused, since food crops are the ones they are able to engineer most easily and so are the preferred media for biopharming.

If the food manufacturers are uncomfortable with biopharming in food plants, they are ready to embrace the closely related practice of engineering "nutraceutical" or "functional" foods containing enhanced levels of health-promoting substances (e.g., lycopenes, omega-3 fatty acids) or reduced levels of unhealthy substances (e.g., trans-fats, saturated fats). Nutritionist Marion Nestle prefers the designation of "techno-foods" for these substances, the development of which she regards as a cynical exercise in the creation of "value-added" products through complex and unnecessary reconstitution and repackaging. She observes that

> from the standpoint of nutrition, techno-foods simply are not necessary. . . . The food marketplace is already glutted with an enormous overabundance of calories and products, and it is not difficult to select a health-promoting diet from this supply at quite low cost. The techno-food approach misses the point that the best health outcomes are associated with dietary *patterns* that follow recommendations, not just eating or avoiding one or another single food. [Nestle 2002:355–356]

The besetting sin of the biotechnology industry seems to be its lack of interest in solving for pattern as opposed to identifying partial, temporary, and commodifiable solutions to narrowly defined problems.

There are two things going on here, I think. The first order of business for the companies is arranging for accumulation as quickly as possible. They therefore push as hard as they can on the limited number of product options now available to them in spite of the uncertainties associated with genetic engineering and the possibility – or even likelihood – of generating environmental and human health externalities. A second order of business is a moment of what I have characterized as a process of permanent primitive accumulation: the separation of producers from their means of production and the creation in consumers of new needs whose satisfaction entails the production of new sorts of commodities. To the extent that genetically engineered Bt crop varieties create Bt resistant populations of pests, the utility of a tool for producing independently of conventional agriculture is undermined and possibly even eliminated. To the extent that biopollution undermines the possibility of organic production, the deployment of GM crops constitutes an expropriation of organic farmers. To the extent that GM substances appear in the food supply, eaters everywhere have imposed on them a system of production and provisioning which is increasingly defined by the parameters selected by a narrow range of commercial interests. Consumers are separated from the means of independent and healthy provisioning just as producers are separated from the means of independent and healthy production. The appropriation of nature becomes a means for the subjugation of the social world. According to food and biotechnology consultant Don Westfall, "The hope of the industry is that over time the market is so flooded [with GMOs] that there's nothing you can do about it, you just sort of surrender" (Laidlaw 2001).

Of course, none of this would be happening absent the historical elaboration of both the technical and social means of stabilizing seeds and DNA as commodities. After all, as economist Lester Thurow (1996:281) notes, "Capitalists do not invest in things they cannot own." As we have seen, plant breeders have long pursued hybrids less for their superior agronomic characteristics than for the "biological patent" that they confer. Given the key role that cytoplasmic male sterility had played in the production of hybrids, researchers were, by the early 1980s, already looking to use biotechnology for the induction of sterility in specified generations of seed parents. As understanding of gene structure and operation advanced during the 1990s, both corporate and public labs explored the ways various functions of a plant – including its fertility – could be switched on and off by application of various chemicals.

In March 1998, U.S. Patent No. 5,723,765 was issued to the USDA and to Delta & Pine Land Co., a leading private seed company that had sponsored the USDA research. Among the claims that were made was one for a method for producing seed that is incapable of germination. Seeds had been engineered so that, when drenched with the antibiotic tetracycline, a toxin gene is activated which does not impede the plant's growth, but renders the seeds it produces infertile. Grain harvested from such a variety could not be saved and used as seed the next year because it would be unable to germinate. Since 1998, such companies as Syngenta, Pharmacia, DuPont, and BASF have developed a variety of similar techniques. Collectively known as "Genetic Use Restriction Technologies," or GURTs, the application of these methods has not yet been commercialized. Their clear purpose is to eliminate once and for all the historical ability of farmers to maintain a degree of independence by short-circuiting the reproduction of capital through the reproduction of their own seed.

In a rhetorical coup, the activists at the ETC Group rechristened the USDA/Delta & Pine patented procedure as "Terminator Technology." The term is appropriate. The technology terminates the reproductive viability of seed. It terminates the long-standing relative autonomy of farmers from the commercial seed market. And in the Third World it could very well terminate people's ability to feed themselves by destroying "the 12,000-year tradition of farmers saving, adapting and exchanging seed in order to advance biodiversity and increase food security" (ETC Group 2002:2).

The subsequent interest of Monsanto - prime exemplar of malign corporate biotechnology - in purchasing Delta & Pine focused additional concern on Terminator Technology. The fact that plants were being rendered sterile touched very fundamental sensibilities about the ethics of manipulating the natural world. The possible spread of Terminator genes to other plants reinforced worries about the unintended ecological effects of genetic engineering. The lack of any agronomic utility to Terminator Technology clearly revealed it as a naked attempt by companies to advantage themselves by limiting the opportunities available to farmers and so highlighted the predatory dimension of concentrating corporate power. The prospect of using Terminator Technology in developing countries seemed especially problematic. While seed companies are anxious to extend their markets in Asia, Africa, and Latin America, the dissemination of varieties incorporating GURTs could have a devastating effect on the food security of the estimated 1.4 billion farmers who depend upon saved seed to grow their own crops for subsistence needs. These factors converged to make Terminator an issue with global visibility and unusual potency, and it has appeared as an iconic referent in anti-GMO and anti-globalization actions and literature worldwide (Kluger 1999; ETC Group 2002).

Opposition to Terminator Technology has not been limited solely to environmental or anti-corporate organizations. In June 1999 Rockefeller Foundation President Gordon Conway delivered a remarkable speech to executives and employees of Monsanto. In it, he exhibited a keen sense of the degree to which the excesses of biotechnology companies, in their rush to market, had created a broad-based backlash against genetic engineering that was threatening to endanger what Conway regards as a useful tool in the effort to develop crop varieties to serve the needs of Third World farmers. He urged Monsanto to consider major changes in its approach, specifically identified Terminator as a strategic point from which a retreat would be useful, and advised disavowal of GURTs as one way to "remove many of the suspicions about abuse of intellectual property to create market domination" (Conway 1999).

Rocked by the inauguration of the European moratorium, the monarch controversy, and intensifying opposition in many quarters, Monsanto CEO Robert Shapiro made a "public commitment not to commercialize sterile seed technologies, such as the one dubbed 'Terminator'" (Shapiro 1999). AstraZeneca made a similar announcement. However, it would be a mistake to think that so fundamental and long-held a structural objective of the seed industry could be so precipitously abandoned. Since 1999, AstraZeneca has been acquired by Syngenta, which along with other companies, has apparently continued Terminator research. The USDA has licensed exclusive rights to its seed sterilization technology to Delta & Pine, which has declared its intention of moving ahead, now with the morphed justification that the technology will be useful as a containment system for transgenes (Rural Advancement Fund International 2001b; Choi 2002).

Moreover, development of Terminator Technology is the mechanism through which valorization of GM seeds can be achieved most effectively in nations in which intellectual property rights are nonexistent or their enforcement is ineffectual. In 2004 Monsanto decided to stop selling its Roundup Ready soybeans in Argentina because half of Argentine GM soy acres were being planted with saved seed for which no royalties or technology fees could be collected (Smith 2004). This situation will likely be reproduced in many developing nations that permit (or tolerate illicit) production of GM crops. The companies may have no choice but to pursue some form of Terminator Technology in order to maintain a market for their seed.

GURT/Terminator appears to bring biotechnology companies much closer to the culmination of their historical quest to pry farmers loose from their central means of production and achieve full commodification of the seed. But since they are not quite there yet technically, the social route to commodification of the seed remains important. Indeed, over the past sixteen years, the companies have relied increasingly on the coercive mechanisms of intellectual property law to keep farmers in the market and to secure valorization of their products. The PVPA had been passed in 1970 with a clause permitting farmers to save, replant, and even sell PVP seed to other farmers. Continuing its policy of incrementally but persistently pursuing its core objectives, the seed industry successfully shepherded through Congress a 1994 revision of the PVPA which revoked that right of sale. However, violators were not aggressively pursued. After the 1985 *Hibberd* decision permitting patenting of plants, PVPA was not a vehicle the companies felt they needed to ride very hard, or very much further.

Patenting opened much broader opportunities for market expansion than did PVPA, for purchase of a patented product carries the right to use but not to make. Farmers would be able to use patented seeds (i.e., grow them) but would presumably be in violation of the law if they made them (i.e., saved and replanted them). GM varieties are, of course, patented. Having spent so much to produce them, having waited so long to release them, and being under so much pressure to produce revenue from them, the companies intended to recover as much from the sale of their seeds as they could. To do so, they would have to change the practices of farmers who, by both tradition and material incentive, were inclined to continue saving seed when possible. Since GM varieties of corn are also hybrids, the battles would be fought principally in soybean and canola fields where seed saving is still common. Since the key GM trait in those crops is Roundup tolerance, Monsanto would be the farmers' chief antagonist.

Certainly, Monsanto worked hard to disabuse farmers of any notion that saving Roundup Ready seed was permissible. In radio spots across the Midwest and in advertisements in magazines such as *Farm Journal,* Monsanto warns that it is "vigorously pursuing growers who pirate any brand or variety of its genetically enhanced seeds" and asks for compliance with the arrangements specified in the Technology Use Agreement (TUA), which any farmer buying their seed must sign. The TUA not only explicitly prohibits the purchaser from replanting seed, it also limits Monsanto's liability for any loss associated with use of the product, specifies how a farmer pursuing a claim must proceed, requires that no herbicide but Roundup be used, and allows the company free access to the farmer's fields. In its own version of "heartland security," Monsanto set up a hotline for farmers to report neighbors who they believe are violating the terms of their agreement. In less than two years, the company had more than 250 cases under investigation in twenty states (Wanamaker 1999). Selective vigorous prosecution has resulted in stiff penalties which have served the purpose of discouraging other farmers from engaging in plant-back. Most of those charged settle out of court.

However, a number of farmers – most prominently Canada's Percy Schmeiser – have defended themselves by claiming that they never planted protected varieties and that the presence of Monsanto's genes in their crops was the result of pollen drift. On May 21, 2004, after six years of legal battles, Canada's Supreme Court held Schmeiser liable for infringing on a Monsanto patent by saving and replanting glyphosate-resistant canola seed even after he had noted that some of his plants were resistant to Roundup. In a split five to four decision, the justices held that how the genes had gotten into Schmeiser's field was immaterial. According to the court, Canadian farmers do not have the right to knowingly "use" patented genes even if they are incorporated into a crop through mechanisms over which the farmer has no control. As University of Guelph agronomist Ann Clark (n.d.) observes, Schmeiser is being "held accountable for something which the seed trade itself cannot do," that is, preventing "adventitious" contamination. Incredibly, the decision puts responsibility for monitoring and reporting the flow of transgenes on the farmer rather than on the company. On the other hand, the court awarded Monsanto no damages since Schmeiser never sprayed Roundup and therefore did not benefit from his use of their genes, nor did the justices choose to require Schmeiser to pay Monsanto's court costs (I-SIS 2004). Monsanto's victory may have been Pyrrhic inasmuch as it may not be worth taking farmers to court for inadvertent infringement if the company does not stand to gain materially.

The Canadian decision on Schmeiser raises more questions than it answers, and it will be especially important to see how such circumstances play out in the U.S. legal system (Hamilton 2003). A trio of North Dakota brothers who find themselves in a situation similar to that of Percy Schmeiser will soon be going to court. In a different iteration of the issue of plant-back of patented materials, Missouri and Illinois farmers have fought back by initiating class action suits against Monsanto claiming that they had never signed the TUA and that payment of the technology fee entitled them to save seed for their own use (Schubert 2001b).

The biotechnology companies would like farmers the world over to traverse a legal path toward what the ETC Group has termed "bioserfdom." But the institutionalization of intellectual property rights remains highly uneven globally. Almost anything is now patentable in the U.S., but this nation is situated at the extreme end of a wide continuum. In 2002, Canada rejected patents on higher forms of life, and in the European Union plants are only patentable under certain conditions. Many nations, especially in the South, do not even have simple Plant Breeders' Rights (PBR) legislation. The creation of the WTO in 1994 marked the coming of age of the project of "globalization," the extension of social rules and arrangements that facilitate the expansion and stability of the commodity form and market-based activity. A key strategy in this process is "harmonization" of regulations, including those encompassed by the Agreement on Trade-Related Aspects of Intellectual Property Rights (TRIPS).

Article 27.3(b) of TRIPS requires nations that are members of WTO to offer some form of intellectual property rights in plants through patenting, PBR, or an "effective *sui generis* system."[18] In theory, the *sui generis* option provides countries with an opportunity to shape legislation to meet their own needs and conditions. In practice, some nations with inadequate resources find it convenient simply to adopt UPOV's PBR framework rather than to develop their own approach. A 1991 change to the UPOV Act made the farmers' "privilege" to save protected seed merely an "optional exception" to the "rights" of the breeder. More worrisome, both the U.S. and the EU nations are using negotiations in bilateral trade agreements to lever their partners not just onto the UPOV path, but beyond its requirements into so-called "TRIPs plus" arrangements which incorporate more restrictive patent-like provisions.

As a result, many countries have established PBR laws that attenuate the farmers' exemption in a variety of ways (GRAIN 2003).[19] A surprising number of Southern nations have banned farmers from using any protected seed other than what they themselves grew on their own land. PBR legislation thus is functioning to constrain the traditional modes of farmer-to-farmer germplasm exchange by which crop biodiversity has been so productively maintained for so long. As the historical example of the U.S. illustrates, the enactment of PBR legislation becomes the platform and justification for the deemphasis of public breeding programs as well as the precursor for the eventual introduction of patents. There can be no doubt that there will be a continuing push beyond PBR and toward patents, for PBR protects only whole plants rather than component parts such as genes, proteins, and DNA sequences. Moreover, PBR arrangements such as the UPOV Act and the U.S. PVPA permit the use of a protected variety in research, whereas a utility patent prohibits such work without a license.

Farmers are therefore not the only ones who find choices about how to perform their work being constrained by the growth of intellectual property rights in general and utility patents in particular. Just as important as the impact on farmers is the impact on plant scientists themselves. Now, in periodicals such as *The Plant Journal* and *The Annual Review of Phytopathlogy* one finds articles with such titles as "Transgenic crops, biotechnology and ownership rights: what scientists need to know" (Kowalski et al. 2002). And they need to know a great deal. The article by Kowalski et al. advises plant scientists to perform a "product deconstruction" before embarking on a particular line of research. In a social analog of the anticipated biological work in the lab, the organisms and methods to be used in the proposed research are dissected into their component parts, and each one is analyzed for the intellectual property rights that might be attached to it. Only after undertaking this risk assessment for patent infringement can scientists know whether they have sufficient "freedom to operate" (or "FTO" as it is now known) that would justify proceeding with the planned work.

And such freedom to operate has, over the past sixteen years, become increasingly hard to obtain. Absent the development of many products, the way to validate scientific effort in the race to cash in on the genetic code was to obtain patents which functioned as proxies for future valorization of current research and so became markers of "success" for attracting investment. Writing in *Nature,* Bobrow and Thomas (2001:763) observe that patent policy

> has more or less evolved through dialogue within a limited circle of participants. Commercial interests, which are well represented to the patent offices, have not been counterbalanced by those who represent the broader public interest. The result has been an innate tendency for the patent system to "creep" in the direction of extending patentability to biotechnology inventions for which the thresholds for novelty, inventiveness and utility have been lowered.

In what is frequently likened to a nineteenth-century style "land grab," vast tracts of the genescape and its products – DNA sequences, exons, introns, individual mutations, expressed sequence tags, single nucleotide polymorphisms, proteins, protein folds, parts of plants, whole organisms, whole classes of organism – are being appropriated via patents. Since the functional premise of biotechnology is the ecological unity of the genome and its component manifestations, it follows that nearly any research or product will draw upon areas covered by multiple patents. This is especially true since not only is genetic material itself being commodified, but the methods and techniques by which it is studied and manipulated are also patentable subject matter, so key enabling technologies such as the "gene gun" are also being subjected to private control.

The result of this proliferation of patenting and subsequent overlapping of property rights in research has been the progressive diminution of "freedom to operate" for both corporate and public scientists. Given the rather narrow range of technical possibility in GM crops, an enormous amount of litigation has taken place around the commercially critical nodes of Bt expression and herbicide resistance. Seven companies have been involved in multiple suits and countersuits over patent infringements and licensing arrangement on Bt corn, and Monsanto has been continually engaged in cases as it tries to negotiate the tricky terrain between enhancing its competitive position on Roundup resistance technologies while stopping short – not necessarily successfully (see Barboza 2004) – of what the Federal Trade Commission might consider to be unfair trade practices (Hayenga 1998). The desire to avoid costly and time-consuming reciprocal infringement suits is one factor that has driven the consolidation trend among biotechnology firms (Mayer 2003).

While the private sector may see problems with the current system, it is they who have principally shaped it and they are unlikely to work for significant reform. The broad scope of what is now patentable entails risks and transaction costs, but the payoffs can be very large. For example, in 1992 the NBF Agracetus received a U.S. patent that gave it rights over all GM soybean varieties, regardless of the manner of transformation. Monsanto, Novartis, and Pioneer Hi-Bred challenged the European version of that patent. However, once Monsanto acquired Agracetus in 1996, it happily reversed its position and, with the May 2003 decision of the European Patent Office to uphold the patent, now stands to profit mightily from its rights in *every* GM soy plant.

Ironically, the rush to patent genetic materials was accelerated by a public agency, the NIH, which in 1995 filed for patents on several thousand partial human DNA sequences even though their function had not been determined. The move was apparently motivated by a desire to preclude acquisition of patents by the private company Celera Genomics. Human Genome Project director Francis Crick explained, “Every piece we get, it’s like saving another block from speculators” (Hayden 2000:51). Whether they like it or not, academic and public researchers find themselves enmeshed in the appropriationist, market-oriented world of corporate biology, and they find they must play by its rules. Increasingly, access to the tools and materials of plant improvement are subject to intellectual property restrictions.

Most critically, patented germplasm – unlike that protected under PVPA - is not available for research purposes and cannot be used in breeding programs except under license. Plant breeding is an additive process in which incremental genetic additions are made to existing cultivars. If such additions – and the techniques to produce them – are patented, there may develop a progressive accumulation of property rights which becomes extremely difficult to manage. Transaction costs balloon. Desirable germplasm may be unavailable or accessible only through acceptance of Materials Transfer

Agreements (MTAs) which specify the conditions under which and for which the material can or cannot be used.[20] Permitting public researchers royalty-free access to germplasm may even be in the interest of a private company since it encourages broad use and distribution of the material without relinquishing the power to set the terms of commercialization (Coffman 1998). Reach Through Licensing Agreements (RTLAs), which involve conferral of rights in future discoveries made with licensed research techniques and tools, are also coming into use (Heller and Eisenberg 1998). According to University of Wisconsin corn breeder William Tracy, "Improvements will slow with the loss of free exchange. That's just basically inevitable" (Pollack 2001:D2).

The prospect of such slowing is beginning to generate broad concern (Dickson 2003). One worry is that the sheer number of patents in functionally related areas will lead to a "tragedy of the anticommons" (Heller and Eisenberg 1998), in which scientific progress is retarded by the legal and managerial difficulties of reconciling and coordinating different owners' property rights. If the "Balkanization" (Byron 2001) of patents in biotechnology is actually impeding the development and application of knowledge, then the fundamental premise for granting them – their ultimate contribution to overall social welfare through technical progress – is called into question. The scope of many biotechnology patents is troubling as well, and one cotton breeder has suggested that Agracetus's claim to all GM cotton plants is like Henry Ford filing for patent protection on all automobiles (van Wijk 1995:15). The declining standard of inventiveness for patents also appears to be permitting the wholesale appropriation of DNA sequences for effort that can be described as "intellectually trivial" (Bobrow and Thomas 2001:764). According to an editorial in the influential journal *The Lancet* (2002), "the current patent system appears to add the fuel of greed and monopoly to a flicker of discovery." The article questions whether DNA sequences ought to be patented at all and calls for a formal rethinking of the relationship between intellectual property rights and the public good.

Equity issues surrounding the patenting of plant genetic material of special importance to poor people and developing nations were highlighted by the sequencing of the rice genome by Syngenta in 2001 and the company's assertion of its intent to withhold some information from the public domain and to patent genes of importance (Pollack and Yoon 2001). If it is increasingly difficult and costly for the public sector in countries such as Britain and the United States to gain access to desired materials, it is even more difficult for the CG system centers to do so, and almost completely prohibitive for national programs of poorer or developing countries. In 2002, the British government's Commission on Intellectual Property Rights released a report which concluded that the global extension of IPRs under the TRIPS framework would generate few benefits for developing nations and would likely impose significant costs, and they specifically warned developing countries against providing patent protection for plants and animals (Commission on Intellectual Property Rights 2002). In a report released the following year and titled *Keeping Science Open,* Britain's Royal Society (2003) echoed that advice and expressed further concern over the increasing commodification of scientific information generally.

A measure of how deep the discomfort over current patenting practice runs, and how critical the situation has become in regard to crop improvement, is reflected in a recent announcement made in the July 11, 2003, issue of *Science.* The presidents of nine prominent public universities joined with the presidents of three leading foundations to announce the establishment of the Public-Sector Intellectual Property Resource for Agriculture (PIPRA). The authors state that "limited or conditional access to a wide range of patented technologies has been identified as a significant barrier to the applications of biotechnology in the development of new crops" (Atkinson et al. 2003:174). The PIPRA consortium proposes to recover their institutions' freedom to operate (FTO) through a collective patent management regime that would retain the right to use licensed technology for specified purposes (such as improvement of subsistence and minor crops) and allow the sharing of new technologies with other public research institutions.

Such arrangements could "begin to overcome the fragmentation of public-sector IP rights and re-establish the necessary FTO in agricultural biotechnology for the public good, while at the same time improving private-sector interactions by more efficiently identifying collective commercial licensing opportunities" (Atkinson et al. 2003:175). This is hardly the uncompromising "no patents on life" position taken by many NGOs. Though it is not without contradictions, the PIPRA initiative is a serious and concrete effort to reclaim at least part of the principle of free exchange and to operationalize collective action among public institutions for the public good. That it has been proposed by organizations with considerable stature and cultural capital is an indicator of the degree to which conditions conducive to the possible reform and restructuring of intellectual property rights are emerging.

Divisions of labor: biotechnologization and freedom to operate

The extension of intellectual property rights has historically been, and continues to be, a powerful tool for private industry to shape the social division of labor in plant improvement. Inasmuch as IPRs facilitate and protect private investment in agricultural research, economists and seed company executives have long suggested that with their implementation public sector scientists ought to be reoriented to work that does not compete with capital. The argument has not changed since the 1950s and is still oft repeated, for example: "Where more effective protection exists for intellectual property rights, the public sector has reallocated public funds away from variety development toward fundamental or pre-technology research" (Fuglie et al. 1996:2). What has changed, however, are the means of enacting that reallocation. Up until recently, its achievement depended largely on the rhetorical and ideological potency of the argument in persuading legislators and deans and research administrators of the propriety of limiting public activity to certain tasks. Now, with expansive private patenting, public breeders are finding that the nature and content of their work are being constrained by the sheer number and scope of the germplasm and techniques that can be – and sometimes are – placed off-limits to them (Coffman 1998; Sears 1998).

There is no question that private industry has increased its commitment to agricultural research in general and that it has moved into a dominant role in plant breeding in particular. While private sector outlays for plant breeding have increased about 7 percent per year since the 1980s, support for similar public sector activity remained flat and has even begun to decline. In the mid-1990s, Iowa State's Kenneth Frey directed a National Plant Breeding Study which surveyed the status of the field. Among his principal findings were that as of 1994 industry accounted for 61 percent of plant breeding expenditures, a similar proportion of scientific person-years, and that while the companies were increasing the number of breeders they employed the number in the public sector was declining (Frey 1996). In a division of labor that he describes as "changing daily," Frey (1998:6) also noted the movement of public sector activity out of cultivar development and into "genepool enrichment," development of methodologies, and the application of basic biology to crop improvement. That, of course, is the established historical dynamic. What is novel, however, is that private industry in effect had also become deeply involved in germplasm enhancement through its pursuit of genetic engineering. In the 1980s, biotechnology dissolved the boundary between basic and applied work and sent industry into the university in pursuit of the expertise it lacked. Having developed that in-house biotechnological capacity, the companies now find themselves once again working on the same terrain as their public counterparts, and Frey (1998:6) sees the need to develop "new formats of cooperation."

Given its early and heavy involvement with biotechnology, industry has certain structural advantages in this current working out of the division of labor. Its principal advantage is its dominant position in regard to intellectual property. In the area of agricultural biotechnology 76 percent of patents have been issued to the private sector (Atkinson et al. 2003:175), and of those privately held patents about three quarters are owned by only five companies:

Pharmacia, DuPont, Syngenta, Dow, and Bayer (Pollack 2001). Control over key enabling technologies is even more tightly concentrated. For example, the gene constructs patented by just four firms are in 96 percent of herbicide tolerant corn varieties, 86 percent of herbicide tolerant soybeans, and 81 percent of herbicide tolerant cotton (Oehmke and Wolf 2002).

In an assessment of the future of public plant breeding, Cornell University's Ronnie Coffman (1998:5, 6) sums up the implications for him and his colleagues:

> With most important enabling technologies controlled by large-scale companies in the private sector, it is difficult to see how public sector programs can continue to be relevant in the production of advanced breeding material unless strong partnerships exist between public sector programs and those holding the enabling technologies in the private sector. Research exemptions are generally available but leave the public sector breeder in the very difficult position of developing technology that s/he may not be able to distribute. . . . More and more of the best science is patented.

To the extent that industry data or technology is nonfungible – e.g., key DNA sequences of a crop plant, or a uniquely powerful technique – public plant scientists may be unable to pursue certain lines of inquiry or may be forced to fit their research to a company's terms simply to gain access to critical means of production.

Industry is still keenly interested in maintaining the sorts of partnerships with public sector researchers referred to by Coffman. Exploring genomics promises to be a complicated and expensive undertaking, and the opportunity to gain preferential access to the insights of university scientists is welcomed. In 1998, Novartis concluded an arrangement with UC-Berkeley's Department of Plant and Microbial Biology that gave Berkeley $25 million and access to Novartis' genomic database in return for a seat on departmental committees and first right to negotiate a license to patents from selected discoveries (Rausser 1999).[21] Although reminiscent of the blockbuster university/industry agreements of the 1980s, such large-scale deals are now uncommon. That scarcity is not a reflection of a lack of interest in university/industry ties, but is indicative of their very ubiquity. The proliferation of such relationships over the last sixteen years has resulted in a routinization of academic/industrial cooperation, and it is the unusual size, rather than the content, of the Berkeley/Novartis partnership that makes it noteworthy. All Land Grant Universities now manage many hundreds of gifts and contracts that are the vehicles for the flow of information and ideas and expertise and property between their researchers and the companies with the financial resources to purchase access to them.

These enhanced ties between universities and corporations are most striking in the life sciences, but are characteristic of developments within the academy as a whole. The emergence of the "entrepreneurial university" (Slaughter and Leslie 1997) and the "commercialization of higher education" (Bok 2003) have occasioned much comment. A major stimulus to the capitalization of academic knowledge was the passage in 1980 of the Bayh-Dole Act allowing universities to retain patents on discoveries made with federal funds and to license them to industry. Over the last decade, universities have rapidly increased the number of patents they hold, and royalties accruing to them more than doubled between 1992 and 1997 to $617 million per year. UC-Berkeley College of Natural Resources dean Gordon Rausser (1999:5) admits that universities are becoming "more like private companies."

With intensified academic/commercial collaboration comes a transfer not only of technology but also of codes and norms that subtly but effectively induce "the university to become isomorphic with its corporate environment" in both attitude and practice (Kleinman and Vallas 2001:466.) Harvard's ex-President Derek Bok concurs: "Certainly many more people in academic life think that you can have all the virtues of that life and be rich at the same time – in fact they think that you ought to" (quoted in Rimer 2003:A16). Most importantly, ideas generated by public funds that would otherwise have remained in the public domain are now privatized.

If the principal normative effect on the university of intensifying cooperation with industry has been an increasingly proprietary approach to the development of knowledge, the principal material effect – in the life sciences generally and in the agricultural sector in particular – has been a profound commitment to the rapid development of biotechnology. During the 1980s, federal funding agencies, state governments, and universities bought into the promised brave new world of biology in a manner that was not much less enthusiastic than Calgene or Monsanto. There is now no LGU without its own biotechnology center. Even in the midst of fiscal straitening, my own state of Wisconsin still found the resources to implement a $317 million BioStar program to construct four state-of-the-art biotechnology research buildings for the University of Wisconsin. Hiring patterns in agronomy and crop science and horticulture and plant pathology departments made complementary shifts.

An associate dean at the University of Wisconsin's College of Agricultural and Life Sciences explained the new dispensation that has been operative since the mid-1980s (Freistadt 1988:10, 29c):

> The plant breeders and plant physiologists, those kinds of applied research programs are being redirected into molecular biology and biotechnology programs. The reason is that there are monies available from both federal and private sources to sponsor molecular research ac-

> tivities. . . . We don't have a hell of a lot of choice. We have to go where the money is.

Where the money has been is evident in what Buttel (1999:1) has described as the "biotechnologization of agricultural R&D."

With reference to the division of labor between public and private plant breeding, biotechnologization has had two principal meanings. First, it has resulted in the continued ebbing of cultivar release by public institutions. This is problematic since the maintenance of public cultivar development is an effective way to provide alternatives to the varieties offered by the seed companies and to supply quality cultivars for minor crops in which markets are too small to elicit investment from private breeders. Industry has unambiguously demonstrated how it intends to develop the crops of the future, and it has been very aggressive in pressuring the public breeders who are still involved in finished variety development to incorporate proprietary genes – for example, Roundup Ready genes in wheat – into their programs. To lose independent public cultivar development is to lose our best vehicle for ensuring that there will be alternatives to the chemically dependent suicide seeds that are the products of corporate laboratories. For example, a soybean breeder at the University of Minnesota has explored making the plants more competitive with weeds, an approach diametrically opposed to herbicide tolerance. Further, the Gene Giants have shown no more willingness to engage in varietal development for "orphan" crops any more than they have been willing to put research into orphan drugs. Despite hand-wringing in the plant science community over the lack of breeders for minor crops (see Frey 1997), the University of Wisconsin is presently allowing its oat breeding program – one of the last of its kind – to slip quietly away as faculty shift to genomics in the lab rather than cultivars in the field.

Second, the growing emphasis of public plant scientists on biotechnology and genomics brings them into a production location similar to corporate knowledge workers. Having successfully levered breeders away from finished cultivars, the companies now confront the need to readjust the division of labor with a new set of public researchers. The marketization of the university, which has proceeded in parallel with the development of biotechnology, for some time obscured this dynamic. With Bayh-Dole providing a novel opportunity to profit from the labor of its biological professoriate, university IPR managers were anxious to license biotechnology inventions to the highest bidder, often not even preserving their own institutions' rights to use the technologies. For instance, although the widely used "gene gun" was developed at Cornell University, that critical enabling technology was licensed to Agracetus and subsequently acquired by Monsanto when it bought Agracetus.

Hence, it was largely their own inexperience and cupidity that brought universities to the point at which the PIPRA institution presidents can complain "Our institutions have found that the public research sector finds itself increasingly restricted when wishing to develop new crops with the technologies it has itself invented, including so-called 'enabling technologies' – the research tools necessary for further experimentation and innovation" (Atkinson et al. 2003:174). The inauguration of PIPRA is a response to the recognition of the difficulties into which universities placed themselves through their embrace of an atomized, market-based approach to knowledge production. Further, it must be understood as a response to the recognition that – as PIPRA points out on its web site (though not in the *Science* article) – "There are only a small number of large multi-national firms that control a large proportion of cutting edge agricultural IP" (PIPRA 2003). Public sector germplasm enhancement and genomics are in danger of being subordinated to industry, as was cultivar development before them. PIPRA is fundamentally an attempt to build a platform from which the public sector can defend and reassert its freedom to operate not just in technological practice but in the social division of labor with a narrow set of Gene Giants.

How that division of labor will be worked out on the terrain of genomics is uncertain. On the one hand, PIPRA participants recognize that serving the "public good" requires them to preserve a sufficient amount of FTO to use the technology they develop "for furthering their goals of achieving food security for the poor and excluded of the world" (Atkinson et al. 2003:175). This appears to mean providing technology preferentially to public breeding programs and small farmers in the developing world, and possibly even to small seed companies in the North. On the other hand, PIPRA's founders simultaneously envision creating "additional opportunities to generate royalty income to support public-sector research by providing convenient one-stop shopping for commercial licensing" (Atkinson et al. 2003:175). As we have seen throughout this book, a balance between the competing goals of public service and institutional profitability is a precarious equilibrium to maintain, especially given industry's constant pressure to shift their pursuits in the direction of commodification. Still, it is significant that a commitment to the public good is being made so explicitly, and that a concrete challenge is being mounted by the public sector to the hegemony of the genomic oligopoly.

Note, however, that the issue of the division of labor that engages PIPRA is couched largely in terms of the use and control of the tools of biotechnology. By "biotechnologization" what is most centrally meant is a general commitment to advanced genetic technologies as the core paradigm for plant improvement. Among public plant scientists, as well as among their private counterparts, there is near universal excitement over the potential of the new knowledge and techniques being developed and deployed. While in the public sector there is frequent criticism of the way biotechnology and genetic engineering are being *used* by companies, condemnation of the technologies themselves is rare. Given the "enormous path dependence" (Buttel 1999:1) that investment in biotechnology has already established, these techniques are going to remain at the core of agricultural research for a long time to come. Happily, there is much interest among public scientists in how biotechnology might be used to achieve a truly sustainable agriculture (e.g., Coffman 1998; Manning 2000).

To the extent that public plant scientists have adopted a corporate understanding of agricultural problematics, that goal will be difficult to achieve.[22] The development of Terminator Technology by the USDA's Agricultural Research Service is a prime example of the ways in which the goals of corporate and public scientists can become isomorphic. Bt corn and Roundup Ready soybeans represent a change in the form of industrial monoculture rather than a shift to a fundamentally new approach. New approaches are available, however. The emergent set of conceptual, technical, and methodological resources known as "agroecology" (see, e.g., Altieri 1996; Giampietro 2004) appears to be particularly promising in that regard. Genomics, with its emphasis on apprehending the interaction of genome and environment from a systems point of view, seems compatible with an agroecological outlook. It is to be hoped that a division of labor between public and private plant science can be maintained that creates for the public sector not only a broad freedom to operate but also encompasses the motivations and objectives for the use of biotechnology that are very different from those of private industry.

How the contest over the shape of the division of labor in the North proceeds will have major ramifications for the scientists, farmers, and eaters of the South. The CGIAR centers and the national agricultural research programs of developing nations have long looked to the public sector plant breeding institutions of the industrialized countries as a source of technology and expertise. If the patent-driven tying up of germplasm and enabling technologies is problematic for the public sector in the North, lack of access to these resources is even more acute in the South. CG system budgets are as flat as those of public sector institutions in the North. With a reduced flow of new knowledge from their accustomed source, the CGIAR has begun to look to the private sector for strategic alliances that will give its research centers access to the information it believes it needs to serve the interests of the world's resource-poor farmers and the hungry.

The CG system leadership and its advisors feel a special urgency since there is among them a widely shared perception that biotechnology is the key to achievement of what Rockefeller Foundation President Gordon Conway has proposed as a "doubly green revolution." CGIAR system chair Ismail Serageldin has similarly called for embracing "Promethean science"

(Serageldin and Persley 2000), by which he means an integration of agroecology, biotechnology, and precision farming. Certainly, genetic engineering as it has been developed by the private sector now has little to offer poor farmers and, indeed, is in many ways a direct threat to their livelihood and wellbeing. The Gene Giants have developed GM cultivars for the farmers who can pay for the seed, and that means corn and soybean varieties to be fed to animals in Iowa feedlots rather than the development of disease resistant varieties of cassava that might actually be of use to subsistence farmers in Zambia. Still, many in the international agricultural research community are convinced that biotechnology *could* make a critical contribution to overcoming the constraints on agricultural production in the Third World if it were applied in response to human needs rather than market signals (Jefferson 1993; Conway and Toenniessen 1999; Victor and Runge 2002).

The problem with such a strategy, just as it is for the public sector in the North, is attaining sufficient freedom to operate in social and technical spaces that are increasingly dominated by capital. The CGIAR and its LDC partners actually face far fewer constraints than does the public sector in the North. Because the countries in which CG centers are located do not now recognize patents on living organisms, germplasm and many enabling technologies patented in the U.S., Europe, or Australia can now be freely used by CG institutions (Nottenburg et al. 2002). Such an approach, however, is being actively discouraged by donor governments who would certainly apply powerful political pressures and might retaliate by restricting funding or information flows. Bilateral trade pacts may already be restricting this option. And the capacity to benefit from access to germplasm or techniques depends critically on the capacity to use them, not simply to possess them. The CGIAR, if not many developing nations, will likely choose cooperation rather than confrontation with those who hold the purse strings and the passwords to the databases. The CGIAR system now feels it has no option but to access the privately held enabling technologies it needs by establishing strategic alliances with industry (Serageldin and Persley 2000). Since the late 1990s, the CG centers have concluded a wide variety of exchanges, contracts, joint ventures, and licensing arrangements with companies such as Pioneer, Monsanto, and Novartis (Manicad 1999). In October 2002, the CGIAR took the unprecedented step of adding the Syngenta Foundation for Sustainable Agriculture as a member. Syngenta now joins the Ford, Kellogg, and Rockefeller Foundations in CG governance and policy activities.

Closer relations between the public and private sectors – in the North and the South – are, unfortunately, founded on fundamental asymmetries of need, resources, and power. The companies are now spending far more money on research than the public sector. When Hoechst and Rhone-Poulenc merged to become Aventis the annual research budget of that company alone was about $3 billion, ten times that of the entire CG system. The private sector has a considerable lead in the number of patents it has acquired and the enabling technologies that it consequently controls. The companies also have substantially more experience in structuring business arrangements than do public sector institutions.

Drawing on his experience from concluding the deal between his own institution and Novartis, UC-Berkeley dean Gordon Rausser suggests that however difficult such alliances are to construct, public institutions in the developing world have no choice but to proceed with them. He advises a "focus on leveraging the complementarities and potential synergies between their knowledge assets and those of the private sector" (Rausser et al. 2000). Berkeley, like many universities in the North, had a considerable stock of intellectual labor power with which to bargain. What do the CG centers and the national research programs of the LDCs have to offer? They cannot offer much intellectual labor power (though what they do have is comparatively inexpensive). What they do have is access to the 80 percent of the world's farmers who do not yet buy seed. And, as Rausser notes, they have something else that is appealing: "germplasm" (Rausser et al. 2000).

Genetic resources and ecoliberalization

My history of plant breeding and improvement is largely the story of how germplasm has been gradually but steadily integrated into the generalized commodity system, which is the defining feature of capitalism. I have argued that the rough border which has marked the division of labor between public and private plant breeding has followed the shifting contours of the commodity form. With their recent foray into patenting, public plant scientists are trying to balance their responsibility for the development of public goods for the public domain with the mandate to make money through the production of commodities for the benefit of the entrepreneurial university.

This is a fundamentally contradictory process, and I fear the tendency over time will be to honor the market rather than the commons. Farmers and indigenous peoples find themselves in an analogous situation with regard to genetic resources. It is thought that elements of the commons can be preserved even as a means of capturing revenue once the provision of access to genetic resources is established. As Camila Montecinos (1996) has pointed out, this has led to a contradictory conjuncture in which western IPRs are rejected, but "'different' forms of IPRs" are embraced. But the market, in all its forms, carries a profoundly incorporative tendency. In Erna Bennet's (2002:5) memorable phrasing, such quasi-property rights may in fact be the "weapons of the enemy" inasmuch as they may ultimately erode the non-market attitudes and practices they are intended to protect.

I myself was insufficiently sensitive to this point sixteen years ago. At that time, I felt that, for reasons of *realpolitik,* the impasse over the FAO Undertaking could not be resolved in favor of "common heritage." As a pragmatic matter, I thought that recognition of genetic resources as national property would be acceptable to the advanced capitalist nations and would elicit among them a willingness to pay for access to genetic information. I hoped that the solidary front that had characterized the countries of the South in regard to the Undertaking could be extended into cooperation on a new, multilateral regime for the equitable management of genetic resources. As it happened, the 1992 Convention on Biological Diversity affirmed the principle of national sovereignty over genetic resources. I was proven correct in my estimation of the outlook of the North. I was shown, however, to be overly sanguine in my anticipation of collective action on the part of the South. What resulted was the emergence of a wide range of bilateral, market-oriented arrangements, while the multilateral FAO Undertaking was relegated to jurisdiction over a narrow range of materials. Although so-called "Farmers' Rights" were recognized, they remain rhetorical constructs, and peasant farmers and indigenous peoples have been subjected to a new round of appropriationist initiatives.

Through most of the 1980s, seed companies were still resisting the prospect of paying for their raw materials. However, given the scientific and commercial excitement over biotechnology, the fiction that genetic resources are the common heritage of mankind proved impossible to maintain. With biotechnology companies selling many millions of dollars worth of stock on the strength of their manipulation and ownership of genes it became unambiguously clear to the states, bureaucrats, scientists, farmers, and indigenous peoples of the South that genetic resources were valuable. This rapid conscientization was accelerated by the growing interest of biotechnology companies in medicinal as well as crop plants. Nations of the South were asked to open their fields and forests to chemical as well as agricultural prospectors (Eisner 1989). Companies came under increasing pressure as activist groups levied charges of "biopiracy." Northern environmental organizations began to lobby for compensation as well, seeing such payments as a means of funding conservation efforts. In his book *The Diversity of Life,* E. O. Wilson (1992:283) described the "unmined riches" available in tropical regions and called for a New Environmentalism that would use bioprospecting to "draw more income from the wildlands without killing them, and so to give the invisible hand of free-market economics a green thumb."

Out of the 1989 FAO Conference came Resolution 4/89, an "agreed interpretation of the Undertaking." This agreed interpretation explicitly recognized the legitimacy of "Plant Breeders' Rights" and the expression of those rights in the legal protection of varieties. Balancing this concession to the developed nations was language specifying that "Farmers' Rights" will also be recognized. Just as plant scientists are entitled to a reward for their labor in creating breeding lines and elite varieties, so farmers have a right to a reward for creating and maintaining landraces and other forms of plant genetic resources. Further, just as the reward for plant breeders is not moral but material, so should farmers be entitled to material reward for use of the fruits of their labor. This *quid pro quo* appeared to offer a way out of the Seed Wars impasse. But Farmers' Rights had no legally binding meaning, and although the FAO created an International Fund for Plant Genetic Resources contributions were voluntary and, consequently, were negligible.

Then, in 1992, the drug company Merck concluded an agreement with Costa Rica's National Biodiversity Institute (INBio) which gave Merck exclusive access to biological materials collected by INBio in return for a lump sum payment of $1 million (Kloppenburg 1992). Two models for managing genetic resources on a compensatory basis were being established. The FAO approach was multilateral and embedded in institutions of international governance. The Merck/INBio approach was a bilateral market transaction between autonomous parties. The FAO and Merck/INBio arrangements marked the end of the discourse of common heritage and represented two alternative paths for shaping a legitimized framework for the collection of genetic resources. If there were to be a new exchange regime, business interests and their political allies preferred that it be based on the familiar foundation of the market. Since then, a process of "ecoliberalization"[23] has been under way in which conventional market mechanisms have increasingly been applied to the acquisition of genetic material.

This emergent ecoliberal, market-oriented regime for genetic resources was given sanction by the Convention on Biological Diversity (CBD) which was passed in 1992. In its Preamble, the Convention explicitly affirms the "sovereign rights of nations" over biodiversity. Second, in Article 15 the Convention provides for access to genetic resources "on mutually agreed terms" (as does the FAO Undertaking), but fails to specify a particular framework for determining those terms. Article 15 itself contains only the vaguest of language regarding "sharing in a fair and equitable way the results of research and development." Article 16 specifies "access to and transfer of technology . . . under fair and most favorable terms" but immediately removes all the teeth from this provision by qualifying it with unambiguous subordination to "effective protection of intellectual property rights." In effect, Articles 15 and 16 of the CBD do little more than affirm that genetic resources may be bought and sold subject to current intellectual property law. And since the Convention is not explicit about the nature of the parties who should come to mutual agreement, the Convention implicitly encompasses transactions made by virtually anyone or any institution.

The indeterminacy of "mutually agreed terms" is further compounded by paragraph 3 of Article 15. The language of that paragraph excludes from the

Convention all *ex situ* collections of genetic resources outside the country of origin and which were acquired prior to the adoption of the Convention. Most notably, this left uncertain the status of the critically important CGIAR gene banks whose holdings were collected originally under the regime of "common heritage." Not only does the Convention not mention Farmers' Rights, it makes only the vaguest noises in the direction of the rights of "indigenous and local communities." Though recognizing the utility of the knowledge and activities of farmers and indigenous peoples, the Convention does no more than "*encourage* the equitable sharing of the benefits arising from the utilization of such knowledge" (Article 8,j; emphasis added). The Convention thus affirms the force of current intellectual property arrangements but fails to seize the opportunity to move effectively toward provision of a symmetrical set of rights for informal or community knowledge production.

As a result of these deficiencies, the Convention did not so much provide for the conservation of PGRFAs as it established the conditions for their commercialization (Athanasiou 1992). And it did so in a way that allowed more or less free play of market forces. Any parties (states, research agencies, companies, communities, individuals) may enter into agreements for the sale of genetic resources on mutually agreed terms as long as they adhere to existing intellectual property arrangements (which do not include Farmers' Rights). The Convention thus gave an official global imprimatur to the commercialization of biodiversity. In doing so, it simply affirmed the commercial status quo, for the Convention countenanced no more than what was already being done (viz. the INBio-Merck deal).

A principal consequence of the generality of the CBD's language is a tendency for the commercialization of genetic resources to be manifested in bilateral rather than multilateral arrangements. And the past decade has seen an enormous proliferation of such arrangements. The government of the United States gave its imprimatur to bilateralism by requiring the bioprospecting initiatives it underwrites through its International Cooperative Biodiversity Grants (ICBG) program to provide compensation for the peoples and communities who host the bioprospectors (NIH 1993). Many scientists and some conservation groups – Conservation International and the New York Botanical Garden, for example – have become directly involved in the search for green gold (e.g., Balick et al. 1994). Lacking the ethnographic skills and sensitivities to efficiently extract culturally embedded information, companies such as Merck, Monsanto, and Eli Lilly have entered into alliances with a variety of universities and conservation organizations whose ethnobiologists provide the interpersonal solvents needed to make the data flow freely (NIH 1993). Scores of companies and many hundreds of academic scientists are now scouring the lands and bodies of farmers and indigenous peoples for commercially or scientifically useful genetic materials and components. While well-known transnational corporations like Monsanto, Pfizer, and Merck focus on the collection of plants and animals, the Human Genome Diversity Project extended bioprospecting to the blood and tissues of indigenous peoples themselves (Harry 1994).

For farmers and indigenous peoples, however, there are a variety of drawbacks to the bilateral, market-oriented approach. The most obvious of these is the difficulty in establishing a price for genetic materials. The utility of any particular accession cannot be determined at the point of collection. It is only after evaluation and extensive research and development that the rare "hit" results in a new drug or an improved plant variety. Moreover, some genetic traits which are apparently useless now may become important only at some time in the future when their latent utility is revealed by changing social or ecological conditions. As a result of this indeterminacy, the "up front" payments that bioprospectors can be expected to offer for access to genetic resources have been quite small. Recourse to royalties or other forms of time-lagged payments linked to future values provides farmers and indigenous peoples with a way to capture financial benefits in such a situation. But playing the royalty game with transnational corporations that possess large stables of accountants and lawyers, who are expert in writing and interpreting contracts in a way that favors their employers, is no simple matter. As Martínez-Alier (1994:81) notes, the poor tend to sell what they have at a low price simply because they are poor, and "If Costa Rica cannot get a good price for bananas, how can it get a good price for biological diversity?"

Perhaps, most fundamentally, western property arrangements were not designed for collective/community innovation and are simply not well suited to the needs of indigenous peoples and farmers or to their landraces. Over the decade of the 1990s there was a flood of literature from progressives and well-meaning ecoliberals (e.g., Hamilton 1993; Kadidal 1993; Reid et al. 1993; Greaves 1994; Brush and Stabinsky 1996) and from organizations representing or allied with farmers and indigenous peoples (e.g., Gupta 1993; Posey and Dutfield 1996; Rothschild 1997) trying their best to figure out how existing IPR and market arrangements might be twisted and stretched and transmogrified to operate so as to protect the rights and reward the inventive activities of community and informal innovation. The provision for *sui generis* arrangements under TRIPS appeared to offer a real opportunity for the design of "traditional resource rights" that might be articulated to existing property rights law in ways that would actually benefit those who produce and reproduce crop landraces and other forms of embodied indigenous knowledge. It may be that when a clear and unambiguous association of crop genetic material with a people or community is possible, bilateral arrangements between a community or a people and an outside entity may be useful and equitable if intelligently crafted.

Certainly the obstacles to be overcome in concluding such agreements are substantial. Indigenous and peasant communities are rarely the homogeneous, solidary, stable entities that some analysts imagine, and different factions may have very different positions on the propriety or conditions of any particular bioprospecting agreement. While the principle of prior informed consent might provide some operational guidance, the farmers and indigenous peoples who are being targeted by the bioprospectors typically do not have very extensive knowledge of what the bioprospectors will do with the information and organisms they collect, or of the legal, scientific, and commercial frameworks into which they are being inserted. Absent such understanding, it is difficult to see how farmers and indigenous peoples can provide informed consent to bioprospecting activities and construct exchange agreements that are adequately sensitive to their own interests. While the literature is replete with examples of abuses (for summaries see especially Rural Advancement Fund International 1995, 2000a), there are few extant models for how hybridization of indigenous and industrial arrangements for intellectual property rights might be successfully accomplished (see especially Cleveland and Murray 1997 and King and Eyzaguirre 1999 for a review of efforts).

This disjuncture between the models and their application seems to point to a fundamental contradiction between the collective, cooperative, multigenerational modes of knowledge production that are frequently characteristic of indigenous and local farm communities and the capitalist property and market institutions with which they are confronted. Over the last two decades, corporate interests have worked very hard indeed to put in place a legal framework of global reach which is designed to allow anything and everything to be privately and individually owned and therefore privately and individually sold. Since all production is social (no Robinson Crusoes on this planet), it ought to be clear that IPRs are actually an attempt to circumvent and obscure the very reality of social production and to subsume the products of social production under private, individual (either literally "individual" or corporate) ownership.

It follows that existing IPRs can be nothing but antagonistic toward social relations founded on collective responsibility and communal or community ownership. Peoples and communities who do choose to engage in bilateral agreements may find themselves drawn inexorably into the web of values, ethics, and activities characteristic of the market in contemporary capitalism (Agrawal 1995; Argumedo 1995). Indeed, it is this tendency that is of particular concern to Camila Montecinos and Erna Bennet (2002). "Regrettably," writes Montecinos (1996:27), "the gradual deviation of the discussion toward alternatives or exceptions inside the existing system has lost us precious time in determining what is really needed, regardless of whether or not it fits into a predetermined definition or regulation. This is especially serious when for various reasons we are being confronted with absolute demands to 'finally' define Farmers' Rights."

Farmers Rights' have indeed recently been given a more determinate form than in the past. In 1994, negotiations were initiated to revise the FAO's International Undertaking on Plant Genetic Resources in order to harmonize it with the Convention on Biodiversity. This revision of the Undertaking was to address significant weaknesses in the CBD. The Convention had failed to deal with plant germplasm already appropriated and stored in gene banks and, even more critically, had also neglected to concretely engage the question of "Farmers Rights." In 1994, an agreement was reached which brought much of the germplasm held in CGIAR gene banks (some 500,000 accessions) under the trusteeship of the FAO, by whom the materials would be held in the public domain – and unpatentable – for the benefit of humanity. Access to that "in trust" germplasm would be on the basis of "facilitated access" rather than "free access." That is, materials would be available to countries adhering to the Undertaking and also agreeing to provide the "benefits sharing" envisioned by the CBD. Further progress was stymied by continued disagreement over what constituted benefits sharing and its relation to Farmers' Rights. Over the next nine years, although it was a signatory neither of the CBD nor of the Undertaking, the United States took the lead in precluding any material expression of Farmers' Rights, which it insisted was merely a concept rather than an actionable mandate requiring financial commitment.

Nevertheless, pressures for some kind of accommodation continued to grow. In the absence of an agreed upon multilateral framework for exchange, breeders the world over complained of the imposition of increasingly onerous restrictions placed on the availability of germplasm by national governments. The sequencing of the rice genome highlighted the potentials that might go unrealized if the flow of plant genetic information were to be subject to continued constraints. Instances of biopiracy in crop plants – patenting of ICRISAT chickpea and lentil varieties in Australia, appropriation of the Indian descriptor "Basmati rice" by the Texas company RiceTec, the patenting of a Mexican bean landrace by an American breeder – underlined the latitude for abuses under existing arrangements.

In November 2001, a revised FAO Undertaking was approved as a legally binding International Treaty on Plant Genetic Resources for Food and Agriculture (FAO 2001). The Treaty created a "Multilateral System of Access and Benefit Sharing" and passed with only two nations abstaining: Japan and the United States.[24] In 2003, the U.S. became a signatory to the Treaty, if for no other reason than to be able to shape how its provisions are implemented. Although the ambivalence of the U.S. is only apparent, it is authentic among those NGOs that have over the last decade worked hardest to encourage the birthing of a truly new international genetic order. For those progressives, the

Treaty can be seen as a glass half empty, or a glass half full, a "trick or treaty," as the ETC Group (2001b) appropriately phrases it. There are five principal "tricks," yet they are hardly covert and will be familiar to anyone who has followed the Seed Wars.

First, the content of Farmers' Rights is still under-determined. While Article 9 of the Treaty accords farmers three important categories of rights, those perquisites are not specified as clearly as might be preferred. Second, most of the important provisions of the Treaty – including Farmers' Rights – incorporate the proviso, "subject to national legislation." Indeed, the Treaty makes it explicit that "responsibility for realizing Farmers' Rights . . . rests with national governments," so there is no multilateral instrument for enforcement.[25] Even more critically, the right to "save, use, exchange, and sell farm saved seed" is also made subject to national law, which is to say, to PBR legislation. Farmers' Rights are effectively made subordinate to plant breeders' rights. Third, although there is a mechanism for sharing the benefits of commercialization of germplasm, the level, form, and manner of payment are yet to be determined, which may well lead to the kind of negotiating that has already resulted in decades of empty coffers for the FAO Gene Fund. Fourth, materials collected from farmers and stored under the auspices of the Multilateral System may have only the thinnest veneer of protection from patenting. The Material Transfer Agreement (MTA) mandated by the Treaty for the exchange of germplasm requires only that intellectual property rights not be claimed on germplasm "*in the form received* from the Multilateral System" (FAO 2001:7, emphasis added). This has been taken by the private sector to mean that any subsequent transformation of the material renders it patentable. Finally, although Article 13.3 suggests that "benefits . . . should flow primarily, directly and indirectly, to farmers," one suspects that "indirectly" through the scientific community will be the dominant route since the Treaty envisions exchange of information, transfer of technology, and capacity building as the main mechanisms for benefits sharing.

Under the Treaty, both public and private plant breeders have free access to the germplasm they want and can formally acquire intellectual property rights to the cultivars they produce. Farmers must respect PBRs and patents and might, some time in the future, be recipients of benefits as they filter down through plant scientists' research programs. In sum, it can be argued that the Treaty has not actually moved the international genetic resources regime much beyond where the Undertaking had it twenty years ago.

While the Treaty is clearly not a qualitative advance over past arrangements, it can be seen to represent meaningful progress toward a more equitable global framework for crop germplasm exchange. There is now a legitimate, formal Multilateral System to manage access to most of the world's most important food crops and to provide a counterbalance to market-based bilateral arrangements. Critically, the rich and invaluable materials stored in CGIAR center gene banks are placed unambiguously in the public domain under the control of an intergovernmental body. Plant genetic materials are available on a "facilitated" (i.e., free but regulated) basis to those nations ratifying the Treaty. Any party requesting germplasm is required to accept a Materials Transfer Agreement that mandates payment of "an equitable share of the benefits" arising from development of any commercial cultivar or product that incorporates material accessed form the Multilateral System.

The wording of the MTA ban on patenting of materials as received from the system *could* be read as disallowing patenting on *any* part of the material. Even if a pro-business reading is ultimately upheld, the provision will at least slow the process of commodification and act to prevent patent claims on existing varieties. Most importantly, though their articulation leaves much room for equivocation, Farmers' Rights appear in the Treaty in a much stronger form than they have in previous agreements. Farmers' Rights are declared to include (FAO 2001:5):

(a) protection of indigenous knowledge relevant to plant genetic resources for food and agriculture;
(b) the right to equitably participate in sharing benefits arising from the utilization of plant genetic resources; and
(c) the right to participate in making decisions, at the national level, on matters related to the conservation and sustainable use of plant genetic resources for food and agriculture.

Operationalizing what are more statements of principle than concrete commitments to specific actions or policies will not be easy, but those provisions are quite solid platforms from which to work.

And work there will be, for even as the Treaty was ratified by the requisite forty nations on June 29, 2004, and became a legally binding document, it is still substantially a work in progress. Genetic Resources Action International (GRAIN), which had lobbied hard for a strong statement of Farmers' Rights and a strict ban on incorporating materials covered by the Multilateral System into any patented cultivar, finds the Treaty a "disappointing compromise" inasmuch as "many of the central issues remain unresolved and open to interpretation" (GRAIN 2001:2). Erna Bennet (2001:7) joins Camila Montecinos in deploring the way Farmers' Rights have taken on some of the institutional and epistemological trappings of intellectual property rights. Of the Treaty, she asks, "What's new about all of this? Nothing. The International Treaty . . . concedes nothing but a few fragments of bracketed text and some '*room for re-opening discussion*' on the '*key issue*' of Farmers' Rights." But it is the very indeterminacy of the treaty – an unstable combination of promise and danger – that provides a continuing opportunity. The seed industry, the U.S. State Department, and the CGIAR system are all as ambivalent about the Treaty as GRAIN and Erna Bennet. How will the phrase "in the form received" be interpreted? Will the CGIAR centers agree to participate? How will benefits sharing be arranged? The very fact that, as GRAIN (2001:3) puts it, "a lot is left up to consensus interpretation and future debate" means that options are still in play and that progressive outcomes are still realizable.

Pat Roy Mooney, Executive Director of the ETC Group, understands the weaknesses of the Treaty as clearly as anyone, but he embraces its indeterminacy as an opportunity: "It will become what we make of it. It is the white elephant turned into the mouse that could roar. We believe it signals a very important breakthrough" (ETC Group 2001b). I agree with the ETC Group, for I believe that a set of social and political and scientific circumstances is now coalescing that makes the achievement of progressive outcomes at least plausible.

Conclusion: revitalizing public plant science

This book has traced the story of how plant breeding and seed production became a means of capital accumulation. A corollary to the rise of private industry and the commodification of the seed has been the relegation of public plant science to areas complementary to rather than competitive with corporate plant science. When I completed *First the Seed* in 1988, the development and deployment of the new biotechnologies was proceeding along these established historical trajectories. Even now, the objectives of the seed industry – transformed into the life industry – remain constant. The Gene Giants want to ensure the continuous circulation and expansion of their capital by encoding in seed the shape and content of agricultural production processes. They want to ensure enlargement of their markets by forcing the farmer to buy seed (and inputs required by the seed) every growing season. They want the assistance of public science in achieving these ends. Moreover, their ambitions have grown. They are seeking now to extend their reach beyond the farm gate and into the supermarket in order to impose a particular mode of consumption on the world's eaters in the same way they have imposed a pattern of production on the farm sector. The influence of industry over the conduct and direction of public science has the effect of foreclosing alternatives even as the need for alternatives has become increasingly urgent.

Happily, one of the most salient features of the story over the last sixteen years has been the emergence and growth of opposition to the manner in which the companies have developed and deployed biotechnology. This opposition has had multiple sources that are widely diffused in both space and in orientation to issues of concern. The environmental community has the deepest roots in resistance. Today, it is a rare environmental organization – local or global, radical or mainstream – that is not critical of the effect of GM crops on the biosphere. Consumers' and food organizations' concerns over biotechnology have grown in proportion to the GM food products available. The initial, narrow pulse of distress that attended the introduction of rBGH in the U.S. in 1993 was revived in America and ballooned in Europe with the widespread planting of GM crops after 1996. As more organisms became candidates for transgene incorporation, religious groups came to question the ethics of such recombination. Animal rights groups similarly balked at the prospect of treating the beings with which we share this earth as objects to be dissected and reconstituted. Farmers' organizations in the North and South objected to the erosion of their ability to save seeds and the possible damage to their ability to sell their crops if consumers rejected GM foods. Groups opposed to corporate globalization found consolidation in the life sciences to be a paradigmatic expression of what they most feared and regarded TRIPS as a prime vehicle for facilitating the imposition of global market structures. Indigenous peoples confronted bioprospecting as a new form of colonization.

Because the issues it touches upon are so wide-ranging and because it manifests itself in different forms in multiple locations over the globe, opposition to the corporate development of biotechnology has taken some time to gain coherence and momentum. But as the underlying connections between apparently disparate issues and locations are revealed and articulated, opposition to genetic engineering in agriculture takes on a unique and unifying global potency. A remarkable range of organizations around the world finds some facet of biotechnology problematic.[26] Concern is not limited to militant organizations such as Greenpeace, but also extends to mainstream organizations such as the Sierra Club and even the relatively staid Audubon Society.

Globally, a steadily growing number of actions have been undertaken by numerous groups at many different scales with the objective of constraining and restraining the activities of the biotechnology industry. Direct action can be fragmentary and covert, as with the destruction of GMO test plots in the United States, Europe, and India. More commonly, it takes the form of demonstrations at points of sale such as supermarkets. In India, the Seed Satyagraha[27] and the Monsanto Quit India! movements have mobilized hundreds of thousands of people (Shiva 2001). Especially, activists have made an effort to organize parallel and alternative conferences wherever biotechnology trade groups hold their annual meetings or wherever CBD or WTO deliberations take place (Tokar 2001).

The opposition looks to find expression in the mechanisms of governance as well in symbolic forms. A variety of municipalities in the U.S. have attempted to declare themselves "GMO-free zones." In March of 2004 Mendocino County, California, became the first locality in the U.S. to succeed in doing so when its residents passed Measure H prohibiting the raising of genetically modified organisms. A principle motivation for the ban is to ensure that crops can continue to be grown organically (Pogash 2004). In Vermont, where seventy-nine towns have passed resolutions opposing the use of GM crops, bills requiring the labeling of GM seed and protecting farmers from liability for the adventitious patent infringement that landed Percy Schmeiser in court were introduced into the state legislature during its 2003–4 session. The "Farmers' Right to Know Act" was passed into law and any GM seed sold in the state must be labeled as such. However, the "Farmer Protection Act" faced stiff opposition from the biotechnology industry because it would require seed manufacturers themselves to indemnify farmers against any liabilities resulting from actual use of GM seed *and* against patent infringement liabilities resulting from inadvertent contamination of crops by GM varieties. In Vermont, it would be the seed companies, not farmers, who are on the hook for any damages resulting from intentional *or* adventitious "use" of GM genes.

Action has been effective on national and international scales as well. When the USDA decided to permit the use of GMOs under its organic rules over 275,000 people responded – more comment than the USDA had ever received on an issue – and forced the Secretary of Agriculture to rescind the proposal. Consumers in the EU collectively refuse to purchase GM foods and popular pressure is the principal reason that the *de facto* moratorium has been maintained. With the leadership of some politically sophisticated and deeply committed NGOs, a signal victory was achieved on the international level with incorporation of the "precautionary principle" into the Cartagena Protocol on Biosafety that was developed as an outcome of the CBD. By allowing the regulation of imports of GMOs even in the absence of scientific certainty about their potential harms, the Protocol gives nations an important measure of control in dealing with WTO-based pressures to open their markets (Egziabher 2003).

Sensitive to growing opposition to its activities, industry has fought back with the considerable political, economic, and cultural resources at its disposal. A principal theme in its discursive efforts to smooth the way for GM varieties has been the benefits such crops would allegedly have for the world's poor and hungry. In addition to its own advertisements and outreach materials (e.g., CBI 2000), the Council for Biotechnology Information recruited ex-President Jimmy Carter as a spokesman. In a *New York Times* op-ed piece titled "Who's Afraid of Genetic Engineering," Carter (1998:A23) criticizes such precautionary latitude as threatening to "leave food rotting on the dock" while people go hungry. He concludes that

> If [GMO] imports like these are regulated unnecessarily, the real losers will be the developing nations. Instead of reaping the benefits of decades of discovery and research, people from Africa and Southeast Asia will

> remain prisoners of outdated technology. Their countries could suffer for years to come. It is crucial that they reject the propaganda of extremist groups before it is too late.

More recently, U.S. trade representative Robert Zoellick criticized the EU moratorium on GMOs as "immoral" inasmuch as it encouraged developing nations to take a similar stance and therefore deprived the hungry of needed – albeit GM – food (Becker 2003:A6). Activists countered that the U.S. had no real interest in feeding hungry people – witness its world-leading stinginess in foreign aid[28] – and that what it was really worried about was making sure the biotech companies were well fed.

Important actors in the international public agricultural research community agree that the hunger card is a red herring and that the biotechnology industry is not motivated by meeting human needs (Conway 1999; Herdt 1999; Paarlberg 2000; Victor and Runge 2002). But where activist critics of the companies see the problem as too much biotechnology, prominent Green Revolutionaries see the problem as too *little* biotechnology. That is, the companies are pushing Bt and herbicide resistant corn and Roundup Ready soybeans when what hungry people in Africa and Asia and Latin America need are disease resistant and insect resistant and locally adapted varieties of corn, chickpeas, cassava, lentils, rice, sorghum millet, and bananas. Though biotechnology could be applied to those crops, it has not been because private industry does not develop seeds for people who cannot pay for them. That responsibility has been assumed by the CGIAR centers and national agricultural research programs. But access by those public institutions to the necessary tools is now being constrained by the narrow range of private interests who own them.

This situation is understood with great clarity within the Rockefeller Foundation. The Foundation's Director for Agricultural Science observes that

> The business plans of the mega-seed companies seem straightforward: control everything from genetic engineering of seed to the selling of seeds to farmers, to marketing plant-grown drugs, modified foods, and industrial products. They aggressively employ patents to claim intellectual property. . . . They can be expected to seek profits from new high-quality seeds for farmers who are capable of paying a premium price. But they have little incentive for producing crops that are important to the poor and disadvantaged farmers who primarily save their own seed. [Herdt 1999:9]

Rockefeller President Gordon Conway summarizes the implications for the public sector:

> Developing-country crop variety development systems are poorly equipped to deal with the rapid changes that are occurring. They have

> depended extensively on international free exchange of germplasm. . . . As plant research in the industrialized world has come to be dominated by private companies who closely guard their proprietary technologies, the process of innovation in the developing countries has slowed. Public-sector plant breeders don't know how to respond, and when they try, they are handicapped by the huge disparity in resources and negotiating power between themselves and the companies. [Conway 1999:2]

Remarkably, there is a growing sense in the international agricultural research community that corporate concentration and the private appropriation of the natural world through intellectual property rights has become *the* major obstacle to equitable agricultural development.

The practical implications of this insight were not lost on Conway. In his extraordinary 1999 speech to the staff of Monsanto, he issued a call to quite radical reform. Voicing his confidence that application of biotechnology holds enormous promise for the "poor and excluded," he proceeded to indict Monsanto, in particular, and the biotechnology industry, in general, for a host of failings including inattention to outcomes in the rush to market, inappropriate patenting of gene sequences, engaging in biopiracy, and abusing market power. He advised Monsanto (and, by implication, the industry generally) to embrace the precautionary principle, abandon utility patents on plants in favor of PVP exclusively, accept labeling of GMOs, phase out the use of antibiotic resistance markers, disavow Terminator Technology, establish a fellowship program for developing country scientists, and agree to license patented enabling technologies in developing nations at no cost. He concluded by calling for development of an agricultural science that employs both "ecological approaches" and "participatory approaches that strengthen farmers' own experimentation and decision-making" (Conway 1999:4).

Note that Conway's position is actually not very far from that of Greenpeace or the ETC Group or the Organic Consumers Association! However, he is speaking not to Greenpeace, but to Monsanto. Although his program is quite radical, his political frame of reference has not kept pace with his structural analysis. His instinct is to reform Monsanto rather than join with unaccustomed allies to change the structures within which industry operates. When he imagines the revitalization of public agricultural science, he anticipates "new forms of public-private collaboration" (Conway and Toeniessen 1999:C56) rather than a public sector emancipated from corporate hegemony. So while the Rockefeller Foundation funds a multi-year initiative to support the emergence of fairer intellectual property policies that "defend the sharing of knowledge, information, and research among scientists, inventors, farmers, and creators" (Rockefeller Foundation 2002), it is also creating the African Agricultural Technology Foundation in cooperation with Monsanto, Syngenta, DuPont, and Dow (Gillis 2003).

One promising development is a proposal originating at Cornell University and involving three CGIAR centers and the public plant science programs of China, Brazil, France, Japan, the UK, and the Netherlands. The "Challenge Program: 'Unlocking Genetic Diversity in Crops for the Resource Poor,'" would fund a "unique *public platform*" for using molecular and traditional means to develop enabling technologies for the breeding of varieties suited to the needs of the poor. A central feature of the program would be a PIPRA-like approach to acquiring intellectual property rights in order to "ensure that research outcomes remain accessible in the public domain" (International Maize and Wheat Improvement Center 2003). Can such initiatives be platforms for the formation of connections with novel partners in civil society? Or will the public sector be drawn back within the orbit of capital?

Parallel developments are occurring in public sector plant science in the United States. A paper presented by Cornell University plant breeder Ronnie Coffman at the 1998 annual meeting of the American Seed Trade Association demonstrates that, like Conway, he is acutely aware of the challenges facing public plant scientists. And like Conway, he clearly locates the source of those challenges in the concentration of corporate power and the expansion of intellectual property rights. Citing a paper by his Kansas State wheat breeder colleague, Rollin Sears (1998), Coffman comments:

> If words were copyrighted, only the few who owned them could communicate and our society would be harmed. Genes are analogous to words in that they allow the creation of new plant cultivars just as words allow the creation of a book. Everyone in society should have the right to use genes. Cultivars (novel genotypes or combinations of genes), not genes should be eligible for patenting. It is now clear that the patenting of genes will result in only two or three companies having a major influence on the food system. [Coffman 1998:8]

Exploring the implications of this tendency, he cites the comments of his Cornell colleague Charles Arntzen:

> With the dramatic shift to proprietary protection for crops, and a massive consolidation of the seed supply industry into vertically integrated food/feed life sciences companies, the future employers of crop breeders are rapidly changing. If the history of corporations involved in this consolidation is any indicator, crop breeders run the risk of becoming "hired hands" working for MBAs and corporate attorneys. [quoted in Coffman 1998:7]

Coffman finds this restrictive division of labor unpalatable and, no less than Conway, is prepared to interrogate the basic assumptions of property and power that have for so long defined the political economy of plant improvement.

For at least some plant scientists, and especially those still involved in cultivar development, questioning has matured into a willingness to directly challenge prevailing patterns. In the summer of 2000, five public plant breeders, one sociologist, and the director of an activist NGO met to explore the prospects for revitalizing public plant science. One of the participants, Washington State University wheat breeder Stephen Jones, was embroiled in controversy over his refusal to incorporate Monsanto and BASF herbicide resistance genes into his wheat cultivars (Schubert 2001b). Jones's stance resonated powerfully with his colleagues, who formed a Caucus on the Future of Public Plant Breeding. The Caucus's vision of that future included opposition to the patenting of naturally occurring genes and cooperation with farmers as central components of "vigorous, creative plant breeding programs, supported by public funds." With the facilitation of the Rural Advancement Fund International-USA, the Caucus initiative took on a more expansive form: in September 2003, some twenty-five public plant scientists met in Washington, D.C., with representatives of selected NGOs and farmers' organizations at a "Summit on Seeds and Breeds for the 21st Century."

Recognizing that public breeders now operate in an environment "where control of elite germplasm has increasingly become proprietary," the goal of the Summit was to "Develop a blueprint or road map for re-invigorating public domain land and animal breeding to meet the needs of a more sustainable agriculture" (Sligh and Lauffer 2004). There are several features of this initiative that closely parallel the concerns outlined by Gordon Conway in his Monsanto address. First, there is a clear and explicit recognition that growing corporate power is the chief force shaping and conditioning the context within which public plant science operates. Any effort at change must necessarily engage the exigencies of political economy, and especially the role of intellectual property rights. Second, there is an explicit commitment to service of the "public interest," which is defined both as the enhancement of "farmer and consumer choice" and as work that is "targeted to support . . . sustainable systems." Third, there is also an explicit commitment to research approaches that incorporate "tie ins with farmers (participatory programs)" and "cooperative problem solving." At both international and domestic levels, there is within the public plant science sector an emergent sense that revitalization of its activities will require confronting corporate dominance, a reorientation of research objectives to sustainability, and inclusion of clients in research processes.

Where this Summit initiative represented a critical and material advance over the Rockefeller Foundation model is in its effort to connect to a different set of social actors in pursuit of solutions to key problematics. Land grant scientists at the Summit were talking not to Monsanto or Syngenta or Dow or DuPont, but to representatives of the Rural Advancement Fund International-USA, the Union of Concerned Scientists, the Organic Farming Research Foundation, the Northern Plains Sustainable Agriculture Society, and the Land Institute. It is in this reworking of the *social* landscape that the Summit broke new ground. The Summit was an opportunity for public sector plant science to withdraw from its accustomed role as a supplicant to private industry and to embrace an alliance with civil society organizations willing to confront the Gene Giants and demand that resources be provided for alternative paths.

Rural Advancement Fund International (2001a) has made the argument that the CGIAR's international research centers should adopt such a strategy:

> The arguments for common cause seem strong. We are both dedicated to ending hunger and achieving food security. We also share a common enemy – the corporate/country driven demand for privatization and the diminution of public research in deference to corporate solutions. We share – to a degree that would shock many – a common vision of "sustainable agriculture" and a mutual antipathy for intellectual property monopoly. All this should be a basis for alliance.

Could it be that this alliance will emerge first in the United States?

The conclusions and policy recommendations that emerged from the summit (see Sligh and Lauffer 2004:171–175) are not themselves particularly remarkable or original. Nor do they adequately reflect the oppositional energy that was so apparent in the meeting's presentations and discussions. What is significant, however, is that public plant scientists found common ground with some quite radical advocacy organizations and established an institutional vehicle for strengthening and reinforcing that relationship. As this book goes to press a follow-up conference is scheduled for September 2004 in order to engage a larger set of stakeholders and to begin mapping concrete strategies for the reinvigoration of public plant and animal breeding.

An obstacle to cooperation among civil society organizations and public plant science may be their differing attitudes to biotechnology. Although there are exceptions (see Altieri 2001; Cox 2002), most public plant scientists anticipate that the new genetic technologies will be useful and even vital adjuncts to conventional breeding. All too often, however, activist and advocacy groups working to oppose the activities of the Gene Giant multinationals have tended to fall into a conceptual shorthand in which "biotechnology" (and especially the subsets "genetic engineering" and "GMOs") is almost completely identified with the companies that are using it. The result is a species of technological determinism in which opposition to "biotechnology" is insufficiently distinguished from opposition to "*corporate* biotechnology."

Failure to make that distinction alienates many scientists - themselves often users of particular biotechnologies - who are otherwise sympathetic to criticisms directed at the manner in which those biotechnologies are being used by commercial interests. It has also led to impoverishment of debate and a discursive climate in which the narrow and dystopian construct of "Frankenfood" confronts the similarly narrow but utopian construct of "Golden Rice." This simplistic dualism obscures important differences not only among technologies and their likely impacts, but between the character of corporate and public research as well (Stone 2002; Ruivenkamp 2003).

A conjoining of activist groups and scientists interested in reclaiming a truly public science would be a very good thing. Achieving such an alliance will likely mean untangling biotechnology from the corporation and developing a more nuanced approach both to the technologies themselves and to distinguishing "between the practice and the practitioner," as wheat breeder Stephen Jones puts it. This should be possible, for the focus of concern and opposition from civil society tends to be on genetic engineering (the transfer of genes from one species to another) and GMOs, rather than on the many other methods and techniques which do not involve transgenes. Moreover, with the cutting edge of research moving to genomics, plant scientists may soon have available information and tools that will allow them to explore the deep reservoirs of diversity that are available within organisms' own genomes (Tanksley and McCouch 1997; Nielsen 2003). Genetic engineering would become "intragenic" rather than "transgenic," and many of the objections to current practice might be obviated.

But to focus too much on the tools rather than on who is using the tools and for what the tools are being used is to misapprehend the problem. What is really needed is not so much the banning of one tool or the approval of another, but a revision of the way in which we develop tools and for what we imagine we are using them. The real problem, as Henry Miller and Gregory Conko (2001:303) point out in the quote that opened this chapter, is that a small authoritarian minority dictates "what kinds of research are permissible and which technologies and products should be available in the marketplace." This is nowhere clearer in the world today than in plant improvement, where a set of five companies dominates the science and commerce of seed production.

That domination has been achieved through the progressive commodification of the seed by legal and technical means, through the relegation of public sector science to complementary activities, and through free access to genetic raw materials. Industry's capacity to maintain those conditions of its hegemony is now in question. The Gene Giants' pursuit of profits led them to press the commercialization of biotechnology more quickly than was socially acceptable. The contradictions embedded in the appropriation of public resources and the erosion of public protections were made palpable to people around the world. Farmers, indigenous peoples, consumers, and scientists all found their choices diminished, their "freedom to operate" – to use the scientific neologism – restricted as a result of corporate activity associated with biotechnology.

Opposition from civil society has accordingly increased substantially over the past decade, especially outside the U.S., and now threatens the commercial viability of GM seeds and foods. In the scientific community, there is a growing perception that the permitted scope of patenting has been stretched beyond its social utility. Public sector plant scientists find themselves uncomfortably confined by corporate property rights even as the field of genomics opens before them. And access to genetic resources is now subject to compensation. A great many circumstances are in flux and, while capital is mobilizing to respond to these challenges, there appears to be considerable opportunity to create space for progressive change.

In particular, public plant scientists are poised to essay a reassertion of their formerly central role in crop improvement. An active and influential core is disturbed by the way industry has used the new biotechnologies, dissatisfied with their appointed role in the sectoral division of labor, frustrated by the difficulties in accessing key means of production, and anxious to fulfill their mandate to serve the public interest by contributing to the development of a sustainable agriculture. Most importantly, the consequences of the massive consolidation of capital that have occurred over the last decade has fostered the maturation of political-economic sensibility that is more accurately attuned to the realities of late capitalism. Some, at least, are ready to seek an alternative future that is relatively autonomous from industry and its straitened objectives

The emerging interest in sustainability and in participatory approaches to plant breeding is especially encouraging. Decades of gradual subordination to industry and the emergence of increasingly marketized research institutions have established powerful scientific and social path dependencies that will not be easily renounced. Organic/sustainable research programs are uncommon, agroecological/alternative method and theory is underdeveloped, and there is little experience with participatory endeavors. But if public scientists are committed to treading some new paths, the time may never be better. Participation can usefully be implemented through connection to some of the many civil society organizations that are working for a just and sustainable food system and through enhanced contacts with farmers. Moreover, plant scientists in public institutions may well find themselves in a position of surprising strength in the near future. Genomics is complicated and expensive and universities could well regain the preeminent standing they had before industry made its massive, and perhaps misplaced, investments in research and patents. Judicious use of patenting may be a useful tactic for the public sector to preclude private appropriation of research results and to preserve free exchange among public entities (i.e., through PIPRA-like arrangements).

If crop production is to be shaped to meet human needs rather than to make profits in the twenty-first century, then public agricultural research institutions must find a way to play a key role in that endeavor. The four principal reasons for maintaining a robust public presence in plant improvement are the same as they were sixteen years ago. First, the public sector can produce public goods that are socially valuable but do not attract private investment because they are not profitable: cultivars for subsistence farmers, varieties in minor crops, cultivars for alternative (e.g., organic) production systems, management oriented input-displacing technologies and practices. Second, by developing and releasing finished cultivars, the public sector can provide a countervailing force to the market power of large companies as well as maintain standards of quality. Third, the public sector can explore options and alternatives and innovations that are not pursued by private industry. Given the many uncertainties associated with global warming and the cumulative and unanticipated effects of human action on the biosphere, it is irresponsible to commit the future to a single path defined by profitability. Fourth, we need the public sector to supply independent and reliable information if we are to have an informed populace capable of making informed decisions in a complex and uncertain world.

In 1988, I concluded this book with a challenge to plant scientists to consider Henry Wallace's call to political duty. I will end this edition with a similar passage from Erna Bennet (2002:10):

> The day is coming when scientists and intellectuals will accept the need to take social action and accept social responsibility as an integral, and not a supplementary part of their scientific responsibility, adding their voice and their actions to those of millions of others. That will be a day of great hope for a direly threatened world.

That day may be near.

Notes

1. Introduction

1 DNA is an acronym for deoxyribonucleic acid, the fundamental genetic material found in all living organisms.

2 "It must be kept in mind that the new forces of production and relations of production do not develop out of *nothing*, nor drop from the sky, nor from the womb of the self-positing Idea; but from within and in antithesis to the existing development of production and the inherited, traditional relations of property" (Marx 1973:278).

3 Robert Merton (1970, 1973) laid the foundations for the "internalist" approach to the social analysis of science. He has emphasized the shaping influence of an ethos subscribed to by all members of the scientific community. Michael Polanyi (1945, 1962) added the notion of the independence and self-governing nature of this "Republic of Science." Science is conceived of as a kind of autonomous handcar powered by scientists who lay track wherever their curiosity leads them and their shared values permit them to go. The model is one of endogenous determination of scientific progress.

4 By "finished" varieties are meant lines to be released for commercial use, as opposed to breeding lines or exotic germplasm.

2. Science, agriculture, and social change

1 Note that Marx conceives of both forces and relations of production in social terms. As he puts it in the *Grundrisse*, they correspond to "two different sides of the development of the social individual" (Marx 1973:706).

2 "Large-scale industry tore aside the veil that concealed from men their own social process of production and turned the various spontaneously divided branches of production into riddles, not only to outsiders but even to the initiated. Its principle, which is to view each process of production in and for itself, and to resolve it into its constituent elements without looking first at the ability of the human hand to perform the new processes, brought into existence the whole of the modern science of technology. The varied, apparently unconnected and petrified forms of the social production process were now dissolved into conscious and planned applications of natural science, divided up systematically in accordance with the particular useful effect aimed at in each case" (Marx 1977:617).

3 Engels goes so far as to suggest that "from the very beginning the origin and development of the sciences has been determined by production." And again,

"Hitherto, what has been boasted of is what production owes to science, but science owes infinitely more to production" (quoted in Rosenberg 1974:716).

4 "It is machines that abolish the role of the handicraftsman as the regulating principle of social production. Thus, on the one hand, the technical reason for the lifelong attachment of the worker to a partial function is swept away. On the other hand, the barriers placed in the way of the domination of capital by this same regulating principle now also fall" (Marx 1977:491).

5 "Thus were the agricultural folk first forcibly expropriated from the soil, driven from their homes, turned into vagabonds, and then whipped, branded and tortured by grotesquely terroristic laws into accepting the discipline necessary for the system of wage labor" (Marx 1977:899).

6 "With the 'setting free' of a part of the agricultural population, therefore, their former means of nourishment were also set free. They were now transformed into material elements of variable capital. The peasant, expropriated and cast from the land had to obtain the value of the means of his subsistence from his new lord, the industrial capitalist, in the form of wages" (Marx 1977:908-9).

7 "The discovery of gold and silver in America, the extirpation, enslavement and entombment in mines of the indigenous population of that continent, the beginnings of the conquest and plunder of India, and the conversion of Africa into a preserve for the commercial hunting of blackskins, are all things which characterize the dawn of the era of capitalist production. These idyllic proceedings are the chief moments of primitive accumulation" (Marx 1977:915).

8 For reviews of the debate over the "articulation of modes of production," see Foster-Carter (1978) and Portes (1981).

9 In the *Grundrisse*, Marx (1973:408) comments: "On the other side, the production of *relative surplus value*, i.e., production of surplus value based on the increase and development of the productive forces, requires the production of new consumption; requires that the consuming circle within circulation expands as did the productive circle previously. Firstly, quantitative expansion of existing consumption; secondly, creation of new needs by propagating existing ones in a wide circle; *thirdly*: production of *new* needs and discovery and creation of new use-values."

10 I define independent commodity production in agriculture as an agricultural enterprise in which (a) there is family ownership of land and other means of production, (b) there is effective entrepreneurial control of the allocation of these factors, (c) labor is provided by family members, (d) commodities are sold in the market, and (e) the family subsists on income from the sale of commodities and the production of use values.

11 In an article entitled "Corporate Farming: A Tough Row to Hoe" (*Fortune* 1972), Dan Cordtz enumerates from a business point of view the reasons for the "disappointing to disastrous" performance of large corporations in farming. Interestingly, he comes up with a list of factors startlingly similar to the set of obstacles discussed here. However, he adds that overproduction and lack of product differentiation capability have also been important.

12 Thus, for example, Mann and Dickinson (1978) criticize Chayanov (1966), P.

H. Mooney (1982) takes Mann and Dickinson (1978) to task, and Mouzelis (1976) critiques Amin and Vergopoulos (1974).

13 I am indebted to Richard Lewontin (1982) for the stylistic approach used in this passage.

14 The NSF tripartite division is somewhat unusual. Generally, "development" is not broken out of "applied," and the discussion that follows will treat the issue in the terms in which it is conventionally addressed, that is, as a dichotomy between "basic" and "applied" research.

15 Hence, industry may invest in what appears to be very "basic" research (e.g., molecular biology) but nevertheless promises to be realized as a product of some sort. Conversely, the state may undertake certain "development" efforts (e.g., the tomato harvester) to which, for a variety of reasons, capital may not have been attracted.

3. The genetic foundation of American agriculture

1 "Thus, the American grower of cash crops in effect became the cat's-paw for our industrial capitalism: he was encouraged to push out his horizons endlessly, to put more and more virgin acres under the plow, to engage in greater and more magnificent livestocks enterprises; so that the railroad manipulators, the steelmasters, and the factory lords could bring their industrial plans to maturity without molestation from foreign competition. Such was the part performed by historical necessity. It was a kind of, perhaps only dimly felt, economic compulsion, but the expansion of American agriculture, during the long period we were a debtor nation, permitted industrial capitalism to triumph" (Hacker 1940:397-8).

2 For example, of over 150 varieties of fruit trees imported by Eben Preble into Boston in 1805, only three varieties, two cherries and a pear, survived in North America (Webber 1900:467).

3 In 1857, a planter by the name of Wray unsuccessfully tried to maintain a monopoly on a set of sixteen particularly good African sorghums. He failed because he could not control their propagation (Klose 1950:50).

4 Many naval officers were farm owners and were already accustomed to collecting seeds of interest at their ports of call. Military activities have historically provided many opportunities for the appropriation of germplasm in foreign lands. Most recently, the dwarfing stock for contemporary peach varieties was brought back from China by an American flyer after World War II. And the germplasm containing the famous Norin 10 dwarfing gene for wheat was sent to the United States from Japan in 1946 by an agricultural adviser working with the army of occupation (Wilkes 1983:141). University of New Hampshire plant breeder Elwyn Meader noticed a gynoecious cucumber plant while serving as horticulturist for the U.S. Army command in Korea in 1946. This germplasm made hybrid cucumbers possible.

5 He continues: "Was it an accident that the founders and patrons of the Royal Society in London – indeed some of the first experimenters in the physical

sciences – were merchants from the City? King Charles II might laugh uncontrollably when he heard that these gentlemen had spent their time weighing air; but their instincts were justified, their procedure was correct: the method itself belonged to their tradition, and there was money in it. The power that was science and the power that was money were, in final analysis, the same kind of power: the power of abstraction, measurement, quantification" (Mumford 1963:25).

6 See Rossiter (1975) for a comprehensive account of Liebig's impact on early agricultural research in the United States.

7 To be sure, eastern farmers joined the industrialists on this issue, for the deterioration of their soil and competition from western farmers made them sympathetic to a more systematic approach to scientific solutions to agricultural problems. That the East was fully settled meant that land-grants for the eastern states would come from tracts located in western states. New York, for example, was ultimately allocated huge acreages of valuable timber land in Wisconsin with which to underwrite the establishment of the New York State College of Agriculture; see Gates (1943).

8 Antagonism in the South was not universal, however. The influential journal *Southern Cultivator* saw promise in an experimental farm "worked by negroes" and an agricultural college to "supply educated overseers and mechanics" for plantation owners (Gates 1960:377).

9 The department would not rate a secretary until raised to cabinet status in 1886.

10 The creation of the SAESs is largely the result of a lobbying effort on the part of land-grant college administrators and scientists who desperately wanted an institutional buffer between the university and the insistent demands of a farm clientele who expected them to, in the words of one farmer, "bring science down out of the sky and hitch it to a plough" (Carstensen 1961:12). See Rosenberg (1976) for an account of the stations' genesis and the difficulties they faced.

11 Some Shaker communities gained a reputation as growers of excellent seed and undertook its production as one of their specialized communal activities. The Shakers were responsible for a number of marketing innovations – the paper packet, for example – that would be taken up by the commercial seed trade (R. F. Becker 1984).

12 The magazine *Horticulture* still persists in this practice. Actually, the journals' principal objections were to the annual report published by the Patent Office and also distributed free to farmers. The 1895 version of this document was 1,376 pages long, and of that total, 1,085 were devoted to agricultural matters. Journals such as *De Bow's Review* saw in these pages "nothing but a cheap and unfair competitor to our statistical and agricultural periodicals" (Rasmussen 1975a:574).

13 It should not be thought, however, that Morton necessarily valued the free dissemination of new ideas any more highly than that of old seeds. With the journal editor's old distaste for government publications he complained of the "vast volume of reading matter given free to all who asked for it" and recom-

mended that departmental publications be "furnished to such citizens only as will pay for their net cost" (USDA 1895b:53).

14 The huckstering of poor seed was nothing new in 1880. Richard Gardiner, in 1603, warned English farmers of "the great and abominable falsehoods of these sortes of people which sell Garden Seedes" (Ferry-Morse 1956:3). The "poor seed evil" never became as serious in the United States as it did in Britain, but the USDA's Chief of Pure Seed Investigation noted in the 1899 *Yearbook of Agriculture* that "gardeners have often been deceived by unscrupulous dealers. They have bought seeds in good faith and have found them worthless; they have thought they were sowing clover, but have reaped thistles" (Pieters 1900:571).

4. Public science ascendant: plant breeding comes of age

1 For example, ASTA president Watson F. Woodruff's comments at the 1909 annual ASTA meeting: "The Government seed shop is working on full time and doing a most flourishing and growing business. An ugly and vicious animal will often hang himself if given rope enough, and it has been my hope that this monster would ultimately sink of its own accord, but alas, on the contrary the powers that be, have elected that this Congressional Pet should be further nursed by increased appropriations . . . The free seed matter has been given such careful thought and study for years by the Association, and the able committee we now have on this subject may have some recommendations. Further than to keep up a perpetual warfare whenever a vulnerable spot shows itself I have nothing new to recommend by way of fighting this monstrous evil" (Woodruff 1909:28-9).

2 For example: Meat Inspection Act (1890), Tea Adulteration Act (1897), Butter Inspection Act (1902), Pure Food and Drug Act (1906).

3 ASTA complained of the USDA: "With one hand it holds up for our inspiration and guidance lofty standards of purity and germination, while with the other hand it distributes broadcast over the land seeds so inferior in quality as to make all of its contentions contemptible" (ASTA 1901:34).

4 "[M]ost farmers were simply not rational businessmen. Highly tradition oriented, the vast majority of farmers grew the same crops year after year, seldom knowing which were profitable, where they were profitable and when they were profitable" (Danbom 1979:67).

5 See especially USDA (1967), Peterson (1975), Pioneer Hi-Bred (1984), and Harlan (1984) for accounts of the contributions of exotic germplasm to disease resistance, pest resistance, plant architecture, environmental tolerances, and sexual characteristics of American crop varieties.

6 This practice was aggressively pursued by cottonseed firms. The USDA's principal cotton breeder, H. J. Webber, became a consultant to Coker's Pedigreed Seed Company and supplied it with a breeding system and its original breeding material. He left public service for a management position with the company in 1920. Two breeders left the Mississippi Agricultural Experiment Station in 1922 to work for Delta and Pine Seed Co., and two other breeders left shortly

thereafter to form their own company, Stoneville Pedigreed Seed Co. All brought public germplasm with them, of course (Ware 1936:668). Cotton is something of an anomaly. There are considerable price differentials for length of staple. Growers began intensive selection for long staple length very early on. But because cotton is open-pollinated, it was difficult to maintain plant type. Growers became accustomed, therefore, to purchasing seed of long-staple varieties every year, thus opening an opportunity for private investment. Most cottonseed companies were once large plantations, and cotton breeding has always had a significant private component.

7 These statistics do not reflect acreage planted to each variety and therefore do not reveal their relative importance. Other data confirm that public varieties produced via hybridization dominated wheat production by 1925. The famous Marquis variety, for example, was planted to 65 percent of wheat acreage in Minnesota and the Dakotas as early as 1918 (Reitz 1954:255).

8 "Agricultural opinion is not made by some undifferentiated group of hard-working farm operators. It is made by the articulate, the prosperous, and the influential both within and without the community of agricultural producers. Educated and more highly capitalized farmers, editors of farm and rural papers, country bankers, insurance agents, merchants, and implement dealers make up the visible agricultural consensus. These were the men active in most farmers' organizations and specialized producers' associations – those who had come to accept scientific knowledge as necessary for successful economic competition" (Rosenberg 1971:176). See also McConnell (1953) for a parallel analysis of the American Farm Bureau Federation. The world's first experiment station, founded in 1851 in Leipzig, was administered by a board of wealthy landowners (Rossiter 1975:130).

9 Said Allgood of an item to appropriate funds to study the role of the barberry as a vector for stem rust in wheat: "Mr. Chairman, I think we could better the condition of the wheat farmer by appropriating some money to disseminate this disease that comes from these bushes so we will cut down the production of wheat. The best friend the cotton farmer ever had was the boll weevil. We did not have enough sense to know it, and we of the South spent millions of dollars of our own hard-earned money and then came to Congress and asked for appropriations to fight the boll weevil. As a result we now have a 2 year supply of cotton and can not sell it . . . It is my opinion that the wheat farmers would prefer asking for an appropriation to propagate this disease rather than asking for an appropriation to help exterminate it" (*Congressional Record* 1932:1045).

10 See especially Kirkendall (1982) for an account of the development of social science within the USDA.

5. Heterosis and the social division of labor

1 For comprehensive overviews of the returns-to-research school, see Fishel (1971), Arndt et al. (1977), and Minnesota Agricultural Experiment Station (1981).

2 For example, testifying before the Senate in response to Hightower's *Hard*

Tomatoes, Hard Times, University of Florida's vice president for research, E. T. York, said: "What are the returns to society from public investment in agricultural research and education? Studies conducted at the University of Chicago (which I would remind you is not a Land Grant institution) have indicated that the annual returns to society on accumulated investment in research on hybrid corn to be about 700% ... Certainly this has been one of the most dramatic breakthroughs in agricultural technology within our country" (York 1978:270).

3 See Wallace and Brown (1956) and Robinson and Knott (1963) for accounts of the corn shows.

4 However, it is important to note that commercial trade in seed corn was still very small. Farmers purchased not large amounts but enough to provide themselves with a population which they could then multiply. Much seed-corn was sold on the ear, unshelled and packed in wooden boxes so that the purchaser could judge how closely it fit show criteria (Crabb 1947:213).

5 Holden's conclusions are particularly interesting because he was not only inbreeding but also crossing his inbred lines, i.e., what would later be called producing hybrid corn. Moreover, he was shortly to go to work for seedsman Gene Funk (Crabb 1947:25). But both Holden and Funk were preoccupied with the production of "fine lines" of commercial corn and failed to see the potential in inbred crosses.

6 By his own account, East's motivation for coming to Connecticut from the performance-oriented station at Illinois was to have the opportunity to pursue studies in fundamental genetics (Hayes 1963:19). Interestingly, the Connecticut station director was willing to permit East this freedom because he was desperately trying to maintain the efficiency of Connecticut farmers in the face of western competition and hoped that basic research might provide a breakthrough (Rosenberg 1976:193). Thus, while Shull's work may have had a strongly serendipitous character, space for East's parallel research was cleared by unambiguously economic forces.

7 By interesting Wallace and Funk in hybrids, the American Breeders Association did, in at least this one very important instance, succeed at doing what its founder W. M. Hays had hoped: bring public agricultural science and private enterprise into direct contact.

8 Though Jones' explanation did much to win acceptance for hybrid vigor, it did not long survive advances in genetics. But then neither has any other explanation of heterosis. In his book *Heterosis*, Frankel (1983:v) admits that "the causal factors for heterosis at the physiological and biochemical level are today almost as obscure as they were 30 years ago."

9 One wonders if "Sibley" instead of "Funk" would now be synonymous with seed corn in Illinois if East's proposal had been accepted. Crabb (1947:90) notes that East did ask for a "modest remuneration," but it is not clear to what extent East was interested in personal gain.

10 Funk was the scion of a family that, like the Sibleys, had accumulated vast acreage of Illinois black bottomland in the first quarter of the nineteenth century. As late as 1941, Eugene Funk still boasted ownership of some 22,000 mostly tenanted acres (Gates 1973:207). He was by no means representative of the

common seedsman and fits solidly in the tradition of the gentleman-farmer with scientific interests and the financial means to pursue them.

11 Simmonds (1979:147-8) explains that inbreeding depression is attributable to the fixation of unfavorable recessives and that heterosis can be thought of as the converse, "unfavorable recessives fixed in one line being covered, so to speak, by dominants from the other... However, random pairs of lines will only rarely have just the right combination of dominants and recessives to complement each other, so the average hybrid will be near to the source population and outstandingly good or bad combinations will be rare."

12 The leaders included the best and the brightest of the new generation of breeders: R. A. Emerson, Cornell University; G. W. Beadle, California Institute of Technology; R. A. Brink, University of Wisconsin; D. F. Jones, University of Connecticut; C. R. Burnham, University of West Virginia; L. J. Stadler, University of Missouri; M. T. Jenkins, USDA; G. F. Sprague, University of Missouri; W. H. Eyster, Bucknell University; E. W. Lindstrom, Iowa State University.

13 Thus, until 1947, all hybrid corn research at Purdue went on in the Department of Botany, because the Department of Agronomy was not willing to abandon its other work (Crabb 1947:94).

14 Jones (1920:78), for example, complained of the shows: "The choosing of seed-corn still proceeds in the primitive way of selection based solely on appearance. The choice can be made on only one side of the family, for no matter how excellent an ear of corn may be, there is no way of judging the qualities of the plants which furnished the pollen to fertilize the seeds on that ear... By propagating elaborate score cards for judging seed corn, agronomists have thrown a cloak of pseudo-science over an antiquated system which is anything but scientific."

15 Pioneer Hi-Bred has in fact pursued this route more thoroughly than any other company, and its great success is generally attributed to this early decision.

16 The ARI's charter members included 31 agribusiness firms (e.g., Standard Oil, Monsanto, John Deere, Funk Seeds, Eli Lilly), 12 professional associations (e.g., American Society of Agronomy), 14 federal agencies (including the Bureau of Plant Industry and the Agricultural Research Administration), and 14 state experiment stations. No farm or consumer group was involved, though the American Farm Bureau Federation has since joined.

17 The year before, in a letter to Zvi Griliches, Wallace had written, "I now wish we could go into the Red River Valley of the North. Both Pioneer and DeKalb have hit southern Wisconsin hard but they have been somewhat handicapped there by the attitude of the state people" (Wallace 1955).

18 Pioneer, 35 percent; DeKalb AgResearch, 13 percent; Funk Seed (Ciba-Geigy), 7 percent; Trojan (Pfizer), 4 percent; Northrup King (Sandoz), 4 percent; O's Gold (Upjohn), 4 percent; Cargill, 3 percent; Golden Harvest, 3 percent (Leibenluft 1981).

19 These varieties were almost exclusively the product of the farmer-breeders, as their names indicate: Leaming, Reid, Lancaster Sure Crop, Hogue's Yellow Dent.

20 Pioneer's John Airy (1951:7) told the ASTA: "Detasseling is both difficult and expensive. The amount of work to be done requires a large number of people who may work for as little as one week to not more than five weeks. The work must be done in fields often widely scattered in communities where the seed producer has located his plant. Workers must be transported to the fields. Supervision is difficult because the workers are scattered over a field of tall corn. The management cannot possibly supervise labor as can be done by the management in a manufacturing plant." In 1951, detasseling costs were about $10 million (U.S. District Court 1970a:9).

21 Prices are computed from data given in various volumes of *Agricultural Statistics* published by the USDA, and using the index of prices paid by farmers for production items in the *Economic Report of the President.*

22 That the CMS system did *not* reduce seed-corn prices for farmers is not widely recognized. Rather, the reverse is conventional wisdom. For example, the National Research Council (1972a:10-1) ingenuously opines that "Both hybrid corn companies and growers assessed this stage of the development [i.e., the CMS system] and were pleased. No longer need the seedsmen hire countless hands to pull tassels from miles of corn rows, and the farmers could obtain the seed at less cost. The farmer could no longer grow his own seed, of course (because the technology was too complex), but he was willing to give up this freedom."

23 Dr. Stanley Becker has studied the controversy and litigation surrounding the Jones patent using the records held by Research Corporation. I am indebted to him for the knowledge he has shared with me. Readers interested in learning more about Donald Jones are referred to Becker (1976a, 1976b).

24 The focus of the antitrust allegations was the Committee to Study The Effects of Patents on Seed Production, a group formed by several seed companies shortly after the patent issued in 1956. Ostensibly an independent organization, the ASTA paid its bills, sent out its notices, and provided meeting facilities. Research Corporation lawyers found that crucial ASTA files regarding the operation of the committee were inexplicably missing. Other irregularities included the ghostwriting of public breeders' opinions for the trial by seed industry executives. Research Corporation attorneys went so far as to suggest that "the ASTA had infiltrated the ASA, a purportedly noncommercial scientific organization, and was using it as a front to carry out its private objectives" (U. S. District Court 1970b:101; see pages 98-102 for allegations regarding violations of the Antitrust Act).

25 The attitude of breeders is illustrated by Jones' (1950:20) response to a questioner at an ASTA meeting who wanted to know Jones' conception of an ideal shank (the point at which the corn cob is attached to the stalk): "There, again, is another question that I think you can answer better than I. This is something you need to have actual experience in the corn-field with corn picking machinery to answer." The machine is taken as a given.

26 The principal impact of hybrid corn on the nutritional status of the American population has been, indirectly, greatly increased consumption of fats and oils. Corn that is used for human consumption has not improved in nutritional quality,

and protein content may actually have declined since the elimination of open-pollinated varieties (Perelman 1977:45). This is not to say that hybrids of improved food quality are not available. A number of hybrids with superior nutritional characteristics have been produced by public breeders concerned with world food problems. But such varieties do not yield as well as standard hybrids and present problems for machine harvest. Hence, farmers will not grow them, and seed companies will not produce them.

27 The following is taken from Fulkerson and Barnes' (1970) article "A Colloquy on Corn Blight," which appeared in the August 1970 issue of the USDA's *Agricultural Science Review*: "Q: Did geneticists know there was a possibility that reliance on T-background [cytoplasm] would cause trouble later on?" "Fulkerson: Yes, geneticists knew it. Pathologists were aware of it too, and at least one of them predicted five years ago that this would happen. Last year three experiment stations held serious discussions about the status of corn blight – similar to the ones we held in USDA last fall. They expressed concern about the potential for increase in this disease in 1969, and several were conducting investigations at that time."

28 An Iowa judge called it "a blizzard of papers" (Supreme Court of Iowa 1977:171).

29 Barnes and Anderson (1975:207) concur: "Does the mechanization of vegetable and fruit harvest imply the development of machines that will duplicate these sophisticated abilities of the human body? The answer is clearly no. The effective approach is to modify the plant to reduce the degree of selectivity required in harvest and to place the harvested parts in a predictable position in relation to the harvest machine."

6. Plant breeders' rights and the social division of labor

1 That is, it encompassed only those species for which budding, grafting, cuttings, and other non-sexual techniques are the means for commercial reproduction. Such crops include most fruits and nuts and many flowers. As of 1985, about 5,000 plant patents have been issued. Some 70 percent of these are for ornamentals, 20 percent for fruits, and 10 percent for trees and shrubs.

2 Though much effort has been given to hybridizing as many species as possible, commercially acceptable hybrids have proved elusive in such important crops as soybeans, wheat, cotton, barley, oats, rice, and peanuts.

3 For treatments of the European seed industry and UPOV, see Sneep et al. (1979), Simmonds (1979), Innes (1984), and Berlan and Lewontin (1986a).

4 It is widely accepted that it is now an opportune time to make initial assessments of the act. Development of new plant varieties generally takes seven to ten years, so the effects of the legislation on the products of breeding should now be evident. And indeed, various analyses have appeared in recent years. These range from frankly partisan accounts from private industry (White 1976; Studebaker and Batcha 1982; Kalton and Richardson 1983) to partisan accounts from opponents of PBR (P. R. Mooney 1983). Allegedly objective analyses from academic investigators include work by Perrin et al. (1983), an ASTA-funded

study by Lesser and Masson (1983), and a congressionally mandated study by Butler and Marion (1985). Accurate empirical analysis has been greatly hampered by the extremely secretive nature of the seed industry. Surveys have been marked by extremely poor response rates (e.g., 35 percent for Butler and Marion) and a great reluctance to divulge financial data even to such investigators as the National Council of Commercial Plant Breeders (Kalton and Richardson 1983:16). The unevenness of the data available makes its placement in historical perspective all the more important and enlightening.

5 In 1965 the index of prices paid by farmers was lower for seed than for any other class of inputs (feed, fertilizer, agricultural chemicals, fuel and energy, etc.). By 1979 the index for seed surpassed those of all other classes of inputs (Lesser and Masson 1983:87). Lee (1985:40) reports that prices for proprietary soybean varieties are more than double those for public lines. He comments, "Do private varieties have sufficient value to repay these higher costs? The answer to the question is unclear."

6 For example, a recent article in *HortScience* was titled "Alternative Methods of Funding Vegetable Research" (Pike 1984). The author specifies what he regards as "key points in obtaining research gifts or grants" from private industry. His principal points are: "Learn the industry needs... Get acquainted with industry people... Establish research goals, research budgets and bring industry and administration together... Provide leadership... Produce results."

7 These seventeen corporations are Royal Dutch Shell, Lafarge-Coppee, Sandoz, Upjohn, Lubrizol, Limagrain, Celanese, International Telephone and Telegraph, Cargill, Ciba-Geigy, Atlantic Richfield Co., Unilever, Rohm and Haas, Monsanto, Pfizer, Stauffer, and Occidental Petroleum.

7. Seeds of struggle: plant genetic resources in the world system

1 Though Marx does not mention it, the spread of tea among the British working class was no less an expression of its pauperization than was its adoption of the potato. Though the aristocracy criticized the "extravagance" of the poor in their fondness for an imported luxury good, the replacement of milk by tea was really a result of the enclosures and the changed patterns of agricultural production they entailed. Hamilton (1948:74) quotes a tract published in 1798 by farm-worker activists: "Wherever the poor can get milk, do they not gladly use it? And where they cannot get it would they not gladly exchange their tea for it? ... Tea-drinking is not the cause, but the consequence of the distresses of the poor."

2 The botanical networks of the imperial powers encompassed the farthest reaches of empire. For example, the Royal Botanic Gardens at Kew near London were, by 1889, linked to subsidiary botanical gardens in Adelaide, Auckland, Bangalore, Barbados, Bombay (3 gardens), Brisbane, British Guiana, Calcutta, Cawnpore, Darjeeling, Dominica, Dublin, Edinburgh, Fiji, Glasgow, Gold Coast (today, Ghana), Granada, Hong Kong, Jamaica (5 gardens), Lagos, Lucknow, Madras, Malacca, Malta, Mauritius (2 gardens), Melbourne, Mungpoo,

Natal (2 gardens), Niger Territories, Penang, Port Darwin, Saharanpur, Singapore, St. Lucia, Tasmania, Trinidad, and Wellington.

3 Lucile Brockway's superb book *Science and Colonial Expansion: The Role of the British Royal Botanic Gardens* details these activities and provides a comprehensive analysis of the place of botanical gardens in the development of the capitalist world system.

4 The chronicles of these expeditions are fascinating reading. The titles of the biographies and autobiographies of the botanist/collectors reflect the peculiar blend of science and imperialist adventure that characterized efforts to collect plant genetic resources. See *Frank N. Meyer: Plant Hunter in Asia* (Cunningham 1984) and *The World Was My Garden: Travels of a Plant Explorer* (Fairchild 1945).

5 The development of high-producing hybrids was also seen as a means of reducing the cost of feeding workers employed by such firms as United Fruit and Goodyear Rubber, which had large plantation operations in Latin America (Crabb 1947:312).

6 For example, the Central American Corn Improvement Project, the Inter-American Food Crop Improvement Program, the Inter-American Potato Improvement Project, the International Wheat Improvement Project (Cleaver 1975:318).

7 The warnings of Sauer and paleobotanist Edgar Anderson were not taken seriously. Corn breeder Paul Mangelsdorf, one of the three plant scientists sent into Mexico to make the original survey for the Rockefeller foundation (and arguably the very sort of "good, aggressive plant breeder" who worried Sauer), summarized the arguments of Sauer and Anderson for Foundation officers: "If the program does not succeed, it will not only have represented a colossal waste of money, but will probably have done the Mexicans more harm than good. If it does 'succeed' it will mean the disappearance of many ancient Mexican varieties of corn and other crops and perhaps the destruction of many picturesque folk ways, which are of great interest to the anthropologist. In other words, to both Anderson and Sauer, Mexico is a kind of glorified ant hill which they are in the process of studying. They resent any attempt to 'improve' the ants. They much prefer to study them as they now are" (quoted in Oasa and Jennings 1982:35).

8 Henry Wallace recognized the linkage of genetic erosion and genetic vulnerability. Explaining the source of his support for the Latin American programs he helped start, he commented: "It seems to me that the whole tendency in modern, commercial hybrid breeding is to narrow the source of germplasm . . . It has been one of the tragedies accompanying the superior power of modern hybrids that so many of the old-fashioned strains have been dropped completely. Who knows which of these so-called inferior sorts may have had just one block of superior genes to contribute at some critical future juncture when the environment may have changed" (Wallace 1956:17). The Third World is now recapitulating the experience of the developed nations.

9 On the one hand, such figures may underestimate the true degree of genetic vulnerability, because, as with the corn blight, a single character can be shared by multiple varieties. For example, all sorghum cultivars now carry the same

cytoplasmic male sterile factor, just as corn did. On the other hand, reliance on the relation between number of varieties and acreage planted may overestimate genetic vulnerability. Dominant varieties may be few, but these are rapidly replaced in the "varietal relay race" that maintains a steady flow of lines into and out of production. Temporal variability replaces spatial variability (Plucknett and Smith 1986); see Duvick (1982c) for an extensive elaboration of the latter position.

10 Contributors to the IBPGR in 1984 were Australia, Austria, Belgium, Canada, China, Denmark, France, Germany, Iceland, India, Ireland, Italy, Japan, Netherlands, Norway, Spain, Sweden, Switzerland, United Kingdom, United Nations Environment Programme, United States Agency for International Development, World Bank (IBPGR 1985:103).

11 For example, recent issues of *Seedsmen's Digest* and *Seed World* carry advertisements in which, against a map of the world made of seed-corn, Funk Seeds asserts it is "Worldwide in the Service of Agriculture," and the Jacklin Seed Co. boasts that "Jacklin Seeds the World."

12 Allegations of genetic protectionism are difficult to corroborate. Only Ethiopia admits to systematically embargoing germplasm as a matter of policy, and the Ethiopian edict is also unique in that it includes all plant genetic material within the nation's borders. The restrictions on exchange imposed by other Third World countries generally relate to one or two cash crops that are of special economic importance to the nation in question. Black pepper, for example, is difficult to obtain from India. Brazil is reluctant to supply rubber germplasm, and Ecuador is uncooperative regarding cocoa. For further information on which countries and crops have been involved in restrictions on the flow of germplasm, see Myers (1981) and P. R. Mooney (1983).

13 Pat Roy Mooney's seminal work (1979, 1983) is unabashedly political in nature, and he has been much involved in organizing opposition to PBR in both the developed and developing nations. Among the public interest groups and nongovernmental organizations that have lobbied against PBR and the inequities of contemporary plant germplasm exchange are the International Coalition for Development Action, Rural Development Fund, Friends of the Earth, International Organization of Consumers Unions, Pesticide Action Network, and a wide variety of similar groups.

14 See the comments of the representatives of Austria, Switzerland, Federal Republic of Germany, New Zealand, Canada, Great Britain, Sweden, United States (FAO 1983c). The testimony of the delegate from the Federal Republic of Germany is indicative of the line of argument used by the advanced capitalist nations: "In the IBPGR we see a body which does useful scientific work, on account of its decentralized form of organization has mobilized significant financial support and brought together successfully a world-wide, growing network of gene banks . . . We cannot agree to the inclusion of breeding material into the Undertaking because of our legal regulations. In addition, this would be a danger for private initiative which is urgently needed in this area . . . We note further that the IBPGR works satisfactorily. We have here no major criticism. It is trusted by the scientific community in my country" (FAO 1983c:10).

15 Forcing the vote was in itself no mean feat. The chairman of the conference was the United States Secretary of Agriculture, John Block. Block was personally chairing the session at which the issue of a vote on the Undertaking came to a head. He used what parliamentary maneuvers he could (albeit unsuccessfully) to avoid a vote.

16 As of this writing, Australia, Canada, and Japan had not provided official responses, but comments by their delegates during the debates made clear their discomfort with the Undertaking.

17 This section quotes extensively from Kloppenburg and Kleinman (1987b). See that article for fuller elaboration of the methodological and substantive points made here.

18 In assigning crops, only primary regions of diversity, where genetic variability is greatest (Harlan 1984), were considered. For the determination of specific assignments, Harlan (1975a), Zhukovsky (1975), Zeven and de Wet (1982), Hawkes (1983), and Wilkes (1983) were consulted. Where there were discrepancies among these sources, majority opinion was followed. The distinction between food and industrial crops is meant to capture an elusive but meaningful distinction. Food crops are defined as those that feed people more or less directly and that are frequently grown by subsistence farmers around the world. Industrial crops are those that feed people indirectly after industrial processing, are often grown on plantations or large-scale farms, or are grown and processed for non-food purposes.

19 Pioneer Hi-Bred International, Inc., the leading American seed firm, now sponsors an annual "Plant Breeding Research Forum" to discuss contemporary issues and problems in breeding. The subject of the 1983 forum was "Conservation and Utilization of Exotic Germplasm to Improve Varieties" (Pioneer Hi-Bred 1984).

20 Ciba-Geigy is not only a leading agrichemical producer, it also owns some 26 seed companies around the world, including Funk Seeds International, a leading U.S. firm (see Table 6.3).

21 He defines domesticated animals as "animals that have undergone modification by means of labour" (Marx 1977:285). Clearly, a similar definition applies for domesticated plants.

22 See also Brown (1985:46) and Arnold et al. (1986:615) for similar formulations.

23 An analogy may help if the point of this analysis is elusive. Computer software has a great deal of utility. It is also easy to copy. By copying software, you are gaining utility without depriving the developer of use of the program. But the developers of Lotus 1-2-3 may well feel "robbed" if you copy that software rather than purchase it from them (and, of course, under American law such an appropriation would in fact be illegal). They are unlikely to be persuaded that your use of their program is legitimate because you have not precluded their continued use of the "resource."

8. Outdoing evolution: biotechnology, botany, and business

1 For a thorough explication of DNA and some of the fundamentals of biology intelligible to laypersons, see Watson and Tooze, *The DNA Story* (1981).

2 Hereafter, I shall use the terms "NBF" (new biotechnology firm) or "start-up" to refer to a set of companies that share certain distinguishing features. Nearly all of these companies have been founded since 1976, they have been established expressly for the commercialization of the new genetic technologies, and their creation is a product of a curious union of university scientists and venture capitalists.

3 See Martin Kenney's *Biotechnology: The University-Industrial Complex* (1986) for the best and most comprehensive treatment of the genesis of the biotechnology industry and emerging patterns of university-industry relations.

4 "One young molecular biologist at the Asilomar Conference, when asked by a reporter to describe the sea urchins whose genetic structure he had been working on for years, responded, 'The only thing I know about sea urchins is what I see under a microscope.' The animal no longer meant anything to the scientist; its chromosomes had become everything" (Hutton 1978:194).

5 For reasonably simple and lucid explanations of these various techniques, see OTA (1981a, 1984), NRC (1984), and Fraley et al. (1986).

6 As of 1984, only a dozen plant genes had been fully sequenced (NRC 1984:27).

7 Such interactions are apparent even in varieties developed by conventional breeding. High-lysine corn is a case in point. It is not grown, because the high protein level is also associated with a 10 percent reduction in yield (NRC 1984:25). Similarly, the inter-specific cross made to incorporate cytoplasmic male sterility into wheat lines used for hybrid production is thought to result in a certain "reduction in fitness" that limits yields of hybrid varieties (Wilson and Driscoll 1983:118).

8 See Carlson et al. (1984) and Ammirato et al. (1984) for further information on somaclonal variation.

9 Monsanto is, in fact, the company most committed to biotechnology. Union Carbide's president noted that "They have bet the company on biotechnology" (quoted in Kenney 1986:331). Pioneer Hi-Bred's biotechnology director comments, "Monsanto hasn't bet on biotechnology, they put money on every square" (Frey 1984).

10 Actually, Agrigenetics has done contract research for both Kellogg and Hoffmann-LaRoche, both of which have held equity interest in the company (see Table 8.2).

11 Financial difficulties appear to have played a role in at least three of these takeovers. International Plant Research Institute was close to bankruptcy when BioRad made its offer. Cetus had recently cut back its agricultural research, and Agrigenetics was planning to go public in an attempt to reduce its considerable debt load. It could be that the long-term nature of plant biotechnology makes it a sector that only companies with very large budgets (or universities) can pursue.

12 The quotable David Padwa (1983:11) has something very similar to say about plant biotechnology: "Rather than gigantic 'breakthroughs' we are likely to continue making gains in crop improvement in an incremental fashion. If we can take the roughly 1 percent annual increase in agricultural productivity which genetics may rightly claim over the last fifty years and increase it over the next fifty years to something as much as or as ambitious as 2 percent we shall have

done something very remarkable and dramatic indeed. I, for one, believe it will be done."

13 The new plane of competition will eventually be based on products, on genetically engineered seeds. However, the mere *promise* of biotechnology is already influencing competitive behavior as expressed in advertising. Seed company advertisements in farm and trade journals are replete with images of high-tech laboratory paraphernalia: test tubes, petri dishes, electrophoresis films, and so on. The text accompanying these images is intended to enhance the appeal of a firm's product by emphasizing the company's commitment to cutting-edge technology; e.g., "At Sakata, research creates the future."

14 Equity interests are sometimes accompanied by a seat on the start-up's board of directors. For example, Hilleshog and Rohm and Haas each have a seat on Advanced Genetic Sciences' board, and Martin-Marietta's investment in Genentech provided for a similar arrangement.

15 Monsanto and Rohm and Haas are leaders in the use of chemical sterilants to produce hybrid wheat varieties.

16 At Illinois, four members of the Department of Agronomy drafted a white paper advising their colleagues not to bother with biotechnology. The resistance of many older breeders to the new techniques recalls the opposition to hybridization research in the 1930s. And, appropriately enough, the historical experience of hybrid corn helped shape subsequent events at Illinois. In response to the white paper, another breeder cautioned the administration, "We once sent East to Connecticut (where he laid the basis for commercial hybridization), let's not do this again" (Laughnan 1984). The point was well taken, and Illinois now has a Genetic Engineering Center.

17 Thus, while heterosis has since 1930 been the basis of all hybrid breeding efforts, there is still no satisfactory explanation of the phenomenon.

18 Could it be that they would exhibit intellectual hybrid vigor?

19 According to Plant Cell Research Institute director Eugene Fox, the needs of the seed companies heavily influence the research goals of the institute (Qualset et al. 1983:476).

20 For details, see Meyerhoff (1982) and Kenney (1986).

21 For example, in addition to $110,000 in consulting fees, Advanced Genetic Science's SAB member, Dr. Milton Schroth, received an $82,302 grant from the company in support of his research at the University of California-Berkeley (Sward 1983:4).

22 The Harvard Project has undertaken three national surveys regarding university-industry relationships in biotechnology. The information collected represents the most comprehensive body of data available on university-industry cooperative research in biotechnology. For published reports of the project, see Blumenthal et al. (1986, "Industrial Support"; 1986, "University-Industry Research"). I cite the project's findings several times. However, the project survey of universities engaged in biotechnology did *not* include agricultural colleges.

23 Clauses specifying that funds from different sources are not to be commingled on the project are very important in establishing property rights in the research.

24 However, Pioneer itself has concluded a $2.5 million research agreement with Cold Spring Harbor national laboratory.

25 The term "biotechnology" is never used, but it is clear that "cutting edge biological science" means just that; see Kenney and Kloppenburg (1983).

26 The two-day meeting was held at the Winrock International Conference Center, Petit Jean Mountain, Morrilton, Arkansas.

27 The participants were Dr. Perry Adkisson (Deputy Chancellor for Agriculture, Texas A&M University System), Dr. James T. Bonnen (Professor of Agricultural Economics, Michigan State University), Dr. Winslow R. Biggs (Director, Department of Plant Biology, Carnegie Institution of Washington), George E. Brown (Chairman, Committee on Agriculture, United States House of Representatives), Dr. Irwin Feller (Director, Institute for Policy Research, Pennsylvania State University), Dr. Ralph Hardy (Director, Life Sciences, E. I. Du Pont), Dr. James B. Kendrick, Jr. (Vice-President, Agriculture and University Services, University of California), Dr. Terry B. Kinney (United States Department of Agriculture), Dr. Lowell Lewis (Director, Agricultural Experiment Station, University of California), Dr. Judith Lyman (Agricultural Sciences Division, The Rockefeller Foundation), Dr. James Martin (President, University of Arkansas), Dr. John Marvel (General Manager, Research Division, Monsanto), Dr. John A. Pino (Director of Agricultural Sciences, The Rockefeller Foundation), Dr. Dennis J. Prager (Assistant Director, Office of Science and Technology Policy, Executive Office of the President), Dr. Peter Van Schaik (Agricultural Research Service, United States Department of Agriculture).

28 Formula funding refers to the manner in which federal appropriations for agricultural research are distributed to the SAES/LGU system. Funds are allocated to states on the basis of a statutory formula that takes into account each state's farm and rural population.

29 Such concerns have received substantial publicity (e.g., *Diversity* 1984; Cochran 1985; Crawford 1986c) even though the United States is acknowledged as the international leader in biotechnology research (Twentieth Century Fund 1984; OTA 1984).

30 A paradigmatic example of opposition can be found in an article in *HortScience* by University of Florida fruit crops professor Norman Childers (1986). Childers complains that "Retiring horticulturists or those leaving for other reasons are being replaced by pure scientists from fields other than horticulture. Grower-oriented research and student teaching are losing priority to 'high-tech' research funded and guided by outside grants... The 'umbilical cord' between the land-grant universities and farmers, growers, and marketing agencies is weakening... The key problem appears to be the federally funded *competitive grants program*, controlled by the 'high-tech' committees out of Washington, D.C.... [A]gricultural deans and directors are 'trapped' and forced to hire these pure scientists with their specialized graduate students to get needed college financing to carry out the research judged important by federal committees... Large industry grants, up to a million dollars, also are influencing administrators to allot space and faculty to projects of special interest to the grantors."

31 See Kloppenburg and Buttel (1987) for more detailed analysis of the political economy of underinvestment in agricultural research.

32 Ironically, the breeders reaching retirement age now are the cohort that replaced the generation that resisted hybrid corn.

33 Privatization is a process that is limited neither to the United States nor to the agricultural sector, as recent events in France and Britain illustrate. However, Britain's treatment of its public plant improvement and seed multiplication capabilities is an archetypal example of the process. Margaret Thatcher's government has decided to sell both the National Seed Development Organisation and the famed Cambridge Plant Breeding Institute to private enterprise. Leading bidders are expected to be Royal Dutch Shell and Imperial Chemical Industries. For further information, see Anderson (1986), Dixon (1986), and *Nature* (1986).

9. Directions for deployment

1 Examples of the danger of such single-character uniformity are not limited to the 1970 corn blight. Hybrid barley has failed because the male sterile factor used is susceptible to ergot (Ramage 1983). In fact, conventional breeding has already created this single-gene uniformity in onions, sunflower, carrot, millet, and sorghum. All hybrids in these species carry the same CMS factor (Peterson 1975; Pioneer Hi-Bred 1984:55). Clearly, we have not learned the lesson of the corn blight. Or, rather, we learned it, but private companies are unwilling to do anything about it. They are unwilling to eliminate CMS in these crops because it would eliminate hybrids altogether, because hand emasculation of onions, sunflowers, carrots, millet, and sorghum is not practical: no CMS, no hybrid.

2 Ten percent of public cotton breeders, 16 percent of public soybean breeders, 33 percent of public wheat breeders, 0 percent of public sorghum breeders, and 7 percent of public maize breeders responded that genetic vulnerability in their crop was a serious problem (Duvick 1982c:40).

3 Seed coating provides additional opportunity for the enhancement of returns via product differentiation. In advertising reminiscent of gasoline commercials, Moew's Seed Company's 1985 Seed Guide talks of the "special additives" incorporated in its "GroZone Coated Seed."

4 Another highly touted promise that probably will turn out to be mythical is the displacement of inorganic fertilizer by plants that fix their own nitrogen. After initial heavy investments, Allied Chemical and ARCO have withdrawn their resources from that area, and activity in general is down (Klausner 1984:774). Sungene executives Thomas Hiatt and Robert Hubbard have, in a very frank interview, denied that genetic engineering will reduce the need for fertilizer, introduce any new crops, or solve the short-term problems of agriculture (*Seedsmen's Digest* 1982:8).

5 The frequent appearance of tobacco in plant biotechnology research is not a reflection of its commercial importance, but its utility as a model system. Many things can be done in tobacco – e.g., regeneration of a whole plant from single cells or fused protoplasts – that cannot now be accomplished in species of greater economic interest.

6 Stanley Cohen (Stanford University) and Nobel laureates Joshua Lederberg (then of Stanford University, now president of Rockefeller University) and James D. Watson (Cold Spring Harbor) opposed the whole proceeding on the final morning. Watson had proposed that the moratorium be lifted and no subsequent guidelines for research be imposed (Watson and Tooze 1981:25).

7 I shall only outline the rDNA controversy here. For a social history of the controversy, see Sheldon Krimsky's *Genetic Alchemy* (1982). James D. Watson and John Tooze's *The DNA Story* (1981) is a collection of documents and articles relating to the controversy and is an invaluable source, spiced by the authors' own recollections.

8 This combination of appeal to national self-interest and global altruism in justifying biotechnology is a curious but common element in the literature. For example, Monsanto's Senior Vice President for Research and Development, Howard Schneiderman (1985:11), writes: "During the next forty years, the world's population will almost certainly double to about nine billion hungry people. Genetic engineering and other new agronomic methods should enable the American farmer to continue to lead the world in agricultural productivity in the next century and to feed a significant number of these nine billion people. None of us want American agriculture to join the steel industry, the automobile industry, and the electronics industries, as industries in which we once *were* the world leaders! However, unless we keep genetic engineering on a fast track with research funding, and unless federal regulations permit the controlled field testing of new crop varieties and beneficial soil microorganisms, America will lose out." See also Godown (1986:28) and U.S. House of Representatives (1982b:185) for other examples.

9 Interviewed ten years after the Asilomar conference, many participants felt that they had overreacted. Says Waclaw Szybalski, "[Asilomar] was a reflection of the Vietnam era and earlier history. Physicists were guilty of the atomic bomb, and chemists were guilty of napalm. Biologists were trying very hard to be guilty of something . . . Now Asilomar is for me an open sore. I'm still embarrassed to have participated" (quoted in J. A. Miller 1985:123).

10 For example, among the signatories of the 1974 Berg letter, David Baltimore founded the NBF Collaborative Research, Herbert Boyer founded Genentech, and Berg himself formed the NBF DNAX, subsequently purchased by Schering-Plough for $29 million (Kenney 1986:100). Frederick Ausubel, a Harvard professor who was in 1975 a member of Science for the People, is now a Senior Research Consultant for the NBF Biotechnica International. And University of California-San Francisco microbiologist David Martin, who in 1977 complained of "capitalism sticking its nose into the lab," joined Genentech as vice president of research in 1982 (Kenney 1986:26).

11 It probably would take a miracle to get a corn plant to grow in the terrain depicted in the advertisement. And none of the products of biotechnology now available has revolutionized much of anything as yet. Insulin harvested from pig and cattle pancreas is still cheaper than the much-touted human insulin produced by genetically engineered microbes (Maranto 1986:58).

12 According to Agracetus research director Winston Brill, "My wife asked me if

we're going to get champagne for the company. I chose not to because this is a historic event, perhaps, but not a commercially important event. It is the first of many, many thousands of plants that will be coming out over the next decade here and around the world" (quoted in Schneider 1986b).

13 Patent applications may thus claim new products, new processes, or both products and processes.

14 But see the dissenting opinion written by Justice Brennan, who declared that he "cannot share the Court's implicit assumption that Congress was engaged in either idle exercises, or mere correction of the public record when it enacted the 1930 and 1970 Acts" (U.S. Supreme Court 1982:320).

15 Breeders are able to cite prior art in a variety of crops ranging from sugar beet to cabbage. One public forage breeder observed of the patent: "It's as ridiculous as anything I have ever heard. My first reaction was one of real anger that anybody would even try this. I mean, the use of clonal lines and crosses--we've got cultivars that are in production now that are developed exactly like they state in the patent, and they have been in production for twenty years." And a vegetable breeder at a major seed company concurs: "Every major brassica producer has used plant tissue culture since the 1970s... No one ever thought to patent what was common knowledge."

16 In congressional testimony, Stanford president Donald Kennedy recalled, "I have been comparing notes with others and there are at least three or four incidents during this past year in which at scientific meetings – at which the traditional valuation of basic research had always been expected to prevail – there were communications in which a scientist actually refused on questioning to divulge some detail of technique, claiming that, in fact, it was a proprietary matter and that he was not free to communicate it. If you are not free to communicate it, then somebody else can't repeat your experiment, and whatever enterprise you are involved in it is not the one that traditionally moves science forward."

17 Jack Harlan (1980:237) provides an example: "[S]oybeans grown in the United States are far removed from a center of origin where they would be challenged by a full array of pests and diseases. Many of these have not even been introduced to the United States from Asia, but our own varieties would be susceptible to them. The introduction of a pest such as rust (*Phakospora pachyrhizi*) could be devastating... Not only are our current varieties of soybean susceptible, we do not even know of a source of resistance in our world soybean collection."

18 For more information on each of these initiatives, see the winter 1986 issue of *Diversity*.

19 In an effort to bolster the United States' position in international trade, the Reagan administration proposes to deny tariff concessions and other aid to countries that do not protect American copyrights, trademarks, and patents (Farnsworth 1986).

20 Such activities are not confined to plants, and private industry is more deeply engaged in collection than is the government. See the April 1985 issue of *Genetic*

Engineering News for several articles describing private collection activities in tropical plants, algae, fungi, and marine organisms.

10. Conclusion

1 The argument summarized here is drawn from Kloppenburg and Kleinman (1986); see also Kloppenburg and Kleinman (1987a, 1987b, 1987c).
2 Occidental Petroleum has recently purchased a collection of rice lines from China, and Zoecon Corporation also bought a set of soybean land races from the Chinese.

11. Still the seed

1 The terms "Terminator Technology" and "Gene Giants" were coined by the exceptionally creative and media-savvy people at the ETC Group (Action Group on Erosion, Technology and Concentration). Previously known as RAFI (Rural Advancement Fund International), the ETC Group has been involved in the issues surrounding plant breeding, biotechnology, and genetic resources since the 1970s. The Group maintains constant and comprehensive oversight of scientific, economic, commercial, and political activity in the broad area of biotechnology and the life sciences. It is a prolific production of position papers, updates, analyses, and commentary that is invaluable for anyone working in this area. Readers are urged to visit the organization's web site www.etcgroup.org.
2 For use of the term "means of reproduction" I am indebted to Dr. Wendy Russell, Department of Biological Sciences, University of Wollagong.
3 "FDA considers a GE [genetically engineered, that is, GM] crop safe if it is substantially the same as its non-GE version and the genetic modification does not cause the crop to produce a substance that is new or used in a new way or that is present in much larger amounts than in currently safe food. If the genetic modification produces a new substance that is not an approved food additive, the safety of the new substance as a food must be proven" (Shoemaker 2001:31).
4 For extensive treatment of corporate resistance to regulation and the politicking that entailed, see especially Marion Nestle's excellent *Safe Food: Bacteria, Biotechnology, and Bioterrorism* (2003).
5 An excellent example is the *Safe Food News* published and distributed by Mothers for Natural Law of the Natural Law Party (2000).
6 See the list at www.truefoodnow.org/home_polls.html. The proportion of respondents to various polls taken between 1997 and 2003 who say they want labeling of GM foods ranges from 68 percent to 94 percent.
7 See *The Non-GMO Source* (Writing Solutions 2003) for a summary of regulations by nation.
8 Says Clive James, head of the International Service for the Acquisition of Agribiotech Applications, "There's piracy going on. These farmers think so much of this technology, they will steal it" (Barboza 2003).

9 Executives recognize this only too well. According to Gene Grabowski of the Grocery Manufacturers of America, "For the price of what it would have cost to market a new breakfast cereal, the biotech industry probably could have saved itself a lot of the struggle that it is going through today" (Eichenwald et al. 2001:C6).

10 Reprints of the advertisements and a list of the groups that are part of the Turning Point Project are available at www.turnpoint.org. Founding members of the CBI were Aventis, BASF, Dow Chemical, DuPont, Monsanto, Novartis, Zeneca, the American Crop Protection Association, and the Biotechnology Industry Organization. Many materials produced by the CBI are available at www.whybiotech.com.

11 See especially Nestle (2003), pages 153–166, for an effective review of the issue.

12 For an exceptionally informative and thought provoking analysis of such "third generation" products and initiatives, see "Biotech's 'generation 3'" (Rural Advancement Fund International 2000b).

13 For example, Sheldon Krimsky, *Biotechnics and Society: The Rise of Industrial Genetics* (1991); Paul Raeburn, *The Last Harvest: The Genetic Gamble that Threatens to Destroy American Agriculture* (1995); Sheldon Krimsky and Roger Wrubel, *Agricultural Biotechnology and the Environment: Science, Policy, and Social Issues* (1996); Vanadana Shiva, *Biopiracy: The Plunder of Nature and Knowledge* (1997); Luke Anderson *Genetic Engineering, Food, and Our Environment* (1999); Martin Teitel and Kimberly A. Wilson, *Genetically Engineered Food: Changing the Nature of Nature* (1999); Richard Manning, *Food's Frontier: The Next Green Revolution* (2000); Daniel Charles, *Lords of the Harvest: Biotech, Big Money, and the Future of Food* (2001); Brian Tokai, *Redesigning Life? The Worldwide Challenge to Genetic Engineering* (2001); Marion Nestle, *Safe Food: Bacteria, Biotechnology, and Bioterrorism* (2003).

14 But see, for example, National Research Council, *Genetically Modified Pest-Protected Plants: Science and Regulation* (2000) and Rissler and Mellon, *The Ecological Risks of Engineered Crops* (1996).

15 According to Annabelle Duncan, chief of molecular science at Australia's Commonwealth Scientific and Industrial Research Organization, "This shows that something we thought was hard – increasing the pathogenicity of a virus – is easy" (Finkel 2001:585).

16 Says one entomologist, "The problem is, we're asking cotton growers to produce insects. They are not interested in growing insects. They want to destroy insects" (Charles 2001:184).

17 The USDA has refused to reveal the pharmaceutical substances the corn was engineered to produce, saying it is confidential business information. "Biopharming" initiatives undertaken by companies such as ProdiGene and Epicyte Pharmaceutical are known to involve research designed to produce vaccines for HIV, cholera, and hepatitis-B, growth hormones, clotting agents, industrial enzymes, human antibodies, contraceptives, spermatocides, immunosuppressant cytokines, and abortion-inducing drugs. The host plant of choice is corn, but possibilities are also being explored using tomato, potato, and banana.

18 *Sui generis* is a "Latin phrase meaning 'of its own kind of class.' This is the sys-

tem for WTO member states that do not allow patents on plants/plant varieties, to provide an effective alternative means of protection. Although not specifically defined in the TRIPs agreement, minimum requirements are indicated" (Kowalski et al. 2002:409).

19 See GRAIN (2003) for a country by country listing of the restrictions.

20 Public breeders report that they regularly receive letters from industry lawyers listing the varieties that cannot be used for anything except testing and mandating that plants be destroyed when testing is completed.

21 An interesting aspect of the Berkeley/Novartis strategic alliance was the manner in which the two parties got together. The agreement was not the result of a one-on-one courtship, but the end result of an auction process in which Novartis was selected from among five different corporate suitors (see Rausser 1999).

22 The degree to which this can occur is illustrated by the comments of public scientists themselves. British breeder Lindsay Innes (n.d.) describes himself as "a so-called public sector plant breeder, though these days I seem to be more of a hybrid between public and private, and at time shave to adopt chameleon-like characteristics." Dan Charles (2001:37) records the acclimatization of Roger Beachy, then-assistant professor at Washington University, to corporate culture. Offended at first by Monsanto's appropriation of his work, Beachy recalls that "after a while I realized this is just the way it is. They are good collaborators. They had to feel that they were owners of it in order to promote the company. And then you say, 'That's what companies do!' You forgive and you go on."

23 The phrase "ecoliberal" is Hammond's (1993).

24 In the throes of security madness, the United States wanted a clause allowing it to refuse to freely exchange germplasm with any nation it considered to be a security threat.

25 Not only is there no mechanism for enforcement, there is not even any mechanism for monitoring. In a clarification of FAO/CGIAR trust arrangement, the CG states that in sending germplasm to those requesting accessions or engaged in regeneration, "in neither case will the source Centre be under an obligation to monitor the compliance of the recipient with these undertakings." The slippage possible in this arrangement is already apparent. In 2001, Jasmine rice from IRRI was shipped to a researcher in the United States without a Materials Transfer Agreement. The situation only came to light through the vigilance of a local Philippine NGO.

26 For example, among the sixty organizations supporting the Turning Point Project's advertisements in *The New York Times* were: the Foundation on Economic Trends, the International Center for Technology Assessment, Greenpeace USA, the Council for Responsible Genetics, Food First/Institute for Food & Development Policy, Friends of the Earth, Humane Society USA, Institute for Agriculture and Trade policy, Organic Consumers association, Pesticide Action Network, Sierra Club, Center for Food Safety, Mothers & Others for a Livable Planet, Earth Island Institute, and the Native Forest Council.

27 *Satyagraha* is the Ghandian term for nonviolent resistance.

28 The United States ranks twenty-second in the percentage of its gross national product devoted to foreign aid, the lowest among the advanced industrial nations.

References

Abrams, Philip (1983) *Historical Sociology*. Ithaca, NY: Cornell University Press.

Advanced Genetic Sciences, Inc. (1983) "Prospectus." September 22. Greenwich, CT: Advanced Genetic Sciences.

Agrawal, Arun (1995) "How to lose your cake without eating it: intellectual property rights and 'indigenous' knowledge." *The Common Property Resource Digest* 36 (December):2-5.

Agricultural Genetics Report (1983) "The new plant genetics – highlights of Teweles' study." *Agricultural Genetics Report* (November/December):2-7.

Agricultural Research Institute (1952) "Proceedings of the first meeting, October 16." Washington, DC: National Research Council.

Agricultural Research Service (1977) Letter from T. W. Edminster, Administrator, Agricultural Research Service, to Richard H. Demuth, chairman, International Board for Plant Genetic Resources.

(1983) "Agricultural Research Service program plan." Miscellaneous publication no. 1429, Agricultural Research Service, January. Washington, DC: USDA.

(1986) "Five year plan for the food and agricultural sciences." Joint Council on Food and Agricultural Sciences, Agricultural Research Service. Washington, DC: USDA.

Agrigenetics Corporation (1981) "Advanced research laboratory." Madison, WI: Agrigenetics Corporation.

(1984) "Preliminary prospectus: issued December 8, 1983." Madison, WI: Agrigenetics Corporation.

Airy, John (1951) "Current problems of detasseling." In *Improved Techniques in Hybrid Seed Corn Production*. Chicago, IL: American Seed Trade Association.

Alexander, Martin (1983) "Statement by Martin Alexander." In *Environmental Implications of Genetic Engineering*. Hearings before the Subcommittee on Investigations and Oversight and the Subcommittee on Science, Research and Technology of the Committee on Science and Technology, U.S. House of Representatives, Ninety-eighth Congress, first session, June 22. Washington, DC: USGPO.

Altieri, Miguel (1996) *Agroecology: The Science of Sustainable Agriculture*. Boulder, CO: Westview Press.

(2001) *Genetic Engineering in Agriculture: The Myths, Environmental Risks, and Alternatives*. Oakland, CA: Food First/Institute for Food and Development Policy.

Alvarez, Lizette (2003) "Consumers in Europe resist gene-altered foods." *The New York Times* (11 February):A3.

American Breeders Association (1905) *Proceedings of the first and second annual meetings of the American Breeders Association, 1903 and 1904.* Washington, DC: American Breeders Association.

(1906) *Proceedings of the Third Annual Meeting of the American Breeders Association, 1905.* Washington, DC: American Breeders Association.

American Medical Association (2000) "Genetically modified crops and foods." Report no. 10 of the Council on Scientific Affairs, www.ama-assn.org/ama/pub/article/2036-3604.html.

American Seed Trade Association (1901) *Proceedings of the American Seed Trade Association, 1901.* Hartford, CT: Hartford Press.

(1909) *Proceedings of the American Seed Trade Association, 1909.* Hartford, CT: Hartford Press.

(1923) *Proceedings of the American Seed Trade Association, 1923.* Hartford, CT: Hartford Press.

(1930) *Proceedings of the American Seed Trade Association, 1930.* Hartford, CT: Hartford Press.

(1934) *Proceedings of the American Seed Trade Association, 1934.* Hartford, CT: Hartford Press.

(1938) *Proceedings of the American Seed Trade Association, 1938.* Hartford, CT: Hartford Press.

(1944) *Proceedings of the American Seed Trade Association, 1944.* Hartford, CT: Hartford Press.

(1970) "Statement of Allenby L. White, Chairman, Breeders' Rights Study Committee, American Seed Trade Association." In U.S. Senate, 1970. Washington, DC: USGPO.

(1983) "1983 membership directory." Washington, DC: American Seed Trade Association.

(1984) "Position paper of the American Seed Trade Association on FAO undertaking on plant genetic resources," May 5. Washington, DC: American Seed Trade Association.

American Society of Agronomy (1956) "Special resolution." *Agronomy Journal* 48:12 (December):603.

(1964) *Plant Breeders' Rights.* American Society of Agronomy special publication no. 3. Madison, WI: American Society of Agronomy.

Amin, Samir, and Kostas Vergopoulos (1974) *La Question Paysanne et le Capitalisme.* Paris: Anthropos.

Ammirato, Philip V., David A. Evans, and William Sharp (1984) "Biotechnology and agricultural improvement." *Trends in Biotechnology* 2:3(May/June):53-58.

Anderson, Alun (1986) "Success breeds privatization." *Nature* 322(7 August):491.

Apfelbaum, Robert S. (1956) "Taking research to the farmer." In William Heckendorn and Joseph Gregory (eds.), *Proceedings of the First Annual Farm Seed Industry-Research Conference.* Washington, DC: American Seed Trade Association.

Arditti, Rita, P. Brennan, and S. Cayrak (eds.) (1980) *Science and Liberation.* Boston, MA: South End Press.

Argumedo, Alejandro (1995) "Protecting and promoting informal innovation systems – indigenous peoples and biodiversity." In Kent Whealy (ed.), *Seed Savers 1995 Harvest Edition.* Decorah, IA: Seed Savers Exchange, Inc.

Arndt, Thomas M., Dana G. Dalrymple, and Vernon W. Ruttan (eds.) (1977) *Resource Allocation and Productivity in National and International Agricultural Research.* Minneapolis, MN: University of Minnesota Press.

Arnold, M. H., et al. (1986) "Plant gene conservation." *Nature* 319(20 February):615.

Ashford, Nicholas A. (1983) "A framework for examining the effects of industrial funding on academic freedom and the integrity of the university." *Science, Technology & Human Values* 8:2 (Spring):16-23.

Association of American Agricultural Colleges and Experiment Stations (1908) "Report of the Commission on Agricultural Research." Washington, DC: AAACES.

Athanasiou, Tom (1992) "After the summit." *Socialist Review* 22:4(October-December):57-92.

Atkinson, Richard C., Roger N. Beachy, Gordon Conway, France A. Cordova, Marye Anne Fox, Karen A. Holbrook, Daniel F. Klessig, Richard L. McCormick, Peter M. McPherson, Hunter R. Rawlings III, Rip Rapson, Larry N. Vanderhoef, John D. Wiley, and Charles E. Young (2003) "Public sector collaboration for agricultural IP management." *Science* 301(11 July):174-175.

Baker, Gladys L., W. D. Rasmussen, V. Wiser, and J. M. Porter (1963) *Century of Service: The First One Hundred Years of the Department of Agriculture.* Washington, DC: USGPO.

Balick, Michael J., Rosita Arvigo, and Leopoldo Romero (1994) "The development of an ethnobiomedical forest reserve in Belize: its role in the preservation of biological and cultural diversity." *Conservation Biology* 8:1(March):316-317.

Banaji, Jarius (1980) "Summary of selected parts of Kautsky's *The Agrarian Question.*" In F. H. Buttel and H. Newby (eds.), *The Rural Sociology of the Advanced Societies.* Montclair, NJ: Allanheld, Osmun and Company.

Barboza, David (2000a) "Industry moves to defend biotechnology." *The New York Times* (4 April):A9.

(2000b) "Ground-level genetics, for the perfect lawn." *The New York Times* (9 July):A1, A18.

(2003) "Development of biotech crops is booming in Asia." *The New York Times* (21 February):A3.

(2004) "Questions linger on price of seeds." *The New York Times* (6 January):A1, C8.

Barnes, K. K., and J. H. Anderson (1975) "If you enjoy eating, thank the machines!" In *That We May Eat: Yearbook of Agriculture, 1975.* Washington, DC: USDA.

Barton, John H. (1982) "The international breeder's rights system and crop plant improvement." *Science,* 216(4 June):1071-1075.

Bateson, William (1902) "Practical aspects of the new discoveries in heredity." In *Proceedings of the International Conference on Plant Breeding and Hybridization, Memoirs of the Horticultural Society of New York, Volume 1.* New York, NY: HSNY.

Batie, Sandra S., and Robert G. Healy (1983) "The future of American agriculture." *Scientific American* 248(February):45-53.

Beachy, Roger (1984) Personal interview, 21 March.

Beard, D. (1966) "Certification for varietal purity only." In John Sutherland (ed.), *Proceedings of the Twelfth Farm Seed Research Conference.* Washington, DC: American Seed Trade Association.

Becker, Elizabeth (2003) "U.S. delays suing Europe over ban on modified food." *The New York Times* (3 February):A6.

(2004) "Europe approves genetically modified corn as animal feed." *The New York Times* (20 July):C11.

Becker, Robert F. (1984) "American vegetable seed industry – a history." *HortScience* 19:5 (October):610-611.

Becker, Stanley L. (1976a) "Donald F. Jones and hybrid corn." Connecticut Agricultural Experiment Station, bulletin 763.

(1976b) "The Jones patent." Unpublished paper, The Research Corporation.

(1984) Personal communication, 25 May.

Bell, Daniel (1973) *The Coming of Post-Industrial Society.* New York, NY: Basic Books.

Benbrook, Charles (1999) *Evidence of the Magnitude and Consequences of the Roundup Ready Soybean Yield Drag from University-Based Varietal Trials in 1998.* AgBio Tech InfoNet Technical Paper no. 1 (13 July).

(2001) *The Farm-Level Economic Impacts of Bt Corn from 1996 Through 2001: An Independent National Assessment.* IATP report, http://www.biotech-info.net/RR_yield_drag_98.pdf (accessed April 22, 2003).

(2003) "Principles governing the long-run risks, benefits, and costs of agricultural biotechnology." Paper presented at the "Conference on Biodiversity, Biotechnology, and the Protection of Traditional Knowledge." Washington University, St. Louis, MO (5 April), http://www.biotech-info.net/biod_biotech.pdf.

Benbrook, Charles M., and Phyllis B. Moses (1986) "Engineering crops to resist herbicides." *Technology Review* 89:8(November/December):54-61, 79.

Bennett, Erna (2002) "The summit-to-summit merry-go-round." *Seedling* (July):3-10.

Bent, Stephen A. (1985) "Protection of plant material under the general patent statute: a sensible policy at the PTO?" *Biotechnology Law Report* 105(March).

Berardi, Gigi, and Charles C. Geisler (eds.) (1984) *The Social Consequences of New Agricultural Technologies.* Boulder, CO: Westview Press.

Berg, Paul, David Baltimore, Herbert W. Boyer, Stanley N. Cohen, Ronald W. Davis, David S. Hogness, Daniel Nathans, Richard Roblin, James D. Watson, Sherman Weissman, and Norton D. Zinder (1974) "Potential biohazards of recombinant DNA molecules." *Science* 185(26 July):303.

Berg, Paul, David Baltimore, Sydney Brenner, Richard O. Roblin III, and Maxine F. Singer (1975) "Asilomar conference on recombinant DNA molecules." *Science* 188(6 June):991-994.

Bergelson, Joy, Colin B. Purrington, and Gale Wichmann (1998) "Promiscuity in transgenic plants." *Nature* 395(3 September):25-26.

Berlan, Jean-Pierre, and Richard Lewontin (1983) "Hybrid corn revisited." Unpublished paper.

(1986a) "Breeders' rights and patenting life forms." *Nature* 322(28 August):785-788.

(1986b) "The political economy of hybrid corn." *Monthly Review* 38:3(July/August): 35-47.

Bernal, J. D. (1939) *The Social Functions of Science.* London: Routledge and Kegan Paul.

(1965) *Science in History: Volume 1, The Emergence of Science.* Cambridge, MA: MIT Press.

Berry, Wendell (1977) *The Unsettling of America: Culture and Agriculture.* San Francisco, CA: Sierra Club Books.

Bingham, E. T. (1983) "Molecular genetic engineering vs plant breeding." *Plant Molecular Biology* 2:222-224.

Biofutur (1984) "Vie des societies." *Biofutur* (November):69.

Biotechnology Newswatch (1984a) "Roche v. Bolar ruling will shield biotechnology patents lawyers assert." *Biotechnology Newswatch* 4:10(21 May):1-2.

(1984b) "Pollen vector expresses anti-rust genes in corn." *Biotechnology Newswatch* 4:11(June 4):2-3.

Bishop, Jerry E. (1985) "Firm's corn plant resistant to class of weed killers." *Wall Street Journal* (28 August):27.
Bliss, Frederick A. (1984) "The application of new plant biotechnology to crop improvement." *HortScience* 19:1(February):43-48.
Blumenthal, David, Michael Gluck, Karen Seashore Louis, and David Wise (1986) "Industrial support of university research in biotechnology." *Science* 231 (17 January):242-246.
Blumenthal, David, Michael Gluck, Karen Seashore Louis, Michael A. Stoto, and David Wise (1986) "University-industry research relationships in biotechnology: implications for the university." *Science* 232(13 June):1361-1366.
Bobrow, Martin, and Sandy Thomas (2001) "Patents in a genetic age." *Nature* 499 (15 February):763-764.
Bok, Derek (2003) *Universities in the Marketplace: The Commercialization of Higher Education.* Princeton, NJ: Princeton University Press.
Bradfield, Richard (1942) "Our job ahead." *Journal of the American Society of Agronomy* 34:12(December):1065-1075.
Braudel, Fernand (1966) *The Mediterranean and the Mediterranean World in the Age of Philip II.* New York, NY: Harper and Row.
(1979) *The Structures of Everyday Life: The Limits of the Possible.* New York, NY: Harper and Row.
Braverman, Harry (1974) *Labor and Monopoly Capital.* New York, NY: Monthly Review Press.
Brill, Winston (1982) "Statements of Dr. Winston J. Brill." In *Potential Application of Recombinant DNA and Genetics on Agricultural Sciences.* Hearings before the Subcommittee on Investigations and Oversight of the Committee on Science and Technology, U.S. House of Representatives: Ninety-seventh Congress, 9 June, 28 July, no. 134. Washington, DC: USGPO.
(1985) "Safety concerns and genetic engineering in agriculture." *Science* 227 (25 January):301-304.
Brink, R. A. (1941) "Closed versus open pedigrees for corn hybrids." In *Report of the Sixth Corn Improvement Conference of the North Central Region.* Washington, DC: USDA.
British Medical Association (1999) "The impact of genetic modification on agriculture, food and health – an interim statement." (May), http://www.bma.org.uk/ap.nsf/Content/Impact+of+genetic+modification+-+%28m%29.
Brockway, Lucile H. (1979) *Science and Colonial Expansion: The Role of the British Royal Botanic Gardens.* New York, NY: Academic Press.
Bronk, D. W. (1953) "Relationship of the Agricultural Research Institute to the National Academy of Sciences." In *Proceedings of the Second Annual Meeting of the Agricultural Research Institute.* Washington, DC: NAS-NRC.
Brooks, Howard J., and Grant Vest (1985) "Public programs on genetics and breeding of horticultural crops in the United States." *HortScience* 20:5(October): 826-830.
Brown, William (1985) "The coming debate over ownership of plant germplasm." In Dolores Wilkinson (ed.), *Proceedings of the 39th Annual Corn and Sorghum Research Conference.* Washington, DC: American Seed Trade Association.
(1986) "The exchange of genetic materials: a corporate perspective on the internationalization of the seed industry." Paper presented at the annual meeting of the American Association for the Advancement of Science, May, Philadelphia.
Brush, Stephen B., and Doreen Stabinsky (1996) *Valuing Local Knowledge: Indigenous People and Intellectual Property Rights.* Washington, DC: Island Press.

Buker, Robert J. (1969) "Public and private plant breeders and their responsibility to agriculture through variety development and release." In *Variety Protection by Plant Patents and Other Means.* Madison, WI: Crop Science Society of America.

Bukharin, Nikolai, et al. (1971) *Science at the Crossroads.* London: Frank Cass and Company.

Busch, Lawrence, and William B. Lacy (1983) *Science, Agriculture, and the Politics of Research.* Boulder, CO: Westview Press.

Business Week (1984a) "Biotech comes of age." *Business Week* (23 January):84-91.

(1984b) "The biotech big shots snapping up small seed companies." *Business Week* (11 June):69-70.

Butler, Leslie J., and Bruce W. Marion (1985) "The impacts of patent protection on the U.S. seed industry and public plant breeding, North Central Region research publication 304, September. Research Division, College of Agriculture and Life Sciences, University of Wisconsin, Madison.

Buttel, Frederick H. (1983) "Beyond the family farm." In Gene Summers (ed.), *Technology and Social Change in Rural Areas.* Boulder, CO: Westview Press.

(1985) "Biotechnology and agricultural research policy: emergent issues." In K. Dahlberg (ed.), *New Directions in Agriculture and Agricultural Research.* Totowa, NJ: Rowman and Allanheld.

(1999) "Agricultural biotechnology: its recent evolution and implications for agrofood political economy." *Sociological Research Online* Vol.-1995, no. 1, http://www.socresonline.org.uk/4/3/buttel.html.

Buttel, Frederick H., Martin Kenney, and Jack Kloppenburg, Jr. (1985) "From green revolution to biorevolution: some observations on the changing technological bases of economic transformation in the Third World." *Economic Development and Cultural Change* 34:1(October):31-55.

Buttel, Frederick H., and Howard Newby (1980) *The Rural Sociology of the Advanced Societies: Critical Perspectives.* Montclair, NJ: Allanheld, Osmun and Company.

Butz, Earl L. (1960) "Agribusiness in the machine age." In *Yearbook of Agriculture, 1960.* Washington, DC: USGPO.

Byron, Janet (2002) "Center proposes solution for ag biotech licensing disputes." *California Agriculture* 55:2:6.

Calgene (1986) "Prospectus." July 8. New York, NY: Hambrecht & Quist, Inc., Paine Webber, Inc., and Piper, Jaffray & Hopwood, Inc.

California Agricultural Lands Project (1982) *Genetic Engineering of Plants.* San Francisco, CA: California Agricultural Lands Project.

Callahan, Patricia, and Scott Kilman (2001) "Seeds of doubt: some ingredients are genetically modified, despite labels' claims." *Wall Street Journal* (5 April).

Canadian Seed Trade Association (1984) *Seeds for a Hungry World: The Role and Rights of Modern Plant Breeders.* Ottawa: Canadian Seed Trade Association.

Caren, J. (1964) "Germ plasm control as it would affect variety improvement and release of vegetables and flowers." In *Plant Breeders' Rights.* ASA special publication no. 3. Madison, WI: Crop Science Society of America.

Carlson, Peter S., Betsey F. Conrad, and Joseph D. Lutz (1984) "Sorting through the variability." *HortScience* 19:3(June):388-392.

Carson, Rachel (1962) *Silent Spring.* Boston, MA: Houghton Mifflin.

Carstensen, Vernon (1961) "Profile of the USDA – first fifty years." In W. D. Rasmussen (ed.), *Lecture Series in Honor of the USDA Centennial Year, 1961.* Washington, DC: USDA Graduate School.

Carter, George S. (1925) "G. S. Carter to Henry Wallace, 29 April, 1925." 56 (277-279). Henry A. Wallace Papers, University of Iowa Libraries, Iowa City, IA.

Carter, Jimmy (1998) "Who's afraid of genetic engineering." *The New York Times* (26 August):A23.

Castells, Manuel (1986) "High technology, economic policies and world development." BRIE working paper 18. Berkeley, CA: Berkeley Roundtable on the International Economy.

Cates, John S. (1929) "The day of super-corn crops has come." *Country Gentleman* 94:3(March):20-21, 130.

Cavalieri, Liebe F. (1981) *The Double-Edged Helix*. New York, NY: Columbia University Press.

Chaney, James W. (1982) "Statement by James W. Chaney." In *University/Industry Cooperation in Biotechnology*. Hearings before the Subcommittee on Investigations and Oversight and the Subcommittee on Science, Research, and Technology of the Committee on Science and Technology, U.S. House of Representatives, Ninety-seventh Congress, Second Session, 16-17 June. Washington, DC: USGPO.

Chang, T. T. (1979) "Crop genetic resources." In J. Sneep and A. J. T. Hendriksen (eds.), and O. Holbek (coed.), *Plant Breeding Perspectives, 1879-1979*. Wageningen, The Netherlands: Centre for Agricultural Publishing and Documentation.

Charles, Daniel (2001) *Lords of the Harvest: Biotech, Big Money, and the Future of Food*. Cambridge, MA: Perseus Publishing.

Chayanov, A. V. (1966) *On the Theory of Peasant Economy* (edited by D. Thorner, B. Kerblay, and R. E. F. Smith). Homewood, IL: American Economic Association.

Childers, Norman F. (1986) "Is there a crisis developing in horticulture?" *HortScience* 21:1(February):8-9.

Chilton, Mary-Dell (1983) "Crown gall gene as a vector for plants." *Bio/Technology* 1:2(April):163-164.

Choi, Charles (2002) "The terminator's back." *Scientific American* 287:3:30.

Christensen, Ken (1957) "Value of applied research to the seed industry." In William Heckendorn and Joseph Gregor (eds.), *Proceedings of the Second Annual Farm Seed Industry-Research Conference*. Washington, DC: American Seed Trade Association.

(1962) "Future of certification as seen from a seedsman's standpoint." Annual report of the International Crop Improvement Association, 1962.

Claffey, Barbara (1981) "Patenting life forms: issues surrounding the Plant Variety Protection Act." *Southern Journal of Agricultural Economics* 13(December):29-37.

Clapp, Stephen (2002) "Food industry alarmed by biopharm contamination incidents." *Food Traceability Report* (14 November).

Clark, Ann E. (n.d.) "The implications of the Schmeiser decision," www.percyschmeiser.com/crime.htm (accessed April 11, 2002).

Clark, J. A. (1936) "Improvement in wheat." In *Yearbook of Agriculture, 1936*. Washington, DC: USGPO.

Cleaver, Harry M., Jr. (1972) "Contradictions of the green revolution." *Monthly Review* (June):80-111.

(1975) *The Origins of the Green Revolution*. Unpublished Ph.D. dissertation, Stanford University.

(1979) *Reading Capital Politically*. Austin, TX: University of Texas Press.

Cleveland, David A. and Stephen C. Murray (1997) "The world's crop genetic resources and the rights of indigenous farmers." *Current Anthropology* 38:4: 477-496.

Cochran, Wendell (1985) "US losing leadership in farm research." *Des Moines Register* (11 September):1.

Cochrane, Willard W. (1979) *The Development of American Agriculture: A Historical Analysis.* Minneapolis, MN: University of Minnesota Press.

Coffey, W. C. (1949) "A challenge to the hybrid seed corn industry." In *What's New in the Production, Storage and Utilization of Hybrid Seed Corn.* Chicago, IL: American Seed Trade Association.

Coffman, W. Ronnie (1998) "Future of plant breeding in public institutions." Paper presented at the Annual Meeting of the American Seed Trade Association, Chicago, IL, 13 November.

Cohen, Geoffrey A. (1978) *Karl Marx's Theory of History: A Defence.* Princeton, NJ: Princeton University Press.

Collins, G. N. (1910) "The value of first-generation hybrids in corn." Bureau of Plant Industry bulletin no. 191. Washington, DC: USGPO.

(1911) "Yield tests as a basis of awarding prizes at corn shows." *American Breeders Magazine* 11:2:103-106.

Colwell, Robert, Elliot A. Norse, David Pimentel, Frances Sharples, and Daniel Simberloff (1985) "Genetic engineering in agriculture." *Science* 229(12 July):111-112.

Commission on Intellectual Property Rights (2002) "Independent commission finds intellectual property rights impose costs on most developing countries and do not help reduce poverty." www.iprcommission.org (accessed September 12).

Congressional Record (1932) *Congressional Record* 76:1:1045.

Consultative Group on International Agricultural Research (1980) "Report of the quinquennial review of the International Board for Plant Genetic Resources." Rome: CGIAR.

Conway, Gordon (1999) "The Rockefeller Foundation and plant biotechnology." (June 24), http://biotech-info.net/gordon_conway.html.

Conway, Gordon G., and Gary Toenniessen (1999) "Feeding the world in the twenty-first century." *Nature* 402(2 December):C55-C58.

Copeland, L. O. (1976) *Principles of Seed Science and Technology.* Minneapolis, MN: Burgess Publishing Company.

Cordtz, Dan (1972) "Corporate farming: a tough row to hoe." *Fortune* 86:2(August): 135-139, 172-175.

Council for Biotechnology Information (2000) *Biotechnology: Good Ideas Are Growing.* Washington, DC: Council for Biotechnology Information.

(2004) "Safety and regulation: biotech foods are safe, say regulators and medical experts," www.whybiotech.com/index.asp?id=1974.

Cox, Stan (2002) "The mirage of genetic engineering." *American Journal of Alternative Agriculture* 17:1:41-43.

Crabb, A. R. (1947) *The Hybrid-Corn Makers: Prophets of Plenty.* New Brunswick, NJ: Rutgers University Press.

Craig, R. (1968) "Implications of the new genetics in horticultural plant breeding." *Horticultural Science* 3:4(Winter):5-249.

Crawford, Mark (1986a) "OSTP ponders plant research initiatives." *Science* 231(17 January):212.

(1986b) "Larger public sector role sought on biotech." *Science* 232(4 April):15-16.

(1986c) "Regulatory tangle snarls agricultural research in the biotechnology arena." *Science* 234(17 October):275-276.

Crill, P., J. W. Strobez, D. S. Burgis, H. H. Bryan, C. A. John, P. H. Everett, J. A. Bartz, N. C. Hayslip, and W. W. Deen (1971) "Florida MH-1, Florida's first machine harvest fresh market tomato." University of Florida, circular S-212.

Crop Science Society of America (1969) *Variety Protection by Plant Patents and Other*

Means. Crop Science Society of America technical series no. 1. Madison, WI: Crop Science Society of America.
Crosby, A. W. (1972) *The Columbian Exchange: Biological and Cultural Consequences of 1492.* Westport, CT: Greenwood Press.
Cummins, Ronnie (2000) *Hazards of Genetically Engineered Foods and Crops: Why We Need a Global Moratorium.* Little Marais, MN: Organic Consumers Association.
Cummins, Ronnie, and Ben Lilliston (2000) *Genetically Engineered Food: A Self Defense Guide for Consumers.* New York, NY: Marlowe.
Cunningham, Isabel S. (1984) *Frank N. Meyer: Plant Hunter in Asia.* Ames, IA: Iowa State University Press.
Daley, Suzanne (1999) "French see hero in war on 'McDomination.'" *The New York Times* (12 October):A1, A4.
Danbom, David B. (1979) *The Resisted Revolution: Urban America and the Industrialization of Agriculture 1900–1930.* Ames, IA: Iowa State University Press.
Davenport, Caroline (1981) "Sowing the seeds – research, development flourish at DeKalb and Pioneer Hi-Bred." *Barron's* 61:9:9–10, 33.
Davis, John E. (1980) "Capitalist agricultural development and the exploitation of the propertied laborer." In F. H. Buttel and H. Newby (eds.), *The Rural Sociology of the Advanced Societies.* Montclair, NJ: Allanheld, Osmun and Company.
Day, Boysie E. (1974) "Support of research by California marketing programs." *HortScience* 9:1(February):49–51.
(1978) "The morality of agronomy." In *Agronomy in Today's Society.* American Society of Agronomy special publication no. 33. Madison, WI: American Society of Agronomy.
Day, Peter R. (1986) "The impact of biotechnology on agricultural research." *Diversity* 9:33–37.
De Janvry, Alain (1980) "Social differentiation in agriculture and the ideology of neopopulism." In F. H. Buttel and H. Newby (eds.), *The Rural Sociology of the Advanced Societies.* Montclair, NJ: Allanheld, Osmun and Company.
Delouche, John C. (1983) "Sea change VIII." *Seedsmen's Digest* 34:8(August):8.
Dickson, David (2003) "The case for a new intellectual property order." *SciDevNet* (12 May), http://www.scidev.net/Editorials/index.cfm?fuseaction=readEditorials&itemid=70&language=1 (accessed June 6, 2003).
Dixon, Bernard (1986) "More dickering over British plant research." *Bio/Technology* 4(7 August):762.
Diversity (1983) "ASTA recommends policy for germplasm and variety release by public breeders." *Diversity* 1:3(November/December):10.
(1984) "Reagan administration zeroes in on US agricultural research." *Diversity* 6(March/April):3–4.
(1986a) "Where have all the plant breeders gone?" *Diversity* 9:14–15.
(1986b) "News in brief." *Diversity* 9:39–41.
DNA Plant Technology Corporation (1983) "Preliminary prospects dated December 19, 1983." Cinnaminson, NJ: DNA Plant Technology Corporation.
Dolan, D. D., and W. Sherring (1982) "Progress in the evaluation and use of plant germplasm in the northeast, 1974–1979." Special report no. 43, New York State Agricultural Station, Geneva, NY.
Donwen, W. J. (1984) "Seed treatment – the tip of the iceberg?" *Seedsmen's Digest* 35:7(July):8–9.
Dorner, Peter (1983) "Technology and U.S. agriculture." In G. Summers (ed.), *Technology and Social Change in Rural Areas.* Boulder, CO: Westview Press.

Dorsey, J. G. (1964) "U.S. laws and regulations on protection of varieties and juridic requirements for breeders' rights." In *Plant Breeders' Rights.* Madison, WI: American Society of Agronomy.

Dowker, B. D., and G. H. Gordon (1983) "Heterosis and hybrid cultivars in onions." In R. Frankel (ed.), *Heterosis.* New York, NY: Springer-Verlag.

Doyle, Jack (1985) *Altered Harvest: Agriculture, Genetics, and the Fate of the World's Food Supply.* New York, NY: Viking.

Duffy, Michael (2001) "Who benefits from biotechnology?" Paper presented at the annual meeting of the American Seed Trade Association, December 5-7, Chicago, http://www.gene.ch/genet2002/Jan/msg00007.html.

DuPuis, Melanie E. (2000) "Not in my body: rBGH and the rise of organic milk." *Agriculture and Human Values* 17:3(September):285-295.

Duvick, Donald N. (1959) "The use of cytoplasmic male sterility in hybrid seed production." *Economic Botany* 13:3(July-September):167-195.

(1977) "Genetic rates of gain in hybrid maize yields during the past 40 years." *MAYDICA* 22:187-196.

(1982a) "Genetic engineering: status and potential for cotton." In *Proceedings of the 1982 Beltwide Cotton Production-Mechanization Conference.* Las Vegas, NV, January 6-7.

(1982b) "Commercial plant breeding and its relationship to publicly supported plant breeding." In *Proceedings and Minutes of the Thirty-first Annual Meeting of the Agricultural Research Institute.* Bethesda, MD: Agricultural Research Institute.

(1982c) "Genetic diversity in major farm crops on the farm and in reserve." *Australasian Plant Breeding and Genetics Newsletter* 32:1-40.

(1983) "Plant breeding with molecular biology," *Plant Molecular Biology* 2:221-222.

(1984) "North American grain production – biotechnology research and the private sector." Paper presented at the Conference on the Future of the North American Granary, St. Paul, MN, June 17-18, 1984.

Earle, Elizabeth D. (1984) "Corn cells in the lab: a new resource for plant breeders." *New York's Food and Life Sciences Quarterly* 15:2:12-13.

East, Edward M. (1907) *The Relation of Certain Biological Principles to Plant Breeding.* New Haven, CT: Connecticut Agricultural Experiment Station.

(1908) "The distinction between development and heredity in inbreeding." *American Naturalist* 43:173-181.

East, Edward M., and H. K. Hayes (1912) "Heterozygosis in evolution and in plant breeding." Bureau of Plant Industry bulletin 243. Washington, DC: USDA.

East, Edward M., and Donald F. Jones (1919) *Inbreeding and Outbreeding: Their Genetic and Sociological Significance.* Philadelphia, PA: J. B. Lippincott Company.

Eckholm, Erik (1986) "Species are lost before they're found." *New York Times* (16 September):17, 20.

Edwards, I. B. (1983) "An assessment of the role of cytoplasmic-genetic systems in hybrid wheat production." *Plant Molecular Biology Reporter* 1:3:139-143.

Egziabher, Tewolde (2003) "When elephants fight over GMOs . . ." *Seedling* (October):1-3.

Eichenwald, Kurt, Gina Kolata, and Melody Petersen (2001) "Biotechnology food: from lab to debacle." *The New York Times* (25 January).

Eisner, Thomas (1989) "Prospecting for nature's chemical riches." *Issues in Science and Technology* 6:2:31-34.

Emmanuel, Arghiri (1972) *Unequal Exchange: A Study in the Imperialism of Trade.* New York, NY: Monthly Review Press.

Engels, Frederick (1940) *Dialectics of Nature.* New York, NY: International Publishers.

Enriquez, Juan (1998) "Genomics and the world's economy." *Science* 282(14 August):925-926.

Ervin, David E., Sandra S. Batie, Rick Welsh, Chantal L. Capentier, Jacqueline I. Fern, Nessa J. Richman, and Mary A. Schulz (2000) *Transgenic Crops: An Environmental Assessment*. Henry A. Wallace Center for Agricultural & Environmental Policy.

ETC Group (2001a) "Globalization, Inc., concentration in corporate power: the unmentioned agenda." *ETC Group Communique* 71(July/August).

(2001b) "Trick or treaty? How the IU became an IOU: is 'The Law of the Seed' a white elephant . . . or the mouse that could roar?" News release, December 6.

(2002) "Defend food Sovereignty: terminate Terminator." January.

Evans, David A. (1983) "Agricultural applications of plant protoplast fusion." *Bio/Technology* 1:3(May):253-261.

Evans, David A., J. E. Bravo, and W. P. Sharp (1983) "Applications of tissue culture technology to development of improved crop varieties." In *Biotech '83: Proceedings of the International Conference on the Commercial Applications and Implications of Biotechnology*. Northwood, Middlesex, UK: Online Publications.

Evenson, Donald D., and Robert E. Evenson (1983) "Legal systems and private sector incentives for the invention of agricultural technology in Latin America." In M. Pineiro and E. Trigo, (eds.), *Technical Change and Social Conflict in Agriculture: Latin American Perspectives*. Boulder, CO: Westview Press.

Everett, H. L. (1984) Personal interview, 7 and 24 August.

Fairchild, David (1945) *The World Was My Garden: Travels of a Plant Explorer*. New York, NY: Charles Scribner's Sons.

Farm Chemicals (1981a) "The DuPont agrichemicals story." *Farm Chemicals* 144:3 (March):13-35.

(1981b) "A look at world pesticide markets." *Farm Chemicals* (September):55-60.

Farm Verified Organic (2000) "GMO contamination in seed corn and other crops." Letter to organic certifiers and other interested parties.

Farnsworth, Clyde H. (1986) "U.S. plans to defend its patents." *New York Times* (7 April):21, 25.

Federal Register (1986) "Coordinated framework for regulation of biotechnology." *Federal Register* 51:23302 (26 June).

Feibleman, James K. (1982) "Pure science, applied science, and technology." In Lynchburg College (ed.), *Science, Technology and Society, Volume 2*. Washington, DC: University Press of America.

Ferry-Morse Seed Company (1956) "The seeds of tomorrow." Detroit, MI: Ferry-Morse Seed Company.

Finkel, Elizabeth (2001) "Engineered mouse virus spurs bioweapon fears." *Science* 291(26 January):585.

Fishel, Walt L. (ed.) (1971) *Resource Allocation in Agricultural Research*. Minneapolis, MN: University of Minnesota Press.

Fisher, Lawrence (1999) "The race to cash in on the genetic code." *The New York Times* (29 August):C1.

Fliegel, Frederick C., and Johannes C. Van Es (1983) "The diffusion-adoption process in agriculture: changes in technology and changing paradigms." In G. Summers (ed.), *Technology and Social Change in Rural Areas*. Boulder, CO: Westview Press.

Flinn, William L., and Frederick H. Buttel (1980) "Sociological aspects of farm size: ideological and social consequences of scale in agriculture." *American Journal of Agricultural Economics* 62:5(December):946-953.

Food and Agriculture Organization of the United Nations (1981) *Animal Genetic Resources: Conservation and Management.* FAO animal production and health paper 24. Rome: FAO.

(1983a) "Plant genetic resources: report of the director general." Document C 83/25, August. Rome: FAO.

(1983b) "Proposal for the establishment of an international genebank and the preparation of a draft international convention for plant genetic resources (conference resolution 6/81)." Document C 83/LIM/2, September. Rome: FAO.

(1983c) "Transcript of the twenty-second session." Document C 8/II/PV/16, November. Rome: FAO.

(1983d) "International undertaking on plant genetic resources." Document C 83/II REP/4 and 5, 22 November. Rome: FAO.

(1984) *Production Yearbook, 1983.* Rome: FAO.

(1985a) "Country and international institutions' response to conference resolution 8/83 and council resolution 1/85." Document CPGR/85/3. Rome: FAO.

(1985b) "Report of the first session of the Commission on Plant Genetic Resources." 11-15 March. Rome: FAO.

(1989) *Report of the Conference of FAO.* 25th Session, 11-1-29 November, C/89/Rep, Rome: FAO.

(2001) *International Treaty on Plant Genetic Resources for Food and Agriculture,* www.fao.org/ag/cgrfa/itpgr.htm (accessed May 8, 2003).

Food and Drug Administration (1992) "Statement of policy: food derived from new plant varieties." *Federal Register* 57:104(29 May):22984-23005.

Food First (ed.) (2003) *Voices from the South: The Third World Debunks Corporate Myths on Genetically Engineered Crops.* Oakland, CA: Food First/Institute for Food and Development Policy.

Ford, K. A. (1942) "Closed versus open pedigrees for corn hybrids: a hybrid seed corn industry point of view." In *Report of the Sixth Corn Improvement Conference of the North Central Region.* Washington, DC: USDA.

Fortmann, Henry R. (1969) "Plant variety protection legislation – the state-federal viewpoint." In John Sutherland and Robert J. Falasca (eds.), *Proceedings of the Fifteenth Annual Farm Seed Conference.* Washington, DC: American Seed Trade Association.

Foster-Carter, Aidan (1978) "The modes of production controversy." *New Left Review* 107:47-78.

Fowler, Cary (1980) "Sowing the seeds of destruction." *Science for the People* (September/October):8-10.

Fox, J. L. (1984) "Patents encroaching on research freedom." *Science* 224(8 June): 1080-1082.

Fraley, Robert T. (1983) "Molecular biology vs plant breeding?" *Plant Molecular Biology* 2:49-50.

Fraley, Robert T., Stephen G. Rogers, and Robert B. Horsch (1986) "Genetic transformation in higher plants." *CRC Critical Reviews in Plant Sciences* 4:1:1-46.

Frankel, Sir Otto H. (1970) "Genetic dangers in the Green Revolution." *World Agriculture* 19:3(July):9-13.

(1974) "Genetic conservation: our evolutionary responsibility." *Genetics* 78(September):53-65.

(1985) "Genetic resources: the founding years." *Diversity* 7(Fall):26-29.

(1986a) "Genetic resources: the founding years – part two: the movement's constituent assembly." *Diversity* 8(Winter):30-32.

(1986b) "Genetic resources: the founding years – part three: the long road to the International Board." *Diversity* 9:30-33.

Frankel, R. (ed.) (1983) *Heterosis: Reappraisal of Theory and Practice.* New York, NY: Springer-Verlag.

Freifeld, Karen (1985) "Seed money." *Forbes* (2 December):219-220.

Freistadt, Margo (1988) "Squeezed out: Traditional ag researchers are watching grant money and students shift to biotechnology." *Agrichemical Age* (October): 10-11, 29.

Frey, Kenneth J. (1985) "Presidential address: the unifying force in agronomy – biotechnology." *Agronomy Journal* 77:2(March/April):187-189.

(1996) *National Plant Breeding Study – I: Human and Financial Resources Devoted to Plant Breeding Research and Development in the United States in 1994.* Iowa Agriculture and Home Economics Experiment Station Special Report 98.

(1997) *National Plant Breeding Study – II: National Plan for Promoting Breeding Programs for Minor Crops in the U.S.* Iowa Agriculture and Home Economics Experiment Station Special Report 100.

(1998) *National Plant Breeding Study – III: National Plan for Genepool Enrichment of U.S. Crops.* Iowa Agriculture and Home Economics Experiment Station Special Report 101.

Frey, Nicholas (1984) Personal interview, 16 March.

Friedland, William, and Amy Barton (1976) "Tomato technology." *Society* 13:6 (September/October):34-42.

Friedland, William H., Amy E. Barton, and Robert J. Thomas (1981) *Manufacturing Green Gold: Capital, Labor, and Technology in the Lettuce Industry.* Cambridge, UK: Cambridge University Press.

Friedmann, Harriet (1980) "Household production and the national economy: concepts for the analysis of agrarian formations." *Journal of Peasant Studies* 7 (January):158-184.

Fuglie, Keith, Nicole Ballenger, Kelly Day, Cssandra Kloz, Michael Ollinger, John Reilly, Utpal Vasavada, and Jet Yee (1996) *Agricultural Research and Development: Public and Private Investments Under Alternative Markets and Institutions.* Agricultural Economic Report 735, Washington, DC: USDA.

Fulkerson, J. F., and J. M. Barnes (1970) "A colloquy on corn blight." *Agricultural Science Review* 8:4:1-10.

Funk, D. J. (1905) "Commercial corn breeding." In *Proceedings of the First Meeting of the American Breeders Association.* Washington, DC: ABA.

Gabelman, Warren H. (1975) "GRAS legislation: viewpoints of a professional horticulturist and plant breeder." *HortScience* 10:3(June):248-250.

Galbraith, John Kenneth (1967) *The New Industrial State.* Boston, MA: Houghton Mifflin.

Galloway, B. T. (1912) "Distribution of seeds and plants by the Department of Agriculture." Bureau of Plant Industry Circular 100. Washington, DC: USGPO.

Garden City Group (2003) "Non-StarLink farmer litigation." Garden City Group, Inc., Merrick, NY, www.non-starlinkfarmerssettlement.com (accessed June 27, 2003).

Gardner, C. M. (1949) *The Grange: Friend of the Farmer.* Washington, DC: The National Grange.

Gardner, C. O. (1983) "Presidential address: reflections." *Agronomy Journal* 75:2 (March/April):157-160.

Gates, Paul W. (1943) *The Wisconsin Pine Lands: A Study in Land Policy and Absentee Ownership.* Ithaca, NY: Cornell University Press.

(1960) *The Farmer's Age: 1815–1960.* New York, NY: Holt, Rinehart and Winston.
(1973) *Landlords and Tenants on the Prairie Frontier.* Ithaca, NY: Cornell University Press.

Gaus, J. M., L. O. Wolcott, and V. B. Lewis (1940) *Public Administration and the United States Department of Agriculture.* Chicago, IL: Committee on Public Administration, Social Science Research Council.

Gebhart, Fred (1984a) "Calgene researchers clone gene for herbicide resistance." *Genetic Engineering News* 4:1(January/February):25.
(1984b) "Sungene completes two field trials of modified corn strains." *Genetic Engineering News* 4:3(April):9.
(1986) "Agri-biotech roundup: current products are few, many more to follow." *Genetic Engineering News* 6:9(October):10, 11.

Gelbart, William M. (1998) "Databases in genomic research." *Science* 282(23 October):659–661.

General Accounting Office (2002) *Genetically Modified Foods: Experts View Regimen of Safety Tests as Adequate, but FDA's Evaluation Process Could Be Enhanced.* GAO-02-566, Washington, DC: United States General Accounting Office.

Genetic Engineering and Biotechnology Monitor (1985) "Patents and intellectual property." *Genetic Engineering and Biotechnology Monitor* 14(September/December): 39–40.

Genetic Engineering News (1982) "Biotechnica and L. William Teweles in plant genetics consulting agreement." *Genetic Engineering News* 2:4(July/August):16.
(1983) "IPRI finds new life in takeover by BioRad." *Genetic Engineering News* 3:4(July/August):14.

Genetic Resources Action International (GRAIN) (2001) "A disappointing compromise." *Seedling* 18:4(December):2–3.
(2003) "Farmers' privilege under attack." (9 June), http://www.grain.org/bio-ipr/?id=105.
(2004) "Confronting contamination: Five reasons to reject co-existence." *Seedling* (April):1–4.

Genter, Clarence (1967) "Inbreeding without inbreeding depression." In John Sutherland (ed.), *Proceedings of the Twenty-second Corn and Sorghum Research Conference.* Washington, DC: American Seed Trade Association.
(1982) "Recurrent selection for high inbred yields from the F2 of a maize single cross." In Harold D. Loden and Dolores Wilkinson (eds.), *Proceedings of the Thirty-seventh Corn and Sorghum Research Conference.* Washington, DC: American Seed Trade Association.

Giamatti, A. Bartlett (1982) "The university, industry, and cooperative research." *Science* 218(24 December):1278–1280.

Giampietro, Mario (2004) *Multi-scale Integrated Analysis of Agroecosystems.* Boca Raton, FL: CRC Press.

Gibb, Jeffrey N. (1986) "OSTP shores up regulatory framework." *Bio/Technology* 4(August):690.

Gillis, Justin (2003) "To feed hungry Africans, firms plant seeds of science." *Washington Post* (11 March):A01.

Glass, B. (1955) "Genetics in the service of man." *Johns Hopkins Magazine* 6:5(February):2–8.

Godden, David (1982) "Plant variety rights in Australia: some economic issues." *Review of Marketing and Agricultural Economics* 50:1(April):51–95.

Godown, Richard D. (1986) "The real question is overregulation." *Genetic Engineering News* 6:5(May):4, 28.

Goodman, Robert M. (2002) Personal communication.
Goodwyn, Lawrence (1980) "The cooperative commonwealth and other abstractions: in search of a democratic presence." *Marxist Perspectives* 3(Summer):8-42.
Goss, Kevin F., Richard D. Rodefeld, and Frederick H. Buttel (1980) "The political economy of class structure in U.S. agriculture: a theoretical outline." In F. H. Buttel and H. Newby (eds.), *The Rural Sociology of the Advanced Societies*. Montclair, NJ: Allanheld, Osmun and Company.
Gracen, Vernon (1984) Personal interview, 22 August.
Gray, Paul E. (1981) "Statement by Paul E. Gray." In *Commercialization of Academic Biomedical Research*. Hearings before the Subcommittee on Investigations and Oversight and the Subcommittee on Science, Research and Technology of the Committee on Science and Technology, U.S. House of Representatives, Ninety-seventh Congress, 8-9 June, 1981, no. 46. Washington, DC: USGPO.
Greaves, Tom (ed.) (1994) *Intellectual Property Rights for Indigenous Peoples: A Source Book*. Oklahoma City, OK: Society for Applied Anthropology.
Gregg, Elizabeth (1982) "The seed industry: perspective and prospects." Research paper. New York, NY: Drexel Burnham Lambert, Inc.
Griffin, Keith (1974) *The Political Economy of Agrarian Change*. Cambridge, MA: Harvard University Press.
Grigg, David B. (1974) *The Agricultural Systems of the World: An Evolutionary Approach*. Cambridge, UK: Cambridge University Press.
Griliches, Zvi (1957) "Hybrid corn: an exploration in the economics of technological change." *Econometrica* 25(October):501-522.
(1958) "Research costs and social returns: hybrid corn and related innovations." *Journal of Political Economy* 66(October):419-431.
(1960) "Hybrid corn and the economics of innovation." *Science* 132(29 July): 275-280.
Gupta, Anil K. (1993) "Editorial." *Honey Bee* 4:4(October-December):1-3.
Hacker, Louis M. (1940) *The Triumph of American Capitalism*. New York, NY: Simon and Schuster.
Hallauer, Arnel R. (1984) Personal interview, 1 March.
Hallauer, Arnel R., and J. B. Miranda (1981) *Quantitative Genetics in Maize Breeding*. Ames, IA: Iowa State University Press.
Hambidge, Gove, and E. N. Bressman (1936) "Foreword and summary." In *Yearbook of Agriculture, 1936*. Washington, DC: USGPO.
Hamilton, Henry (1948) *England: A History of the Homeland*. New York, NY: Norton.
Hamilton, Neil D. (1993) "Who owns dinner? Evolving legal mechanisms for ownership of plant genetic resources." *Tulsa Law Journal* 28:587-657.
(2003) "Forced feeding: Inventorying new legal issues in the biotechnology policy debate." Unpublished paper.
Hammond, Edward Hopkins, III (1993) "More on the rainforest harvest." Unpublished paper.
Handelsman, J., and R. M. Goodman (1991) "Banning biotechnology." Paper presented at Rural Wisconsin's Economy and Society, The Influence of Policy and Technology: A Conference on How Policies and Technologies Shape Wisconsin's Economy and Society, April 10-11, Agricultural Technology and Family Farm Institute, Madison, WI.
Hanson, C. H. (1974) *The Effect of FDA Regulations (GRAS) on Plant Breeding and Processing*. Crop Science Society of America special publication no. 5. Madison, WI: Crop Science Society of America.
Hanway, D. G. (1965) "Role of public agencies in hybrid research." In John Suther-

land (ed.), *Proceedings of the Nineteenth Hybrid Corn Industry-Research Conference, 1964*. Washington, DC: American Seed Trade Association.

(1978a) "Agricultural experiment stations and the Plant Variety Protection Act: part 1." *Crops and Soils Magazine* 30:5(February):5-6.

(1978b) "Agricultural experiment stations and the Plant Variety Protection Act: part 2." *Crops and Soils Magazine* 30:6(March):5-7.

Hardy, Ralph W. F., and David J. Glass (1985) "Our investment: what is at stake?" *Issues in Science and Technology* 1:3(Spring):69-82.

Harlan, H. V., and M. L. Martini (1936) "Problems and results in barley breeding." In *Yearbook of Agriculture, 1936*. Washington, DC: USGPO.

Harlan, Jack R. (1971) "Agricultural origins: centers and noncenters." *Science* 174 (29 October):468-474.

(1975a) *Crops and Man*. Madison, WI: American Society of Agronomy.

(1975b) "Our vanishing genetic resources." *Science* 188(9 May):618-621.

(1980) "Crop monoculture and the future of American agriculture." In Sandra S. Batie and Robert G. Healy (eds.), *The Future of American Agriculture as a Strategic Resource*. Washington, DC: The Conservation Foundation.

(1984) "Gene centers and gene utilization in American agriculture." In C. W. Yeatmann et al. (eds.), *Plant Genetic Resources: A Conservation Imperative*. Boulder, CO: Westview Press.

Harry, Debra (1994) "The Human Genome Diversity Project: implications for indigenous peoples." *Abya Yala* 8:4(Winter):13-15.

Harsanyi, Zsolt (1981) "Statement by Zsolt Harsanyi." In *Commercialization of Academic Biomedical Research*, Hearings before the Subcommittee on Investigations and Oversight and the Subcommittee on Science, Research, and Technology of the Committee on Science and Technology, U.S. House of Representatives, Ninety-seventh Congress, 8-9 June, no. 46. Washington, DC: USGPO.

Hartley, C. P. (1905) "Corn-breeding work in the United States." In *Proceedings of the First Meeting of the American Breeders Association*. Washington, DC: ABA.

Harvard Business School (1978) "Pioneer Hi-Bred International, Inc." Harvard Business School case study. Boston, MA: Harvard University.

(1982) "Advanced Genetic Sciences, Inc." Harvard Business School case study. Boston, MA: Harvard University.

Hawkes, J. G. (1983) *The Diversity of Crop Plants*. Cambridge, MA: Harvard University Press.

Hayden, Thomas (2000) "A genome milestone." *Newsweek* (3 July):51.

Hayenga, Marvin L. (1998) "Structural change in the biotech seed and chemical industrial complex." *AgBioForum* 1:2:43-55.

Hayes, H. K. (1957) "A half-century of crop breeding research." *Agronomy Journal* 49:12(December):626-631.

(1963) *A Professor's Story of Hybrid Corn*. Minneapolis, MN: Burgess Publishing Company.

Hayes, H. K., and R. J. Garber (1919) "Synthetic production of high protein corn in relation to breeding." *Journal of the American Society of Agronomy* 11:308-318.

Hays, Willet M. (1900) "Breeding staple food plants." *Journal of the Royal Horticultural Society* 24:257-265.

Hayter, Edward W. (1968) *The Troubled Farmer, 1850-1900*. DeKalb, IL: Northern Illinois University Press.

Hazell, Peter B. R. (1984) "Sources of increased instability in Indian and U.S. cereal production." *American Journal of Agricultural Economics* 304(August):302-311.

Heilbroner, Robert (1966) *The Limits of American Capitalism.* New York, NY: Harper and Row.

Heisey, Paul W. (2002) "Public-sector plant breeding in privatizing world." *Agricultural Outlook* (January-February):26-29.

Heisey, Paul W., C. S. Srinivasan, and Colin Thirtle (2001) *Public Sector Plant Breeding in a Privatizing World.* Agriculture Information Bulletin no. 772, Resource Economics Division, Economic Research Service, Washington, DC: United States Department of Agriculture.

Heller, Michael A., and Rebecca S. Eisenberg (1998) "Can patents deter innovation? The anticommons in biomedical research." *Science* 280:698-701.

Herdt, Robert A. (1999) "Enclosing the global plant genetic commons." Paper presented at the China Center For Economic Research, May 24.

Hess, Charles E. (1986) "Avoiding conflict of interest." *Agricultural Engineering* 66:2(February):17-18.

Hessen, Boris (1971) "The social and economic roots of Newton's *Principia,*" in N. Bukharin et al. (eds.), *Science at the Crossroads.* London: Frank Cass and Company.

Hightower, Jim (1973) *Hard Tomatoes, Hard Times.* Cambridge, MA: Schenkman Publishing Company.

Hobsbawm, Eric (1981) "Looking forward: history and the future." *New Left Review* 125(January/February):3-19.

Hodges, R. D., and A. M. Scofield (1983) "Agricologenic disease." *Biological Agriculture and Horticulture* 1:269-325.

Hoffman, W. (1971) *David: Portrait of a Rockefeller.* New York, NY: L. Stuart.

Hoppe, Richard (1986) "Why biotech can't wait to be regulated." *Business Week* (5 May):29.

Horsfall, James G. (1975) "The fire brigade stops a raging corn epidemic." In *That We May Eat: Yearbook of Agriculture, 1975.* Washington, DC: USDA.

(1979) "Iatrogenic disease: mechanisms of action." In J. G. Horsfall and E. B. Cowling (eds.), *Plant Disease: An Advanced Treatise.* New York, NY: Academic Press.

Hougas, R. W. (1983) "Comments." In Pioneer Hi-Bred International, Inc., *Report of the 1982 Plant Breeding Research Forum.*" Des Moines, IA: Pioneer Hi-Bred International, Inc.

House, C. (1981) "Litigation – the ultimate weapon." In Harold D. Loden and Dolores Wilkinson (eds.), *Proceedings of the Eleventh Soybean Seed Research Conference.* Washington, DC: American Seed Trade Association.

Howard, N. (1982) "Genetic engineering's manpower problem." *Dun's Business Month* (January):92-94.

Hubbard, R. (1976) "Introductory essay: the many faces of ideology." In H. Rose and S. Rose (eds.), *Ideology of/in the Natural Sciences.* Cambridge, MA: Schenkman Publishing Company.

Huey, J. R. (1962) "Opportunities for public relations in the corn industry." In M. Clare Bechtold (ed.), *Proceedings of the Sixteenth Hybrid Corn Industry-Research Conference, 1961.* Washington, DC: American Seed Trade Association.

Hutton, R. (1978) *Biorevolution.* New York, NY: New American Library.

icda Seedling (1984) "IBPGR's birthday party." *icda Seedling* (October):1-2.

Iltis, Hugh H. (1981-2) "Discovery of no. 832: an essay in defense of the National Science Foundation." *Desert Plants* 3:4(Winter):175-192.

(1983) "From teosinte to maize: the catastrophic sexual transmutation." *Science* 222(25 November):886-894.

Ingersoll, F. (1983) "Private financing of public research." *Seed World* (February): 19-20.

Innes, N. L. (1982) "Patents and plant breeding." *Nature* 298(26 August):786.

(1984) "Public and private plant breeding of horticultural food crops in western Europe." *HortScience* 19:6(December):803-808.

(n.d.) "Plant breeding and intellectual property rights," www.agric-econ.uni-kiel.de/Abteilungen/II/forschung/file5.pdf.

International Board for Plant Genetic Resources (1984) "Institutes conserving crop germplasm: the IBPGR global network of genebanks." Rome: IBPGR.

(1985) "International Board for Plant Genetic Resources Annual Report 1983." Rome: IBPGR.

International Maize and Wheat Improvement Center (2003) *Challenge Program: Unlocking Genetic Diversity in Crops for the Resource Poor.* A proposal for a CGIAR Challenge Program, 6 February.

Iowa State University (2001) *Meeting Summary: Strategies for Coexistence of GMO, Non-GMO, and Organic Crop Production.* (December) Ames, IA: Iowa State University.

I-SIS (2004) "Questions over Schmeiser's ruling." Accessed at www.i-sis.org.uk.QOSR/phb.

Jacobs, Jane (1969) *The Economy of Cities.* New York, NY: Random House.

Jain, H. K. (1982) "Plant breeders' rights and genetic resources." *Indian Journal of Genetics* 42:2(July):121-128.

James, Clive (2003) *Preview: Global Status of Commercialized Transgenic Crops: 2003.* ISAAA Brief No. 30, Ithaca, NY: ISAAA.

Jantsch, Erich (1972) *Technological Planning and Social Futures.* New York, NY: John Wiley and Sons.

Japanese Association for Plant Tissue Culture (1982) *Proceedings of the Fifth International Congress of Plant Tissue and Cell Culture.* Tokyo: JAPTC.

Jefferson, Richard A. (1993) "Beyond model systems: new strategies, methods, and mechanisms for agricultural research." *Biotechnology R&D Trends,* 700:53-73.

Jenkins, Merle T. (1936) "Corn improvement." In *Yearbook of Agriculture, 1936.* Washington, DC: USGPO.

(1978) "Maize breeding during the development of hybrid maize." In D. B. Walden (ed.), *Maize Breeding and Genetics.* New York, NY: John Wiley and Sons.

Jennings, C. (1978) "What's going on in the private sector?" In Harold D. Loden and Dolores Wilkinson (eds.), *Proceedings of the Eighth Soybean Seed Research Conference.* Washington, DC: American Seed Trade Association.

Jennings, P. R. (1974) "Rice breeding and world food production." *Science* 185 (20 December):1085-1088.

Johnson, Virgil (1983) "Relationship between public and private breeding research as viewed by a publicly supported research program director." In Pioneer Hi-Bred International, Inc., *Report of the 1982 Plant Breeding Research Forum.* Des Moines, IA: Pioneer Hi-Bred International, Inc.

(1984) Personal interview, 19 March, Columbia, MO

Joly, Pierre-Benoit, and Stéphanie Lemarié (1998) "Industry consolidation, public attitude and the future of plant biotechnology in Europe." *AgBioForum* 1:2:85-90.

Jones, C. C. (1957) "The Burlington Railroad and agricultural policy in the 1920s." *Agricultural History* 31:4(October):67-74.

Jones, Donald F. (1917) "Dominance of linked factors as a means of accounting for heterosis." *Genetics* 2:466-479.

(1918) "The effects of inbreeding and outbreeding upon development." Connecticut Agricultural Experiment Station bulletin 207.

(1919) "The basis for corn betterment." *Wallace's Farmer* 44:2514-2515.

(1920) "Selection in self-fertilized lines as the basis for corn improvement." *Journal of the American Society of Agronomy* 12:3(March):77-100.

(1950) "Changes in hybrid seed corn production in the future." In William Heckendorn and Mabel Anderson (eds.), *Progress in Corn Production.* Chicago, IL: American Seed Trade Association.

Jones, D. F., and H. L. Everett (1949) "Hybrid field corn." Connecticut Agricultural Experiment Station bulletin 532.

Jones, H. A., and A. E. Clarke (1943) "Inheritance of male sterility in the onion and production of hybrid seed." *Proceedings of the American Society for Horticultural Science* 43:189-194.

Joravsky, David (1970) *The Lysenko Affair.* Cambridge, MA: Harvard University Press.

Journal of Commerce (1984) "Grace, Cetus form agricultural venture." *Journal of Commerce* (21 August):22B.

Judd, R. W. (1979) "Soybeans in the future." In Harold D. Loden and Dolores Wilkinson (eds.), *Proceedings of the Ninth Soybean Seed Research Conference.* Washington, DC: American Seed Trade Association.

Judson, H. F. (1979) *The Eighth Day of Creation: Makers of the Revolution in Biology.* New York, NY: Simon and Schuster.

Jugenheimer, R. W. (ed.) (1976) *Corn: Improvement, Seed Production, and Uses.* New York, NY: Wiley Interscience.

Kadidal, Shayana (1993) "Plants, poverty, and pharmaceutical patents." *Yale Law Journal* 103:1(October):223-258.

Kahn, Joseph (2002) "The science and politics of super rice." *The New York Times* (22 October):C1, C12.

Kalton, Robert R. (1963) "Breeders' rights." In William Heckendorn and John Sutherland (eds.), *Proceedings of the Ninth Annual Farm Seed Industry-Research Conference.* Washington, DC: American Seed Trade Association.

(1984) "Who is going to breed the 'minor' crops and train the next generation of plant breeders?" *Seedsmen's Digest* 35:12(December):8, 12.

Kalton, Robert R., and P. Richardson (1983) "Private sector plant breeding programs: a major thrust in U.S. agriculture." *Diversity* 5(November/December): 16-18.

Kane, R. J. (1964) "Populism, progressivism and pure food." *Agricultural History* 38:4(October):161-166.

Kapuscinski, Anne R., Robert M. Goodman, Stuart D. Hann, Lawrence R. Jacobs, Emily E. Pullins, Charles S. Johnson, Jean D. Kinsey, Ronald L. Krall, Antonio G. M. La Viña, Margaret G. Mellon, and Vernon W. Ruttan (2003) "Making 'safety first' a reality for biotechnology products." *Nature Biotechnology* 21:6: 599-601.

Keeler, Kathleen H. (1985) "Implications of weed genetics and ecology for the deliberate release of genetically-engineered crop plants." *Recombinant DNA Technical Bulletin* 8:4(December):165-172.

Kennedy, Donald (1981) "Statement by Donald Kennedy." In *Commercialization of Academic Biomedical Research.* Hearings before the Subcommittee on Investigations and Oversight and the Subcommittee on Science, Research, and Technology of the Committee on Science and Technology, U.S. House of Representatives, Ninety-seventh Congress, 8-9 June, no. 46. Washington, DC: USGPO.

(1982) "The social sponsorship of innovation." *Technology in Society* 4:4:253-265.

Kennedy, W. K. (1963) "Cooperative agricultural research programs – pros and cons – discussion." In *Proceedings of the Eleventh Annual Meeting of the Agricultural Research Institute.* Washington, DC:N4 NAS-NRC.

Kenney, Martin (1986) *Biotechnology: The University-Industrial Complex.* New Haven, CT: Yale University Press.

Kenney, Martin, Frederick H. Buttel, and Jack Kloppenburg, Jr. (1985) "Understanding the socioeconomic impacts of plant tissue culture technology on Third World countries." *ATAS Bulletin* 1(November):48-51.

Kenney, Martin, and Jack Kloppenburg, Jr. (1983) "The American agricultural research system: an obsolete structure?" *Agricultural Administration* 14:1-10.

Kent, James (1986) "The driving force behind the restructuring of the global seed industry." *Seed World* 124:7(June):25-26.

Keppel, B. (1984) "Monsanto shifting its emphasis to embrace new product fields." *The Columbia State* (25 November):3G.

Kidd, Charles V. (1959) "Basic research – description versus definition." *Science* 129(13 February):368-371.

Kidd, George H., and L. William Teweles (1986) "Restructuring the international seed & plant biotechnology industries." *Seedsmen's Digest* 37:2(December):44-47.

Kimmelman, Barbara A. (1983) "The American Breeders' Association: genetics and eugenics in an agricultural context, 1903-13." *Social Studies of Science* 13:2 (May):163-204.

Kimpel, Janice A. (1999) "Freedom to operate: intellectual property protection in plant biology and its implications for the conduct of research." *Annual Review of Phytopathology* 37:29-51.

King, Amanda B. and Pablo B. Eyzaguirre (1999) "Intellectual property rights and agricultural biodiversity: literature addressing the suitability of IPR for the protection of indigenous resources." *Agriculture and Human Values* 16:41-49.

King, Jonathan (1978) "New diseases in new niches." *Nature* 276(2 November):4-7.

(1982) "Patenting modified life forms: the case against." *Environment* 24:6 (July/August):38, 40-41, 57-58.

Kinney, Terry B. (1983) Remarks by Terry B. Kinney, Jr., administrator, Agricultural Research Service, USDA, before the Organization of Professional Employees of the Department of Agriculture, Beltsville, MD, November 21.

Kinsell, R. (1981) "Plant variety protection: is it working, used, enforced, needed?" In Harold D. Loden and Dolores Wilkinson (eds.), *Proceedings of the Eleventh Soybean Seed Research Conference.* Washington, DC: American Seed Trade Association.

Kirkendall, Richard S. (1982) *Social Scientists and Farm Politics in the Age of Roosevelt.* Ames, IA: Iowa State University Press.

Klausner, Arthur (1984) "Agriculture and supply attract biotech start-ups." *Bio/Technology* 2:9(September):774-775.

(1986) "U.S. to screen marine organisms for drugs." *Bio/Technology* 4(April):684.

Kleinman, Daniel Lee (1986) "Ideological struggle and the biotechnology controversy." Unpublished master's thesis, Department of Sociology, University of Wisconsin, Madison, WI.

Kleinman, Daniel Lee, and Steven P. Vallas (2001) "Science, capitalism, and the rise of the 'knowledge worker': the changing structure of knowledge production in the United States." *Theory and Society* 30:451-492.

Kloppenburg, Jack, Jr. (1984) "The social impacts of biogenetic technology in agriculture: past and future." In G. Berardi and C. Geisler (eds.), *The Social Consequences and Challenges of New Agricultural Technologies.* Boulder, CO: Westview Press.

(1992) "Conservationists or corsairs?" *Seedling* 9:2 & 3(June/July):12-17.

(1993) "Planetary patriots or sophisticated scoundrels?" *Biotechnology and Development Monitor* 16(September):24.

Kloppenburg, Jack, Jr., and Frederick H. Buttel (1987) "Two blades of grass: the contradictions of agricultural research as state intervention." In Richard G. Braungart (ed.), *Research in Political Sociology, Volume 3.* Greenwich, CT: JAI Press.

Kloppenburg, Jack, Jr., and Charles C. Geisler (1985) "The agricultural ladder: agrarian ideology and the changing structure of U.S. agriculture." *Journal of Rural Studies* 1:1:59-72.

Kloppenburg, Jack, Jr., and Daniel Lee Kleinman (1986) "Seed wars: common heritage, private property, and political strategy." Paper presented at the Union for Radical Political Economics regional conference, Ann Arbor, MI, 24-26 October.

(1987a) "Seeds of struggle: the geopolitics of plant genetic resources." *Technology Review* 90:2(February/March):48-56.

(1987b) "The plant germplasm controversy." *BioScience* 37:3(March):190-198.

(1987c) "Seeds and sovereignty." *Diversity* 10:29-33.

Klose, Norman (1950) *America's Crop Heritage: The History of Foreign Plant Introduction by the Federal Government.* Ames, IA: Iowa State College Press.

Kluger, Jeffrey (1999) "Suicide seeds." *Time* (1 February):44-45.

Kohl, Danny (2001) "GM food – another view." *The Nation* (16 April):7, 23.

Kolata, Gina (1985) "How safe are engineered organisms?" *Science* 229(5 July):34-39.

(2001) "Researchers find grave defect risk in cloning animals." *The New York Times* (25 March).

Kosuge, T., C. P. Meredith, and A. Hollaender (1983) *Genetic Engineering of Plants: An Agricultural Perspective.* New York, NY: Plenum Press.

Kowalski, Stanley P., Reynaldo V. Ebora, R. David Kryder, and Robert H. Potter (2002) "Transgenic crops, biotechnology and ownership rights: what scientists need to know." *The Plant Journal* 31:4:407-421.

Krimsky, Sheldon (1982) *Genetic Alchemy: The Social History of the Recombinant DNA Debate.* Cambridge, MA: MIT Press.

(1983) "Biotechnology and unnatural selection: the social control of genes." In G. Summers (ed.), *Technology and Social Change in Rural Areas.* Boulder, CO: Westview Press.

Kuhn, Thomas (1962) *The Structure of Scientific Revolutions.* Chicago, IL: University of Chicago Press.

Kuznets, Simon (1966) *Modern Economic Growth.* New Haven, CT: Yale University Press.

Laidlaw, Stuart (2001) "StarLink fallout could cost billions, future of modified crops thrown in doubt, report says." *The Toronto Star* (9 January), www.mindfully.org/GE/StarLink-Fallout-Cost-Billions.htm.

Lamberts, H., and J. Sneep (1984) "Board of Eucarpia disapproves of patents on breeding methods." *Euphytica* 33:1-2.

Lancet, The (2002) "DNA patents: putting an end to 'business as usual.'" *The Lancet* 360:9330(3 August), www.thelancet.com/journal (accessed August 5, 2002).

Lander, Eric S. (1996) "The new genomics: global views of biology." *Science* 274 (25 October):536–539.

Landes, David S. (1969) *The Unbound Prometheus.* Cambridge, UK: Cambridge University Press.

Langer, William L. (1975) "American foods and Europe's population growth 1750–1850." *Journal of Social History* (Winter):51-66.

Laufer, B. (1929) "The American plant migration." *Scientific Monthly* 28(March): 239-251.

Laughnan, John (1984) Telephone interview, 13 April.

Lawrence, Robert H. (1983) "The scientific background of genetic engineering: current technologies and prospects for the future." In *Genetic Engineering and Plant Breeding.* Geneva, Switzerland: UPOV.

Lawton, M. A., and M. Chilton (1984) "Agrobacterium Ti plasmids as potential vectors for genetic engineering," *HortScience* 19:1(February):40-42.

Leath, M. N., L. H. Meyer, and L. D. Hall (1982) "U.S. corn industry." Agricultural economic report no. 479, Economic Research Service. Washington, DC: USDA.

Lee, Gary J. (1985) "A farmer's view of seed certification." In M. B. McDonald and W. D. Pardee (eds.), *The Role of Seed Certification in the Seed Industry.* CSSA special publication no. 10. Madison, WI: Crop Science Society of America.

Leenders, Hans W. (1986) "Reflections on 25 years of service to the International Seed Trade Federation." *Seedsmen's Digest* 37:5(May):8–9.

Leffel, Robert C. (1981) "The future of public soybean improvement programs." In Harold D. Loden and Dolores Wilkinson (eds.), *Proceedings of the Eleventh Soybean Seed Research Conference.* Washington, DC: American Seed Trade Association.

Leibenluft, Robert F. (1981) "Competition in farm inputs: an examination of four industries." Office of Policy Planning, Federal Trade Commission. Washington, DC: National Technical Information Service.

Lemmon, K. (1968) *Golden Age of Plant Hunters.* London: Phoenix House.

Lenin, V. I. (1939) *Imperialism, the Highest Stage of Capitalism.* New York: International Publishers.

(1967) "New data on the laws governing the development of capitalism in agriculture." In *On the United States of America.* Moscow: Progress Publishers.

Lepkowski, Will (1982) "Shakeup ahead for agricultural research." *Chemical and Engineering News* (22 November):8-16.

Lesser, William H. (1986) "Patenting seeds: what to expect." Unpublished manuscript.

Lesser, William H., and Robert T. Masson (1983) *An Economic Analysis of the Plant Variety Protection Act.* Washington, DC: American Seed Trade Association.

Levidow, Les, and Bob Young (eds.) (1981) *Science, Technology and the Labour Process: Marxist Studies, Volume 1.* London: CSE Books.

Levine, Andrew, and Erik Olin Wright (1980) "Rationality and class struggle." *New Left Review* 123(September/October):47-68.

Levins, Richard, and Richard Lewontin (1985) *The Dialectical Biologist.* Cambridge, MA: Harvard University Press.

Lewis, W. J., J. C. vanLenteren, S. C. Phatak, and J. H. Tumlinson (1997) "A total systems approach to pest management." *Proceedings of the National Academy of Sciences* 94:12243-12248.

Lewontin, Richard (1982) "Agricultural research and the penetration of capital." *Science for the People* 14:1(January/February):12-17.

(2001) "Genes in the food!" *The New York Review of Books* (21 June).

Lewontin, Richard, and Richard Levins (1976) "The problem of Lysenkoism." In H. Rose and S. Rose (eds.), *Ideology of/in the Natural Sciences.* Cambridge, MA: Schenkman Publishing Company.

Lianos, T. P., and Q. Paris (1972) "American agriculture and the prophecy of increasing misery." *American Journal of Agricultural Economics* 54:4(November): 570-577.

Lipton, Michael, and Richard Longhurst (1986) "Modern varieties, international agricultural research, and the poor." CGIAR study paper no. 2. Washington, DC: World Bank.

Little, W. R. (1958) "Relationship of seedsmen and certified seed growers." *Annual Meeting of the International Crop Improvement Association, 1958.* ICIA.

Litzow, M. E., and J. L. Ozbun (1979) "Seventy-five years of research on growth and development of vegetable crops." *HortScience* 14:3(June):350-354.

Loden, Harold D. (1963) "Research in the seed industry." *Annual Report of the International Crop Improvement Association, 1963.* ICIA.

Losey, J. E., L. S. Rayor, and M. E. Carter (1999) "Transgenic pollen harms monarch larvae." *Nature* 399:214.

Lovett, J. V. (1982) "Agro-chemical alternatives in a future agriculture." *Biological Agriculture and Horticulture* 1:1:15-27.

Lower, Richard L. (1984) "Genetic engineering: the relationship between industry, academia, and plant sciences." *HortScience* 19:1(February):49-51.

McAuliffe, Sharon, and Kathleen McAuliffe (1981) *Life for Sale.* New York, NY: Coward, McCann and Geoghegan.

McCalla, Alex F. (1978) "Politics of the agricultural research establishment." In D. F. Hadwiger and W. P. Browne (eds.), *The New Politics of Food.* Lexington, MA: Lexington Books.

McConnell, Grant (1953) *The Decline of Agrarian Democracy.* New York, NY: Atheneum.

McCormick, Douglas (1986) "Tunnel vision." *Bio/Technology* 3:12(December):1045.

McDonnell, John (1986) "Helping corporations become green giants." *Corporate Report Wisconsin* 1:10(June):42-43.

McElroy, W. D. (1977) "The global age: roles of basic and applied research." *Science* 196(15 April):267-270.

McGaughey, William H., and Mark E. Whalon (1992) "Managing insect resistance to *Bacillus thuringiensis* toxins." *Science* 258(27 November):1451-1455.

McHughen, Alan (2000) *Pandora's Picnic Box: The Potential and Hazards of Genetically Modified Foods.* London: Oxford University Press.

MacKenzie, Deborah (1984) "Law of the sea flounders in Geneva." *New Scientist* (30 August):3.

McNeil, Donald G., Jr. (2000) "Protests on new genes and seeds grow more passionate in Europe." *The New York Times* (14 March):A1, A10.

McNeill, W. (1974) *The Shape of European History.* New York, NY: Oxford University Press.

Macy, L. K., L. E. Arnold, and E. C. McKibben (1938) "Changes in technology and labor requirements in crop production: corn." Report no. A-5, National Research Project on Reemployment Opportunities and Recent Changes in Industrial Techniques. Philadelphia, PA: Works Progress Administration.

Magdoff, Harry (1982) "Imperialism: a historical survey." In H. Alavi and T. Shanin (eds.), *Introduction to the Sociology of the Developing Societies.* New York, NY: Monthly Review Press.

Malcolm, Andrew H. (1986) "Scientists try to create forest of cloned trees." *New York Times* (4 February):20.

Manchester, Alden C. (1985) "Agriculture's links with U.S. and world economies." Agriculture information bulletin no. 496. Washington, DC: USDA.

Mandel, Ernest (1978) *Late Capitalism.* London: Verso Editions.

Mangelsdorf, Paul C. (1951) "Hybrid corn." *Scientific American* 185:2(August):39-47.

(1974) *Corn: Its Origin, Evolution and Improvement.* Cambridge, MA: Harvard University Press.

Manicad, Gigi (1999) "CGIAR and the private sector: public good versus proprietary technology in agricultural research." *Biotechnology and Development Monitor* 37(March):8-13.

Mann, Susan, and James Dickinson (1978) "Obstacles to the development of capitalist agriculture." *Journal of Peasant Studies* 5(July):466-481.

Manning, Richard (2000) *Food's Frontier: The Next Green Revolution.* San Francisco, CA: North Point Press.

Maranto, Gina (1986) "Genetic engineering: hype, hubris, and haste." *Discover* (June):50-64.

Marcuse, Herbert (1964) *One Dimensional Man.* Boston, MA: Beacon Press.

Margaronis, Maria (1999) "The politics of food." *The Nation* (27 December):11-12, 15-16.

Martínez-Alier, Joan (1994) "The merchandising of biodiversity." *Etnoecológica* 2:3:69-86.

Marx, Karl (1973) *Grundrisse.* New York, NY: Vintage Books.

(1977) *Capital, Volume 1.* New York, NY: Vintage Books.

(1981) *Capital, Volume 2.* New York, NY: Vintage Books.

Marx, Karl, and Frederick Engels (1970) *The German Ideology.* New York, NY: International Publishers.

(1978) "Manifesto of the Communist Party," in Robert C. Tucker (ed.), *The Marx-Engels Reader.* New York, NY: W. W. Norton & Company.

Massey, C., and A. Catalano (1978) *Capital and Land.* London: Edward Arnold.

Massey, Jim (2001) "UW professor calls for biotech moratorium." *The Country Today* (14 February):1A.

May, Edward (1949) "The development of hybrid corn in Iowa." In *Plant Research in the Tropics.* Iowa Agricultural Experiment Station research bulletin 371.

Mayer, Andre, and Jean Mayer (1974) "Agriculture, the island empire." *Daedalus* 103(Summer):83-95.

Mayer, Jorge (2003) "Intellectual property rights and access to agbiotech by developing countries." *AgBiotechNet* 5(March):1-5.

Mayr, Ernst (1982) *The Growth of Biological Thought.* Cambridge, MA: Belknap Press.

Medawar, P. B. (1977) "The DNA scare: fear and DNA." *New York Review of Books* 24:15(27 October); reprinted in J. D. Watson and J. Tooze, (eds.), *The DNA Story: A Documentary History of Gene Cloning.* San Francisco, CA: W. H. Freeman and Company.

Medvedev, Zhores A. (1969) *The Rise and Fall of T. D. Lysenko.* New York, NY: Columbia University Press.

Merton, Robert K. (1970) *Science, Technology and Society in 17th Century England.* New York, NY: Harper and Row.

(1973) *The Sociology of Science: Theoretical and Empirical Investigations.* Chicago, IL: University of Chicago Press.

Meyerhoff, Albert H. (1982) "Statement by Albert H. Meyerhoff." In *University/Industry Cooperation in Biotechnology.* Hearings before the Subcommittee on Investigations and Oversight and the Subcommittee on Science, Research, and Technology of the Committee on Science and Technology, U.S. House of Representatives, Ninety-seventh Congress, second session, 16-17 June. Washington, DC: USGPO.

Miller, Henry I., and Gregory Conko (2001) "Precaution without principle." *Nature Biotechnology* 19(April):302-303.

Miller, Jon D. (1985) "The attitudes of religious, environmental and science policy leaders toward biotechnology." DeKalb, IL: Public Opinion Laboratory, Northern Illinois University.

Miller, Julie Ann (1985) "Lessons from Asilomar." *Science News* 127(February 23): 122-123, 127.

Minnesota Agricultural Experiment Station (1981) *Evaluation of Agricultural Research.* Miscellaneous publication 8-1981.

Mintz, Jerome (1984) "Biotechnology – the biological revolution will change how we think about life itself." *Venture* (February):38-58.

Monsanto Company (n.d.) "Genetic engineering: a natural science." St. Louis, MO: Monsanto Company.

Montecinos, Camila (1996) "*Sui generis* – a dead end alley?" *Seedling* 13:4(December): 19-28.

Mooney, Patrick H. (1982) "Labor time, production time and capitalist development in agriculture." *Sociologia Ruralis* 22:3/4:279-291.

(1983) "Toward a class analysis of Midwestern agriculture." *Rural Sociology* 48:4 (Winter):563-584.

Mooney, Pat Roy (1979) *Seeds of the Earth: A Private or Public Resource?* Ottawa: Inter Pares.

(1983) "The law of the seed." *Development Dialogue* 1-2:1-172.

Moran, Katy, Steven R. King, and Thomas J. Carlson (2001) "Biodiversity prospecting: lessons and prospects." *Annual Review of Anthropology* 30:505-526.

Morgan, Dan (1979) *Merchants of Grain.* New York, NY: Penguin Books.

Mothers for Natural Law of the Natural Law Party (2000) *Safe Food News.* Fairfield, IA: Mothers for Natural Law of the Natural Law Party.

Mouzelis, Nicos (1976) "Capitalism and the Development of Agriculture." *The Journal of Peasant Studies* 3:4(July):483-492.

Mulkay, Michael (1979) *Science and the Sociology of Knowledge.* London: George Allen and Unwin.

Mumford, Lewis (1963) *Technics and Civilization.* New York, NY: Harcourt, Brace and World, Inc.

Munger, Henry M. (1952) "Future of F_1 hybrids." American Seed Trade Association informational pamphlet no. 5:42-47. Chicago, IL: American Seed Trade Association. (1984) "Public/private responsibilities in using horticultural genetic resources." *Diversity* 6(March/April):20, 22.

Murphy, Sheldon R. (1982) "Statements of Dr. Sheldon R. Murphy." In *Potential Application of Recombinant DNA and Genetics on Agricultural Sciences.* Hearings before the Subcommittee on Investigations and Oversight of the Committee on Science and Technology, U.S. House of Representatives: Ninety-seventh Congress, 9 June, 28 July, no. 134. Washington, DC: USGPO.

Murray, J. R. (1982) "Biological diversity and genetic engineering." In U.S. Department of State, *Proceedings of the U.S. Strategy Conference on Biological Diversity.* Washington, DC: U.S. Department of State.

Myers, Norman (1979) *The Sinking Ark: A New Look at the Problem of Disappearing Species.* Oxford, UK: Pergamon Press.

(1981) "The exhausted earth." *Foreign Policy* 42(Spring):141-155.

(1983) *A Wealth of Wild Species.* Boulder, CO: Westview Press.

Myers, W. M. (1964) "Germ plasm control as it would affect variety improvement and release of field crops." In *Plant Breeders' Rights.* Madison, WI: American Society of Agronomy.

Myerson, Allen R. (1997) "Breeding seeds of discontent: cotton growers say strain cuts yields." *The New York Times* (19 November):C1, C5.

Nash, J. Madeline (2000) "Grains of hope." *Time* (July 31):38-46.

National Agricultural Research and Extension Users Advisory Board (1983) "Agricultural research, extension, and education recommendations – report to the Secretary." July. Washington, DC: USDA.

National Association of State Universities and Land Grant Colleges (1983) "Emerging biotechnologies in agriculture: issues and policies, progress report II." Division of Agriculture, Committee on Biotechnology, NASULGC.

National Institutes of Health (1993) "First five year awards are announced under Interagency Biodiversity Program." *NIH News* (7 December).

National Research Council (1972a) *Genetic Vulnerability of Major Crops.* Washington, DC: National Academy of Sciences.

(1972b) "Report of the Committee on Research Advisory to the U.S. Department of Agriculture." Washington, DC: National Academy of Sciences.

(1984) *Genetic Engineering of Plants.* Washington, DC: National Academy Press.

(1985) *New Directions for Biosciences Research in Agriculture: High Reward Opportunities.* Washington, DC: National Academy Press.

(2000) *Genetically Modified Pest-Protected Plants: Science and Regulation.* Washington, DC: National Academy Press.

National Science Board (1986) *Science Indicators, 1985.* Washington, DC: National Academy of Sciences.

Nature (1986) "Selling the family silver." *Nature* 322(7 August):485-486.

Neagley, C. H., D. D. Jeffery, and A. R. Diepenbrock (1984) "Genetic engineering patent law trends affecting development of plant patents." *Genetic Engineering News* 4:3(April):10-11.

Neal, Norman (1983) "The Wisconsin hybrid corn program 1923-1967." Unpublished paper, July.

Nelson, W. L. (1970) "Improving our agronomic image." *Agronomy Journal* 62:1 (January/February):1-4.

Nestle, Marion (2002) *Food Politics.* Berkeley, CA: University of California Press.

(2003) *Safe Food: Bacteria, Biotechnology, and Bioterrorism.* Berkeley, CA: University of California Press.

New Scientist (1982) "An Eldorado of plants that cure, feed and fuel." *New Scientist* 96(21 October):158.

(1983) "Foreign fields save western crops – free of charge." *New Scientist* 96 (28 October):218.

Newlin, Owen J. (1986) "Support for plant breeding research." *Seed World* 123:12 (November):40-50.

Newman, S. A. (1982) "The 'scientific' selling of RNA." *Environment* 24:6(July/August):21-23, 53-57.

New York Times, The (1981) "Talking business: interview with Thomas Urban." *The New York Times* (May 5):D2.

(1982) "The worm in the bud." *The New York Times* (21 October):A30.

(1986a) "A novel strain of recklessness." *The New York Times* (6 April):22.

(1986b) "Toying with genes and Argentina." *The New York Times* (November 12):22.

Nielsen, Kaare M. (2003) "Transgenic organisms – time for conceptual diversification?" *Nature Biotechnology* 21(March):227-228.

Nilan, Robert (1981) "Genetic engineering in plants." In Harold D. Loden and Dolores Wilkinson (eds.), *Proceedings of the Twenty-seventh Farm Seed Conference.* Washington, DC: American Seed Trade Association.

Noble, David N. (1977) *America by Design.* New York, NY: Oxford University Press.

(1980) "Corporate roots of American science." In R. Arditti, P. Brennan, and S. Caurak (eds.), *Science and Liberation.* Boston, MA: South End Press.

(1983) "Academia incorporated." *Science for the People* 15:1(January/February): 7-11, 50-52.

(1984) *Forces of Production: A Social History of Industrial Automation.* New York, NY: Alfred A. Knopf.

Norman, Colin (1981) *The God that Limps.* New York, NY: W. W. Norton and Company.

Norse, Elliot A. (1986) "Statement of Dr. Elliot A. Norse." Hearings before the Committee on Science and Technology Subcommittee on Investigations and Oversight, 23 July; reprinted in *Recombinant DNA Technical Bulletin* 9:3(September):172-177.

Nottenburg, Carol, Philip G. Pardey, and Brian D. Wright (2002) "Accessing other people's technology for non-profit research." *Australian Journal of Agricultural and Resource Economics* 46:3:289-416.

NOVA (1985) "Seeds of tomorrow." NOVA (television broadcast), 15 October.

Oasa, Edmund K., and Bruce Jennings (1982) "Science and authority in international agricultural research." *Bulletin of Concerned Asian Scholars* 14:4(October/December):30-44.

O'Brien, Kelly (1985) "What makes a good researcher?" *Agricultural Age* 27:4 (April):30, 35.

O'Brien, Patrick (1982) "European economic development: the contribution of the periphery." *Economic History Review* 35:1(February):1-18.

O'Connor, James (1975) "The twisted dream." *Monthly Review* (March):41-54.

Oehmke, James F. and Christopher A. Wolf (2002) "Is concentration affecting biotechnology industry R&D?" *Choices* (Spring):11-15.

Office of Technology Assessment (1979) *Pest Management Strategies in Crop Protection, Volume One.* Washington, DC: USGPO.

(1981a) *Impacts of Applied Genetics.* Washington, DC: USGPO.

(1981b) *An Assessment of the United States Food and Agricultural Research System.* Washington, DC: USGPO.

(1984) *Commercial Biotechnology: An International Analysis.* Washington, DC: USGPO.

Oheim, C. L. (1954) "Industry's stake in the ARI." In *Proceedings of the Third Annual Meeting of the Agricultural Research Institute.* Washington, DC: NAS-NRC.

Orton, Thomas J. (1985) Personal communication.

Paarlberg, Don F. (1964) *American Farm Policy.* New York, NY: John Wiley and Sons.

(1978) "A new agenda for agriculture." In D. Hadwiger and W. Browne (eds.), *The New Politics of Food.* Lexington, MA: Lexington Books.

Paarlberg, Robert (2000) "The global food fight." *Foreign Affairs* 79:3(May/June): 24-38.

Padwa, David J. (1982) "Future impact of genetic engineering." In *Proceedings of the International Conference on Genetic Engineering, Volume 5*. Reston, VA: Battelle Memorial Institute.

(1983) "Genetic engineering: a new tool for plant breeders." In *Genetic Engineering and Plant Breeding*. Geneva, Switzerland: UPOV.

Pardee, W., D. Phillipson, J. Billings, and R. Kalton (1981) "Panel discussion – public vs private research." In Harold D. Loden and Dolores Wilkinson (eds.), *Proceedings of the Twenty-seventh Annual Farm Seed Conference*. Washington, DC: American Seed Trade Association.

Paul, Bill (1984) "Third World battles for fruit of its seed stocks." *Wall Street Journal* (June 15):1.

Pearse, Andrew (1980) *Seeds of Plenty, Seeds of Want*. New York, NY: Oxford University Press.

Pearson, O. H. (1983) "Heterosis in vegetable crops." In R. Frankel (ed.), *Heterosis*. New York, NY: Springer-Verlag.

Perelman, Michael (1977) *Farming for Profit in a Hungry World*. Montclair, NJ: Allanheld, Osmun and Company.

Perrin, R. K., K. A. Kunnings, and L. A. Ihnen (1983) "Some effects of the U.S. Plant Variety Protection Act of 1970." August. Economics research report 46, Department of Economics and Business, North Carolina State University.

Perrow, Charles (1984) *Normal Accidents: Living With High Risk Technologies*. New York, NY: Basic Books.

Peterson, Clinton E. (1975) "Plant introductions in the improvement of vegetable cultivars." *HortScience* 10:6(December):575-579.

(1984) Personal communication.

Pfeffer, Max John (1982) "The labor process and capitalist development of agriculture." *Rural Sociologist* 2:2:72-80.

Phillips, Robert L. (1983) "Genetic engineering of plants: some perspectives on the conference, the present, and the future." In T. Kosuge et al. (eds.), *Genetic Engineering of Plants: An Agricultural Perspective*. New York, NY: Plenum Press.

Pieters, A. J. (1900) "Seed selling, seed growing, and seed testing." In *Yearbook of Agriculture, 1899*. Washington, DC: USGPO.

Pike, Leonard M. (1984) "Alternative methods of funding vegetable research." *HortScience* 19:6(December):802-803.

Pimentel, David, L. E. Hurd, A. C. Bellotti, M. J. Forster, I. N. Oka, O. D. Sholes, and R. J. Whitman (1973) "Food production and the energy crisis." *Science* 182 (2 November):443-449.

Pimentel, David, and Lois Levitan (1986) "Pesticides: amounts applied and amount reaching pests." *BioScience* 36:2(February):86-90.

Pioneer Hi-Bred International, Inc. (n.d.) "Biotechnology: science or alchemy?" Des Moines, IA: Pioneer Hi-Bred International, Inc.

(1983) *Report of the 1982 Plant Breeding Research Forum*. Des Moines, IA: Pioneer Hi-Bred International, Inc.

(1984) "Conservation and utilization of exotic germplasm to improve varieties." In *Report of the 1983 Plant Breeding Research Forum*. Des Moines, IA: Pioneer Hi-Bred International, Inc.

Plant Disease (1983) "Industry news." *Plant Disease* (September):1051.

Plant Variety Protection Office (1984) *Plant Variety Protection Office Journal* 12:2 (April-June).

Plucknett, Donald L., and Nigel J. H. Smith (1982) "Agricultural research and third world food production." *Science* 217(16 July):215-220.

(1986) "Sustaining agricultural yields." *BioScience* 36:1(January):40-45.

Plucknett, Donald L., N. J. H. Smith, J. T. Williams, and N. M. Anishetty (1983) "Germplasm conservation and developing countries." *Science* 220(8 April): 163-168.

Poehlman, John M. (1968) "Testimony of John M. Poehlman." In U.S. Senate, 1968.

Pogash, Carol (2004) "California county debates use of gene-altered foods." *The New York Times* (2 March):A17.

Polanyi, Michael (1945) *The Logic of Liberty.* London: Routledge and Kegan Paul.

(1962) "The republic of science." *Minerva* 1:1:54-73.

Pollack, Andrew (2000a) "Kraft recalls taco shells with bioengineered corn." *The New York Times* (23 September):B1.

(2000b) "Bioengineered corn found in European food." *The New York Times* (7 November):C2.

(2001) "The Green Revolution yields to the bottom line." *The New York Times* (15 May):D1-D2.

(2002) "Spread of gene-altered pharmaceutical corn spurs $3 million fine." *The New York Times* (7 December):A15.

(2003a) "Widely used crop herbicide is losing weed resistance." *The New York Times* (14 January):C1.

(2003b) "2nd cancer is attributed to gene used in F.D.A. test." *The New York Times* (17 January):A20.

(2003c) "RNA trades bit part for starring role in the cell." *The New York Times* (21 January):D1, D4.

(2003d) "Report says more farmers don't follow biotech rule." *The New York Times* (19 June):C5.

(2004a) "No foolproof way is seen to contain altered genes." *The New York Times* (21 January):A10.

(2004b) "So the fish glow. But will they sell?" *The New York Times* (25 January):B5.

Pollack, Andrew, and Carol Kaesuk Yoon (2001) "Rice genome called a crop breakthrough." *The New York Times* (27 January):A8.

Pollan, Michael (1998) "Playing god in the garden." *The New York Times Magazine* (25 October):44-51, 62, 82, 92.

Porter, C. R. (1961) "Certified seed in a merchandising program." In *Annual Report of the International Crop Improvement Association.* ICIA.

Portes, Alejandro (1981) *Labor, Class, and the International System.* New York, NY: Academic Press.

Posey, Darrell A. and Graham Dutfield (1996) *Beyond Intellectual Property: Toward Traditional Resource Rights for Indigenous Peoples and Local Communities.* Ottawa: IDRC.

Powell, F. W. (1927) *The Bureau of Plant Industry: Its History, Activities, Organization.* Service Monographs of the U.S. Government. Baltimore, MD: Johns Hopkins University Press.

Pramik, Mary Jean (1982) "Atlantic Richfield builds research team for agricultural applications in biotechnology." *Genetic Engineering News* 2:3(May/June):14-15.

Pramik, Mary Jean, and John Sterling (1986) "Advanced Genetic Sciences accused of illegal experiment." *Genetic Engineering News* 6:3(March):1, 10.

Price, Don K. (1954) *Government and Science.* New York, NY: New York University Press.

Proebsting, Edward L., Jr. (1984) "Life in the slow lane." *HortScience* 19:5(October): 615-617.

Public Sector Intellectual Property Resource for Agriculture (2003) "Background," www.pipra.org/background.php (accessed July 24, 2003).

Pursell, Carroll W. (1968) "The administration of science in the Department of Agriculture, 1933-1940." *Agricultural History* 42(July).

Qaim, Matin, and David Zilberman (2003) "Yield effects of genetically modified crops in developing countries." *Science* 299(7 February):900-902.

Qualset, Calvin O., A. Hollaender, M. Clutter, D. N. Duvick, J. E. Fox, R. L. Garcia, E. G. Jaworski, and R. H. Lawrence, Jr. (1983) "Roundtable discussion on research priorities." In T. Kosuge et al. (eds.), *Genetic Engineering of Plants: An Agricultural Perspective.* New York, NY: Plenum Press.

Quisenberry, K. S., N. J. Volk, and F. L. Winter (1956) "Panel discussion on industry-government relations: plant breeding." In *Proceedings of the Sixth Annual Meeting of the Agricultural Research Institute.* 15-16 October. Washington, DC: NAS-NRC.

Quist, David and Ignacio H. Chapela (2001) "Transgenic DNA introgressed into traditional maize landraces in Oaxaca, Mexico." *Nature* 414(29 November): 541-543.

Ramage, R. T. (1983) "Heterosis and hybrid seed production in barley." In R. Frankel (ed.), *Heterosis.* New York, NY: Springer-Verlag.

Rasmussen, Wayne D. (1975a) *Agriculture in the United States: A Documentary History, Volumes 1-4.* New York, NY: Random House.

(1975b) "Experiment or starve: the early settlers." In *That We May Eat: The Yearbook of Agriculture, 1975.* Washington, DC: USGPO.

Rausser, Gordon (1999) "Public/private alliances." *AgBioForum* 2:1:5-10.

Rausser, Gordon, Leo Simon, and Holly Amden (2000) "Agricultural biotechnology R&D in developing countries: public-private research partnerships." Paper presented at the Fourth International Conference on the Economics of Agricultural Biotechnology, August 24-28, Ravello, Italy.

Reichert, W. (1982) "Agriculture's diminishing diversity." *Environment* 24:9(November):6-11, 39-43.

Reid, Walter V., Sarah A. Laird, Carrie A. Meyer, Rodrigo Gámez, Ana Sittenfeld, Daniel H. Janzen, Michael A. Gollin, and Calestous Juma (eds.) (1993) *Biodiversity Prospecting: Using Genetic Resources for Sustainable Development.* Washington, DC: World Resources Institute.

Reitz, L. P. (1954) "Wheat breeding and our food supply." *Economic Botany* 8:3(July-September):251-268.

(1962) "The improvement of wheat." In *After a Hundred Years: Yearbook of Agriculture, 1962.* Washington, DC: USGPO.

Reitz, L. P., and J. C. Craddock (1969) "Diversity of germplasm in small grain cereals," *Economic Botany* 23:315-323.

Richey, Frederick D. (1937) "Why plant research?" *Journal of the American Society of Agronomy* 29:12(December):967-977.

Rifkin, Jeremy (1983) *Algeny.* New York, NY: The Viking Press.

(1986) "Biotechnology parallels nuclear industry by playing ecological roulette with environment." *Genetic Engineering News* 6:6(June):4, 29.

Rigl, Ted C. (1986) "Field test of gene altered pest resistant plant begins." *Genetic Engineering News* 6:9(October):1, 33.

Rimer, Sara (2003) "A caution against mixing commerce and academics." *The New York Times* (16 April):A16.

Riordan, Teresa (1999) "Patents: bioluminescent champagne." *The New York Times* (6 December):C8.

Rissler, Jane and Margaret Mellon (1996) *The Ecological Risks of Engineered Crops.* Cambridge, MA: MIT Press.

Roberts, Thomas H. (1979) "Public and private roles in seed improvement." *Seedsmen's Digest* 30:9(September):42-47.

Robinson, J. L., and O. A. Knott (1963) *The Story of the Iowa Crop Improvement Association and its Predecessors.* Ames, IA: Iowa Crop Improvement Association.

Rockefeller Foundation (2002) "Rockefeller Foundation initiative to promote intellectual property (IP) polices fairer to poor people." Press release, 4 November.

Rockefeller Foundation and the Office of Science and Technology Policy (1982) "Science for agriculture: report of a workshop on critical issues in American agriculture." New York, NY: The Rockefeller Foundation.

Rogers, Everett M. (1962) *Diffusion of Innovations.* New York, NY: Free Press.

Rogers, Michael (1975) "The Pandora's box conference." *Rolling Stone* 189(19 June); reprinted in J. D. Watson, and J. Tooze, (eds.), *The DNA Story: A Documentary History of Gene Cloning.* San Francisco, CA: W. H. Freeman and Company.

Rose, Hilary, and Steven Rose (eds.) (1976a) *Ideology of/in the Natural Sciences.* Cambridge, MA: Schenkman Publishing Company.

(1976b) "The problematic inheritance: Marx and Engels on the natural sciences." In H. Rose and S. Rose (eds.), *Ideology of/in the Natural Sciences.* Cambridge, MA: Schenkman Publishing Company.

(1976c) "The incorporation of science." In H. Rose and S. Rose (eds.), *Ideology of/in the Natural Sciences.* Cambridge, MA: Schenkman Publishing Company.

Rose, P. S. (1935) "Pathway to the future." *Country Gentleman* 105:9(September):20.

Rose, Steven (ed.) (1982) *Against Biological Determinism.* New York, NY: Allison and Busby.

Rosenberg, Charles E. (1976) *No Other Gods: On Science and American Social Thought.* Baltimore, MD: Johns Hopkins University Press.

Rosenberg, Nathan (1974) "Karl Marx on the economic role of science." *Journal of Political Economy* (July/August):713-728.

Rossiter, Margaret W. (1975) *The Emergence of Agricultural Science: Justus Liebig and the Americans 1840-1880.* New Haven, CT: Yale University Press.

Rossman, P. (1984) "More productive crops and more effective means of crop protection." *Ciba-Geigy Journal* 1:84.

Rothschild, David (ed.) (1997) *Protecting What's Ours: Indigenous Peoples and Biodiversity.* Oakland, CA: South and Meso American Indian Rights Center.

Royal Society, The (2003) *Keeping Science Open: The Effects of Intellectual, Property Policy on the Conduct of Science,* www.royalsoc.ac.uk.

Ruivenkamp, Guido (2003) "Monitoring biotechnological developments: looking back for finding new perspectives." *Biotechnology and Development Monitor* 50 (March):2-5.

Rural Advancement Fund International (1995) "Biopiracy update: a global pandemic." *RAFI Communique* (September/October).

(2000a) "Biopiracy – RAFI's sixth annual update." *RAFI Communique* (May/June).

(2000b) "Biotech's 'generation 3'." *RAFI Communique* 67(November/December).

(2001a) "In search of common ground II: CMDT – can dinosaurs make teammates?" *RAFI Communique* 70(May/June).

(2001b) "USDA says yes to terminator." News release, 3 August.

Russell, D. (1983) "The marketing of genetic science." *Amicus Journal* 5:1(Summer):14-23.

Russell, H. L. (1931) "Commercial support for agricultural research." In *Proceedings of the Forty-fifth Annual Convention of the Association of Land Grant Colleges and Universities.* Burlington, VT: Free Press.

Russell, W. A. (1974) "Comparative performance for maize hybrids representing different eras of maize breeding." In Dolores Wilkingson (ed.), *Proceedings of the Twenty-ninth Annual Corn and Sorghum Research Conference.* Washington, DC: American Seed Trade Association.

Ruttan, Vernon W. (1980) "Bureaucratic productivity: the case of agricultural research." *Public Choice* 35:529-547.

(1982a) "Changing role of public and private sectors in agricultural research." *Science* 216(2 April):23-29.

(1982b) *Agricultural Research Policy.* Minneapolis, MN: University of Minnesota Press.

(1983) "Agricultural research policy issues." *HortScience* 18:6(December):809-818.

(1999) "Biotechnology and agriculture: a skeptical perspective." *AgBioForum* 2:1: 54-60.

Ruttan, Vernon, and W. B. Sundquist (1982) "Agricultural research as an investment: past experience and future opportunities." In Pioneer Hi-Bred International, Inc., *Report of the 1982 Plant Breeding Research Forum.* Des Moines, IA: Pioneer Hi-Bred International, Inc.

Ryan, Bryce (1948) "A study in technological diffusion." *Rural Sociology* 13(September):273-285.

Ryan, Bryce, and Nathan Gross (1943) "The diffusion of hybrid seed corn." *Rural Sociology* 8(March):15-24.

(1950) "Acceptance and diffusion of hybrid corn seed in two Iowa communities." Iowa Agricultural Experiment Station research bulletin 372.

Ryder, Edward J. (1984) "The art and science of plant breeding in the modern world of research management." *HortScience* 19:6(December):808-811.

Ryerson, K. A. (1933) "The history and significance of plant introduction work of the United States Department of Agriculture." *Agricultural History* 7:2(April): 110-128.

Salomon, Jean-Jacques (1973) *Science and Politics.* Cambridge, MA: MIT Press.

Saloutos, Theodore (1982) *The American Farmer and the New Deal.* Ames, IA: Iowa State University Press.

Sanford, R. (1984) "Monsanto planting seeds of the future." *St. Louis Post-Dispatch* (18 March):G1, G6.

Sanger, David E. (2003) "Bush links Europe's ban on bio-crops with hunger." *The New York Times* (23 May):A3.

Satheesh, P. V. (2003) "Finally I have seen Bt cotton." *Seedling* (January):1-3.

Schapaugh, William T. (1985) "ASTA/FAS trip report: meeting of FAO Commission on Plant Genetic Resources, March 11-15, 1985." Washington, DC: American Seed Trade Association.

Schell, Josef, and Marc Van Montagu (1983) "The Ti plasmid as natural and as practical gene vectors for plants." *Bio/Technology* 1:2(April):175-180.

Schmeck, Harold M. (1986) "Government panel urges relaxation of gene splicing rules." *New York Times* (30 September):22.

Schmitz, Andrew, and David Seckler (1970) "Mechanized agriculture and social welfare: the case of the tomato harvester." *American Journal of Agricultural Economics* 52(November):569-577.

Schneider, Keith (1986a) "U.S. quietly approved the sale of genetically altered vaccine." *New York Times* (4 April):1, 9.

(1986b) "Wisconsin scientists plant gene-altered tobacco outdoors." *The New York Times* (31 May).
(1986c) "Live vaccine test angers Argentina." *New York Times* (11 November): 1, 9.
(1986d) "A test: do microbes and politics mix?" *New York Times* (12 November):10.
(1986e) "Biotech's stalled revolution." *New York Times Magazine* (16 November):43, 45-48, 66, 102, 108.
Schneiderman, Howard A. (1982) "Statement by Howard A. Schneiderman." In *University/Industry Cooperation in Biotechnology.* Hearings before the Subcommittee on Investigations and Oversight and the Subcommittee on Science, Research, and Technology of the Committee on Science and Technology, U.S. House of Representatives, Ninety-seventh Congress, second session, 16-17 June. Washington, DC: USGPO.
(1985) "Genetic engineering in agriculture – will it pay?" Reprint of remarks to the Ohio Grain and Feed Association and Ohio Fertilizer and Pesticide Association, OhioAgroExpo, St. Louis, MO: Monsanto Company.
Schneiderman, Howard A., and Will D. Carpenter (1990) "Planetary patriotism: sustainable agriculture for the future." *Environmental Science and Technology* 24:4:466-473.
Schrag, P. (1978) "Rubber tomatoes: the unsavory partnership of research and agribusiness," *Harper's Magazine* 256(June):24-25, 27-29.
Schubert, Robert (2001a) "Monsanto still suing Nelsons, other growers." *CropChoice* (21 May), http://www.cropchoice.com/leadstry.asp?recid=326.
(2001b) "Wheat breeder avoids the GMO path." *CropChoice* (30 July), http://www.cropchoice.com/leadstry.asp?recid=386.
Schultz, Theodore W. (1983) "An unpersuasive plea for centralised control of agricultural research: on a report of the Rockefeller Foundation." *Minerva* 21:1 (Spring):141-143.
Science (1932) "Appropriations for the Department of Agriculture." *Science* 75 (6 May):10.
Scott, E. W. (1971) "Industry needs in relation to genetics, plant breeding and variety selection." *HortScience* 6:5(October):468-469.
Scott, R. V. (1962) "Railroads and farmers: educational trains in Missouri 1902-1914." *Agricultural History* 36:1(January):3-15.
Sears, E. R. (1947) "Genetics and farming." In *Science in Farming: Yearbook of Agriculture, 1943-1947.* Washington, DC: USDA.
Sears, Rollin G. (1998) "Status of public wheat breeding: 1998," www.wheatimprovement.org/Forum/1/Sears.htm (accessed May 16, 2003).
Seed Certification Officials, Foundation Seedstocks Personnel, and Extension Agronomists, North Central States (1982) *Proceedings of the Thirty-second Annual Meeting.* Madison, WI.
Seed Leader (1982) "Meet the new seed leader . . ." *Seed Leader* 1:1(August):1.
Seed World (1983) "A.S.T.A. – in the beginning." *Seed World* 121:7(June):30-32.
Seedsmen's Digest (1982) "Genetically engineered seed." *Seedsmen's Digest* 33:10 (October):8.
Seely, Ron (2002) "Want to know food's pedigree?" *Wisconsin State Journal* (17 November):A1, A8.
Seidel, George (1982) "Statements of Dr. George Seidel." In *Potential Application of Recombinant DNA and Genetics on Agricultural Sciences,* Hearings before the Subcommittee on Investigations and Oversight of the Committee on Science and Technology, U.S. House of Representatives: Ninety-seventh Congress, 9 June, 28 July, no. 134. Washington, DC: USGPO.

Seifert, Robert (1984) Personal interview, 16 March.

Serageldin, Ismail, and G. J. Persley (2000) "Promethean science: agricultural biotechnology, the environment, and the poor." CGIAR Background Brief, May 2, Washington, DC: CGIAR Secretariat.

Shabecoff, Philip (1986) "Pesticide control finally tops the E.P.A.'s list of most pressing problems." *New York Times* (6 March):16.

Shands, Henry (1986) "A synopsis of the U.S. National Germplasm System." Unpublished manuscript, July. Washington, DC: USDA.

Shapiro, Robert B. (1999) "Open letter from Monsanto CEO Robert B. Shapiro to Rockefeller Foundation President Gordon Conway." (October 4), http://www.biotechinfo.net/Monsanto_letter.pdf.

Shiva, Vandana (2001) "Seed Satyagraha: a movement for farmers' rights and freedoms in a world of intellectual property rights, globalized agriculture, and biotechnology." In Brian Tokar (ed.) *Redesigning Life: The Worldwide Challenge to Genetic Engineering,* 351-360. New York, NY: Zed Books.

Shoemaker, Robbin (ed.) (2001) *Economic Issues in Agricultural biotechnology.* Agriculture Information Bulletin no. 762, Economic Research Service, Washington, DC: United States Department of Agriculture.

Shull, George H. (1908) "The composition of a field of maize." In *Annual Report of the American Breeders Association, Volume 4,* pp. 296-301. Washington, DC: American Breeders Association.

(1909) "A pure-line method in corn breeding." In *Annual Report of the American Breeders Association, Volume 5,* pp. 51-59, Washington, DC: American Breeders Association.

(1946) "Hybrid seed corn." *Science* 103(3 May):547-550.

(1952) "Beginnings of the heterosis concept." In J. Gowen (ed.), *Heterosis.* Ames, IA: Iowa State University Press.

Silk, Leonard (1960) *The Research Revolution.* New York, NY: McGraw-Hill.

Simmonds, Norman W. (1979) *Principles of Crop Improvement.* New York, NY: Longman.

(1983a) "Plant breeding: the state of the art." In T. Kosuge et al. (eds.), *Genetic Engineering of Plants: An Agricultural Perspective.* New York, NY: Plenum Press.

(1983b) "Conference review: genetic engineering of plants." *Tropical Agriculture* 60:1(January):66-69.

Simpson, James R., and Donald E. Farris (1982) *The World's Beef Business.* Ames, IA: Iowa State University Press.

Singer, Maxine, and Dieter Soll (1973) "Guidelines for DNA hybrid molecules." *Science* 181(21 September):114.

Simon, J. Y. (1963) "The politics of the Morrill Act." *Agricultural History* 37:2 (April):103-111.

Sink, K. C. (1984) "Protoplast fusion for plant improvement." *HortScience* 19:1(February):33-37.

Sinsheimer, Robert (1975) "Troubled dawn for genetic engineering." *New Scientist* 68(16 October):148-151.

Sirkin, B. (1984a) "Princeton to break ground this spring for major new biotech facility." *Genetic Engineering News* 4:2(March):14-15.

(1984b) "Boston venture fund focuses on agriculture in selection of biotechnology investments." *Genetic Engineering News* 4:3(April):1, 23, 24.

Skocpol, Theda S., and Kenneth Finegold (1982) "State capacity and economic intervention in the early New Deal." *Political Science Quarterly* 97:255-278.

Slaughter, S., and L. Leslie (1997) *Academic Capitalism: Politics, Policies, and the Entrepreneurial University.* Baltimore, MD: Johns Hopkins University Press.

Sligh, Michael, and Laura Lauffer (2004) *Summit Proceedings: Summit on Seeds and Breeds for 21st Century Agriculture.* Pittsboro, NC: Rural Advancement Foundation International.

Smith, Tony (2003) "Farmers help deliver modified crops to Brazil." *The New York Times* (14 October):A1, A7.

(2004) "Argentina soy exports are up, but Monsanto is not amused." *The New York Times* (21 January):W1.

Sneep, J., and A. J. T. Hendriksen (eds.), and O. Holbek (coed.) (1979) *Plant Breeding Perspectives.* Wageningen, The Netherlands: Centre for Agricultural Publishing and Documentation.

Sneep, J., B. R. Murty, and H. F. Utz (1979) "Current breeding methods." In J. Sneep and A. J. T. Hendrickson (eds.) and O. Holbek (coed.), *Plant Breeding Perspectives.* Wageningen, The Netherlands: Centre for Agricultural Publishing and Documentation.

Snow, Allison A., and Pedro Morán Palma (1997) "Commercialization of transgenic plants: potential ecological risks." *BioScience* 47:2:86–96.

Snow, C. P. (1961) *Science and Government.* Cambridge, MA: Harvard University Press.

Sommers, Charles E. (1986) "More pests unfazed by chemicals." *Successful Farming* (January):23.

Soth, Lauren (1957) *Farm Trouble.* Princeton, NJ: Princeton University Press.

Sprague, George F. (1960) "Agronomic research and development." *Agronomy Journal* 52:12(December):675-677.

(1967) "Agricultural production in the developing countries." *Science* 157(18 August):774-778.

(1971) "Genetic vulnerability in corn and sorghum." In John Sutherland and Robert J. Falasca (eds.), *Proceedings of the Twenty-sixth Annual Hybrid Corn Industry-Research Conference.* Washington, DC: American Seed Trade Association.

(1980) "The changing role of the private and public sectors in corn breeding." In Harold Loden and Dolores Wilkinson (eds.), *Proceedings of the Thirty-fifth Annual Corn and Sorghum Research Conference.* Washington, DC: American Seed Trade Association.

(1983) "Heterosis in maize: theory and practice." In O. H. Frankel (ed.), *Heterosis.* New York, NY: Springer-Verlag.

Sprague, George F., D. E. Alexander, and J. W. Dudley (1980) "Plant breeding and genetic engineering: a perspective." *BioScience* 30:1(January):17-21.

Stakman, E. C., R. Bradfield, and P. C. Mangelsdorf (1967) *Campaigns Against Hunger.* Cambridge, MA: Belknap Press.

Staub, W. J., and M. G. Blase (1971) "Genetic technology and agricultural development." *Science* 173(9 July):119-123.

Steele, Leonard (1978) "The hybrid corn industry in the United States." In D. B. Walden (ed.), *Maize Breeding and Genetics.* New York, NY: Wiley Interscience.

Stevens, M. Allen (1974) "Varietal influence on nutritional value." In P. L. White and N. Selvey (eds.), *Nutritional Quality of Fresh Fruits and Vegetables.* Mt. Kisco, NY: Futura.

Stone, Glenn Davis (2002) "Both sides now: fallacies in the genetic modification wars, implications for developing countries, anthropological perspectives." *Current Anthropology* 43:4:611-630.

Strosneider, Robert E. (1984) "A private seed company's views of the roles of public and private breeders – cooperation and support." *HortScience* 19:6(December):800-802.

Studebaker, J. A. (1982) "Fifty years with breeders' rights." *Seedsmen's Digest* 33:5(May):22-27.

Studebaker, J. A., and J. A. Batcha (1982) "A chronicle of plant variety protection." Kalamazoo, MI: Asgrow Seed Company.

Sun, Marjorie (1986a) "The global fight over plant genes." *Science* (7 February): 231:445-447.

(1986b) "Local opposition halts biotechnology test." *Science* 231(14 February): 667-668.

Sundquist, W. B., K. M. Menz, and C. F. Neumeyer (1982) *A Technology Assessment of Commercial Corn Production in the United States.* Minnesota Agricultural Experiment Station research bulletin 546.

Supreme Court of Iowa (1977) "Lucas v. Pioneer Inc." 256 N.W. 2d 167. In *North Western Reporter.* St. Paul, MN: West Publishing Company.

Swanson, Timothy, and Timo Goeschl (2000) "Genetic use restriction technologies (GURTS): impacts on developing countries." *International Journal of Biotechnology* 3:56-84.

Sward, S. (1983) "UC acts on 5 possible faculty conflicts of interest." *San Francisco Chronicle* (23 August):4.

Sylvester, E. J., and L. C. Klotz (1983) *The Gene Age: Genetic Engineering and the Next Industrial Revolution.* New York, NY: Charles Scribner's Sons.

Szybalski, Waclaw (1985) "Genetic engineering in agriculture." *Science* 229 (12 July): 112-113.

Tanksley, Steven D., and Susan R. McCouch (1997) "Seed banks and molecular maps: unlocking genetic potential from the wild." *Science* 277(27 August):1063-1066.

Ten Eyck, A. M. (1910) "Breeding, multiplying, and distributing improved seed grain by the Kansas Experiment Station." In *Proceedings of the American Society of Agronomy, Volume 1,* 1907, 1908, 1909, Washington, DC: American Society of Agronomy.

Teweles, L. William (1976) "Some plain talk about seed company mergers and acquisitions." In Harold D. Loden and Dolores Wilkinson (eds.), *Proceedings of the Twenty-Second Annual Farm Seed Conference.* Washington, DC: American Seed Trade Association.

(1983) "Bringing the new plant biotechnologies to the marketplace." In *Biotech '83: Proceedings of the International Conference on the Commercial Applications and Implications of Biotechnology.* Northwood, Middlesex, UK: Online Publications.

Thurow, Lester C. (1996) *The Future of Capitalism.* New York, NY: Penguin Books.

Tokar, Brian (2001) "Resisting the engineering of life." In Brian Tokar (ed.) *Redesigning Life: The Worldwide Challenge to Genetic Engineering,* 320-336. New York, NY: Zed Books.

Toner, Mike (2002) "Designer crops already abundant on grocery shelves." *The Atlanta Journal-Constitution* (19 May):A1, A13.

True, A. C. (1900) "Agricultural education in the United States." *Yearbook of Agriculture, 1899.* Washington, DC: USGPO.

Tucker, Robert C. (1978) *The Marx-Engels Reader.* New York, NY: W. W. Norton & Company.

Tudge, Colin (1984) "Drugs and dyes from plant cell cultures." *New Scientist* (12 January):25.
Turning Point Project (1999) "Who plays God in the 21st century?" *The New York Times* (11 October):A11.
Twentieth Century Fund Task Force on the Commercialization of Scientific Research (1984) *The Science Business.* New York, NY: Priority Press.
Underwood, Wayne (1978) "Hotline – farm seeds in a world market." In Harold D. Loden and Dolores Wilkinson (eds.), *Proceedings of the Twenty-fourth Annual Farm Seed Conference.* Washington, DC: American Seed Trade Association.
(1984) "Crop improvement in developing nations." *Seed World* (June):38–39.
United States Agency for International Development (1985) "Private sector seed development in the Thai seed industry." AID evaluation study no. 23, June. Washington, DC: United States Agency for International Development.
United States Department of Agriculture (1888) *Report of the Commissioner of Agriculture, 1887.* Washington, DC: USGPO.
(1894) *Report of the Secretary of Agriculture, 1893.* Washington, DC: USGPO.
(1895a) *Report of the Secretary of Agriculture, 1894.* Washington, DC: USGPO.
(1895b) *Report of the Secretary of Agriculture, 1895.* Washington, DC: USGPO.
(1896) *Report of the Secretary of Agriculture, 1896.* Washington, DC: USGPO.
(1899) *Foreign Seeds and Plants: Inventory No. 1.* Washington, DC: USGPO.
(1940) *Technology on the Farm.* Special report by an interbureau committee and the Bureau of Agricultural Economics. Washington, DC: USDA.
(1967) *The National Program for Conservation of Crop Germ Plasm.* Athens, GA: University of Georgia Press.
(1978) "Changes in farm production and efficiency, 1977." Statistical bulletin no. 612, Economics, Statistics and Cooperatives Service. Washington, DC: USDA.
(1981) *A Time to Choose: Summary Report on the Structure of Agriculture.* Washington, DC: USDA.
(1986) "Research progress in 1985." Agricultural Research Service, May. Washington, DC: USDA.
United States Department of State (1985) "U.S. position on FAO undertaking and Commission on Plant Genetic Resources." March. Washington, DC: Department of State.
United States District Court (1970a) "Findings of fact and conclusions of law proposed on behalf of plaintiff." In Civil Action No. 63 C 597, Research Corporation v. Pfister Associated Growers, Inc., et al., v. Paul C. Mangelsdorf. U.S. District Court, Northern District of Illinois, Eastern Division, Chicago, IL: Twentieth Century Press, Inc.
(1970b) "Plaintiff's post trial brief." In Civil Action No. 63 C 597, Research Corporation v. Pfister Associated Growers, Inc., et al., v. Paul C. Mangelsdorf. U.S. District Court, Northern District of Illinois, Eastern Division. Chicago, IL: Twentieth Century Press, Inc.
(1983) "Delta and Pine Land Company v. Peoples Gin Company, no. GC 81-68-WK-O, U.S. District Court, N.D. Mississippi, Greenville Division. March 3, 1982." *Federal Supplement* 546:939–945. St. Paul, MN: West Publishing Company.
United States General Accounting Office (1977) "Management of agricultural research: need and opportunities for improvement." Gaithersburg, MD: USGAO.
(1981) "The Department of Agriculture can minimize the risk of potential crop failure." April 10. Gaithersburg, MD: USGAO.

(1985a) "Agriculture overview: U.S. food/agriculture in a volatile world economy." GAO/RCED-86-3BR, November. Gaithersburg, MD: USGAO.

(1985b) "Biotechnology: the U.S. Department of Agriculture's biotechnology research efforts." GAO/RCED-86-39BR. Gaithersburg, MD: USGAO.

(1986) "Biotechnology: agriculture's regulatory system needs clarification." GAO/RCED-86-59. Gaithersburg, MD: USGAO.

United States House of Representatives (1906) Arguments before the Committee on Patents of the House of Representatives on H.R. 18851, to amend the laws of the United States relating to patents in the interest of the originators of horticultural products, 17 May. Fifty-ninth Congress, first session. Washington, DC: USGPO.

(1922) *Endowment of Agricultural Experiment Stations,* Hearings before the Committee on Agriculture, Sixty-seventh Congress, second session, 26 January, 23-25 February. Washington, DC: USGPO.

(1930) *Plant Patents,* Hearings before the Committee on Patents, House of Representatives, Seventy-first Congress, second session, on H.R. 11372, a bill to provide for plant patents, 9 April. Washington, DC: USGPO.

(1951) "Research and related services in the United States Department of Agriculture." Report prepared for the Committee on Agriculture, Eighty-first Congress, second session, 21 December, Volume 1. Washington, DC: USGPO.

(1981) *Commercialization of Academic Biomedical Research,* Hearings before the Subcommittee on Investigations and Oversight and the Subcommittee on Science, Research, and Technology of the Committee on Science and Technology, U.S. House of Representatives, Ninety-seventh Congress, 8-9 June, no. 46. Washington, DC: USGPO.

(1982a) *University/Industry Cooperation in Biotechnology,* Hearings before the Subcommittee on Investigations and Oversight and the Subcommittee on Science, Research, and Technology of the Committee on Science and Technology, U.S. House of Representatives, Ninety-seventh Congress, second session, 16-17 June. Washington, DC: USGPO.

(1982b) *Potential Applications of Recombinant DNA and Genetics on Agricultural Sciences,* Hearings before the Subcommittee on Investigations and Oversight of the Committee on Science and Technology, U.S. House of Representatives, Ninety-seventh Congress, 9 June, 28 July, no. 134. Washington, DC: USGPO.

(1984) *Agricultural Research.* Hearings before the Subcommittee on Department Operations, Research and Foreign Agriculture of the Committee on Agriculture, Ninety-eighth Congress, first session, 22-23, 28-29 June. Washington, DC: USGPO.

United States Patent Office (1956) "United States patent 2,753,663: production of hybrid seed corn." Washington, DC: U.S. Patent Office.

(1980) "United States patent 4,237,224: process for producing biologically functional molecular chimeras." Washington, DC: U.S. Patent Office.

United States Senate (1968) *Patent Law Revision.* Hearings before the Subcommittee on Patents, Trademarks, and Copyrights of the Committee on the Judiciary, United States Senate, Ninetieth Congress, second session, pursuant to S. Res. 37, Part 2, 30-31 January and 1 February. Washington, DC: USGPO.

(1970) *Plant Variety Protection Act.* Hearings before the Subcommittee on Agricultural Research and General Legislation of the Committee on Agriculture and Forestry, United States Senate, Ninety-first Congress, second session, on S.3070, June 11. Washington, DC: USGPO.

United States Supreme Court (1982) "Diamond, Commissioner of Patents and Trademarks v. Chakrabarty." *United States Reports* 447:303-322. Washington, DC: USGPO.

van Wijk, Jeroen (1995) "Broad biotechnology patents hamper innovation." *Biotechnology and Development Monitor* 25(December):15-17.

Vavilov, Nikolai I. (1951) "The origin, variation, immunity and breeding of cultivated plants." *Chronica Botanica* 13:1-364.

(1971) "The problem of the origin of the world's agriculture in the light of the latest investigations." In H. Bukharin et al. (eds.), *Science at the Crossroads*. London: Frank Cass and Company.

Vest, Grant (1984) "The vegetable seed industry and public plant breeding: some concerns." *HortScience* 19:2(April):167-168.

Victor, David G., and C. Ford Runge (2002) "Farming the genetic frontier." *Foreign Affairs* 81:3:107-121.

Vogel, Gretchen (2001) "Infant monkey carries jellyfish gene." *Science* 291(12 January):226.

Vogeler, Ingolf (1981) *The Myth of the Family Farm*. Boulder, CO: Westview Press.

Wade, Nicholas (2000a) "Genetic code of human life is cracked by scientists." *The New York Times* (27 June):A1.

(2000b) "Now the hard part: putting the genome to work." *The New York Times* (27 June):D4-D5.

(2003a) "Genome's riddle: few genes, much complexity." *The New York Times* (13 February):D1, D4.

(2003b) "Double helix leaps from lab into real life." *The New York Times* (25 February):D1, D6-D7.

Wallace, Henry A. (1919) "Corn show evolution." *Wallace's Farmer* 44:51(December 19):2509, 2524.

(1934a) "Give research a chance." *Country Gentleman* 104:9(September):5-6, 34, 36.

(1934b) "The year in agriculture: secretary's report to the president." In *Yearbook of Agriculture*. Washington, DC: USGPO.

(1936) "Report of the secretary of agriculture." In *Yearbook of Agriculture, 1936*. Washington, DC: USGPO.

(1955) "H. A. Wallace to Zvi Griliches, 28 December, 1955." 50(584), Henry A. Wallace Papers, University of Iowa Libraries, Iowa City, IA.

(1956) "Public and private contributions to corn – past and future." In William Heckendorn and Joseph Gregory (eds.), *Proceedings of the Tenth Annual Hybrid Corn Industry-Research Conference, 1955*. Chicago, IL: American Seed Trade Association.

(1961) "The department as I have known it." In W. D. Rasmussen (ed.), *Lecture Series in Honor of the USDA Centennial Year*. Washington, DC: USDA Graduate School.

Wallace, Henry A., and William L. Brown (1956) *Corn and Its Early Fathers*. East Lansing, MI: Michigan State University Press.

Wall Street Journal (1983) "Upjohn agrees to buy 61% of O's Gold Seed for total $16.8 million." *Wall Street Journal* (6 September):21.

Walsh, John (1984) "Seeds of dissension sprout at FAO." *Science* 223(13 January): 147-148.

Wanamaker, John (1999) "Avast, ye seed pirates." *Wisconsin Agriculturist* (January): 27.

Ward, Bill (1986) "Seed: companies faced with tough transition." *Agri-Marketing* (May):108, 110.

Ware, J. O. (1936) "Plant breeding and the cotton industry." In *Yearbook of Agriculture, 1936*. Washington, DC: USDA.

Warren, G. F. (1969) "Mechanized growing and harvesting of vegetable crops in the eastern United States." *HortScience* 4:3(Autumn):237-239.

Waters, H. J. (1910) "The function of land-grant colleges in promoting collegiate and graduate instruction in agriculture." In *Proceedings of the Twenty-third Annual Convention of the Association of American Agricultural Colleges and Experiment Stations, 1909*. Washington, DC: USGPO.

Watson, James D., and John Tooze (1981) *The DNA Story: A Documentary History of Gene Cloning*. San Francisco, CA: W. H. Freeman and Company.

Webb, R. E., and W. M. Bruce (1968) "Redesigning the tomato for mechanized production." In *Science for Better Living: Yearbook of Agriculture, 1968*. Washington, DC: USDA.

Webber, H. J. (1900) "Progress of plant breeding in the United States." In *Yearbook of Agriculture, 1899*. Washington, DC: USGPO.

Weiss, M. G. (1969) "Public-industry position on plant variety protection." *HortScience* 4:2(Summer):84-86.

Whetzel, H. H. (1945) "History of industrial fellowships in the Department of Plant Pathology at Cornell University." *Agricultural History* 19:2(April):99-104.

Whitaker, T. W. (1979) "The breeding of vegetable crops: highlights of the last seventy-five years." *HortScience* 14:3(June):359-363.

White, Allenby L. (1959) "Research problems as observed by the seed industry." In William Heckendorn and Benjamin R. Blankenship, Jr. (eds.), *Proceedings of the Fourth Annual Farm Seed Industry Research Conference*, Washington, DC: American Seed Trade Association.

(1969) "Plant variety protection – what it means to the seedsman." In John Sutherland and Robert J. Falasca (eds.), *Proceedings of the Fifteenth Farm Seed Conference*, Washington, DC: American Seed Trade Association.

(1976) "Plant variety protection update." In Harold D. Loden and Dolores Wilkinson (eds.), *Proceedings of the Sixth Soybean Seed Research Conference*, Washington, DC: American Seed Trade Association.

Wiley, John P., Jr. (1986) "Phenomena, comment and notes." *Smithsonian* 17:8(November):42, 44, 46-48.

Wilkes, Garrison (1977) "The world's crop plant germplasm – an endangered resource." *Bulletin of the Atomic Scientists* 33:8-16.

(1983) "Current status of crop germplasm." *Critical Reviews in Plant Sciences* 1:2:133-181.

Williams, C. G. (1928) "The responsibility of the agricultural experiment station in the present agricultural situation." *Science* 67(25 May):519-522.

Williams, Sidney B. (1984) "Protection of plant varieties as intellectual property." *Science* 225(6 July):18-23.

Wilson, E. O. (1992) *The Diversity of Life*. New York, NY: W. W. Norton & Co.

Wilson, P., and C. J. Driscoll (1983) "Hybrid wheat." In R. Frankel (ed.), *Heterosis*. New York, NY: Springer-Verlag.

Witt, Steven C. (1986) "FAO still debating germplasm issues." *Diversity* 8(Winter):24-25.

Wittwer, Sylvan W. (1973) "Potentials for improving production efficiency of fruits and vegetables." In *Proceedings of the Twenty-first Annual Meeting of the Agricultural Research Institute, 1972*. Washington, DC: Agricultural Research Institute.

(1985) "The new horticulture: programming the future." *HortScience* 20:5(October):815-821.

Wolf, Edward C. (1985) "Conserving biological diversity." In Linda Starke (ed.), *State of the World 1985*. New York, NY: W. W. Norton and Company.

Wolfenbarger, L. L., and P. R. Phifer (2000) "The ecological risks of genetically engineered plants." *Science* 290:2088-2093.

Writing Solutions, Inc. (2003) "Update of worldwide GM food labeling rules." *The Non-GMO Source* 3:6(June):4-5.

Yarwood, C. E. (1970) "Man-made plant diseases." *Science* 168(10 April):218-220.

Yeatmann, Christopher W., David Kafton, and Garrison Wilkes (1984) *Plant Genetic Resources: A Conservation Imperative.* Boulder, CO: Westview Press.

Yoon, Carol Kaesuk (1999) "Altered corn may imperil butterfly, researchers say." *The New York Times* (20 April):A1, A20.

York, E. T. (1978) "Statement before the Senate. Subcommittee on Migratory Labor." In R. Rodefeld et al. (eds.), *Change in Rural America.* St. Louis, MO: C. V. Mosby Company.

Young, J. A., and J. M. Newton (1980) *Capitalism and Human Obsolescence.* Montclair, NJ: Allanheld, Osmun and Company.

Yoxen, E. (1983) *The Gene Business: Who Should Control Biotechnology.* London: Pan Books.

Zeven, A. C., and J. M. J. de Wet (1982) *Dictionary of Cultivated Plants and their Regions of Diversity.* Wageningen, The Netherlands: Centre for Agricultural Publishing and Documentation.

Zhukovsky, P. M. (1975) *World Gene Pool of Plants for Breeding: Mega-gene-centers and Micro-gene-centers.* Leningrad: USSR Academy of Sciences.

Zirkle, Conrad (1969) "Plant hybridization and plant breeding in eighteenth-century America." *Agricultural History* 53:1(January):25-38.

Zuber, M. S., and L. L. Darrah (1980) "1979 U.S. corn germplasm base." In Harold D. Loden and Dolores Wilkinson (eds.), *Proceedings of the Thirty-fifth Annual Corn and Sorghum Research Conference.* Washington, DC: American Seed Trade Association.

Index

Science and Technology in Society

DANIEL LEE KLEINMAN
Impure Cultures: University Biology and the World of Commerce

DANIEL LEE KLEINMAN, ABBY J. KINCHY, AND
JO HANDELSMAN, EDITORS
Controversies in Science and Technology: From Maize to Menopause

JACK RALPH KLOPPENBURG, JR.
First the Seed: The Political Economy of Plant Biotechnology, second edition